The Earthquake Observers

The Earthquake Observers

Disaster Science from Lisbon to Richter

DEBORAH R. COEN

The University of Chicago Press
Chicago and London

The University of Chicago Press, Chicago, 60637
The University of Chicago Press, Ltd., London
© 2013 by The University of Chicago
All rights reserved. Published 2013
Paperback edition 2014
Printed in the United States of America

22 21 20 19 18 17 16 15 14   2 3 4 5 6

ISBN-13: 978-0-226-11181-0 (cloth)
ISBN-13: 978-0-226-21205-0 (paper)
ISBN-13: 978-0-226-11183-4 (e-book)
DOI: 10.728/chicago/9780226111834.001.0001

Library of Congress Cataloguing-in-Publication Data

Coen, Deborah R.
 The earthquake observers: disaster science from Lisbon to Richter /
Deborah R. Coen. pages ; cm
 Includes bibliographical references and index.
 ISBN-13: 978-0-226-11181-0 (cloth: alkaline paper)
 ISBN-10: 0-226-11181-4 (cloth: alkaline paper)
 ISBN-13: 978-0-226-11183-4 (e-book)
 ISBN-10: 0-226-11183-0 (e-book)
1. Seismology—History.   2. Earthquakes—Observations—History—19th century.
I. Title
 QE539.C64 2013
 551.209'034—dc23

                                                          2012015828

For Paul

CONTENTS

Introduction

No tremor is felt, no liquid is spilled, without the direction in which it occurred being reported to the paper the next morning. . . . Day in, day out, today, tomorrow, eternally. Until the world truly comes crashing down.

—Karl Kraus, "The Earthquake" (1908)[1]

I should admit from the outset that I have never felt an earthquake. What's more, unlike many of the nineteenth-century witnesses that people the following pages, I have no longing to experience one. Nor was I fascinated by seismology when I began to investigate its history; that came later. Instead, this book began with a metaphor: an earthquake of the mind.[2]

Nietzsche used the metaphor of the intellectual earthquake to describe the consequences of the ruthless empiricism of the nineteenth century. "As cities collapse and grow desolate when there is an earthquake, and man erects his house on volcanic land only in fear and trembling and only briefly, so life itself caves in and grows weak and fearful when the concept-quake [*Begriffsbeben*] caused by science [*Wissenschaft*] robs man of the foundation of all his rest and security, his belief in the enduring and eternal."[3] The nineteenth century's relentless scientific spirit had shaken traditional beliefs like an earthquake. The result was a sense of radical contingency and disorientation. Nietzsche may have judged this condition a "sickness," but not all his contemporaries agreed. The version of the metaphor that first piqued my curiosity appeared in a didactic allegory by the Austrian physicist Franz Serafin Exner. A young man sets out to explore the world and encounters novelties that contradict the wisdom inherited from his elders: "As in an earthquake, what begins to shake is just what we have since childhood grown used to regarding as the one solid thing in the world, and so in an instant our previous faith proves to be false and one-sided."[4] For Exner, I found, this crisis was a requisite step on the way to intellectual maturity: acquiring a scientific mindset meant embracing the predicament of contingency with the help of probabilistic reasoning. So why, I wondered, did he resort to what appeared to be a symbol of nihilism? Similarly, in the 1930s, when Kurt Gödel demonstrated that the dream of a complete logical system could never be achieved, two of his most eminent central European colleagues likened the impact to an earthquake. "The work on formally undecidable propositions was received like an earthquake" wrote Karl Popper, while Karl Menger reflected that the "various edifices [of mathematics] are not secure against the earthquake of a contradiction."[5] As one commentator has suggested, the earthquake metaphor implies that the theorem caused "widespread confusion and despair."[6] By all accounts, however, philosophers like Popper and Menger readily embraced Gödel's idea and its radical implications. Why then the violent metaphor?

With further research, I began to suspect that the metaphor was more than metaphorical. There was a concrete sense in which earthquakes were losing the apocalyptic associations of an earlier age. Nineteenth-century scientists were enlisting ordinary people to record seismic events, from barely perceptible tremors to catastrophic shocks. Popular writers were discussing earthquakes as elements of a natural landscape with which people had to learn to live—much as Exner, Popper, and Menger believed that philosophy had to adapt to the instability of the modern intellectual landscape. And the notion of an "intellectual earthquake" was becoming vivid, as the human sciences probed the psychic effects of seismic events. Earthquake observers, in short, were confronting head-on the crisis Nietzsche identified: the mental plunge into the uncertain universe of modern science.

What I learned, in short, is that the world became shakier in the nineteenth century. By that I do not mean simply that political revolutions and industrialization were perceived in terms of precipitous, dizzying change. I mean that the earth was caught trembling more often than before. Reports of jerks, bumps, rumbles, and thuds proliferated. They passed from hand to hand in newspapers, letters, scientific transactions, and medical case studies. Certain countries and even towns acquired reputations for impressive instability. Travelers looked forward to the thrill of a first earthquake. People learned to suspect that the sound of a distant wagon rolling over cobblestones was the trembling of the earth itself—and they could check the morning papers to confirm it. "One could read ten times in a row that Herr Jonas Blau, Frau Isole Klümpel, Herr Isidor Käsler, Herr Wotan Kohn and

Herr Leo Kohn or Fräulein Sprinzel-Kohary were just sitting down to dessert when the glasses started to clatter."[7]

And then, sometime around the Great War, the trembling apparently subsided. The newspapers turned to larger cataclysms and more "worldly" matters. The din of Herr Blau and Frau Klümpel's glasses faded into silence.

Disaster Science

Disaster is, by definition, that which "cannot be comprehended *exactly*."[8] It is a hopelessly hybrid entity: inextricably entangling the natural and the social, freighting objectivity with subjectivity, and binding global science to local contingencies. This book shows how the earthquake as disaster—as an interface between nature and society—was made and unmade as a scientific object. Earthquakes furnished ideal conditions for geophysical research: they gave access to the planet's hidden structure and to the forces that had shaped the earth throughout its history. The problem, in an age before reliable seismographs, was how to turn an instant of panic and confusion into a field for the production of scientific evidence. This was the achievement of the permanent networks of seismic observers organized in the late nineteenth century. The result was a natural experiment at the nexus of human behavior and planetary physics.

In the nineteenth century, a scientific description of an earthquake was built of stories—stories from as many people, in as many places, in as many different situations as possible. "It will be obvious," wrote one nineteenth-century geologist, "that no single person can, from his own observations, estimate the area agitated by an earthquake, though much may be accomplished by the combined observations of many."[9] "Likely in no other field," admitted another, "is the researcher so completely dependent on the help of the non-geologist, and nowhere is the observation of each individual of such high value as with earthquakes."[10] In the chapters that follow, we will encounter the nineteenth century's most eloquent seismic commentators, including Alexander von Humboldt, Charles Darwin, Mark Twain, Charles Dickens, Ernst Mach, John Muir, Gertrude Atherton, and William James. We will also meet countless others whose names are long forgotten: citizen-observers, many of whom were women. Sometimes their stories told of fear and devastation, sometimes of excitement and curiosity, sometimes of wonder, incomprehension, disbelief, or uncertainty. As an observer in New Jersey put it when reporting a possible tremor to a local scientist, "You are interested in manifestations, which, otherwise unexplainable, are referred to earthquakes." In this case, the observer had been reading peacefully when he was suddenly disturbed by a shaking, which "seemed to result from some giant having braced his feet well apart on the room overhead and then made an effort to sway the building." Others in the house had also noticed something odd; one assumed that a Mr. W. was chasing his wife through the house, or perhaps she was chasing him. "Had I known what was coming I should have been prepared for some careful observations as to time. . . . As I had never had a similar experience I did not realize what I was missing until too late."[11] Observers had to learn how to report such ambiguities, and seismologists had to learn how to work with them. Many of these earthquake reports have been preserved in archives, and they are a window onto an unusual dialogue. "Observe conscientiously," instructed an 1879 guide to earthquake observing, "but have no fearfulness with respect to us in the transmission of observations. All that is genuinely observed is welcome, even what is perceived in uncertainty, as long as it is marked as such."[12]

These were stories, above all, about individuals and communities and their relationships to the land they lived on. As a late nineteenth-century antiquarian from the Swiss village of Fleurier noted of the earthquake reports of a generation past, "These notes . . . do not have the dryness and banality of a newspaper report; they [stand out] for their candidness, originality, and local flavor, and it is not the least of their merits that they transport us to the simple and rustic Fleurier of old."[13] Quaint as they may at first appear, earthquake reports based on human observations ("felt reports") hold a rich ore of information. As we will see, it was on the basis of felt reports that earthquakes came to be understood as the result of horizontal movements of the earth's crust. It was also on this basis that scientists learned that earthquakes can be triggered by human activities (a phenomenon now known as induced seismicity). Felt reports profit from the familiarity of local observers with the normal state of their surroundings: locals are in the best position to recognize anomalies such as variations of groundwater levels, unusual weather, remarkable animal behavior, or changes in the surface of the land.[14] Their observations bear clues to the spatial variability of the impacts of earthquakes, which is a complex function of factors such as tectonic structure, soil type, and building style. Maps based on felt reports ("macroseismic intensity maps") shed light on the geography of seismic risk; they can help locate faults and guide reconstruction in order to mitigate future damage. What's more, such sources reveal how earthquakes have been perceived and interpreted in different times and places. Like recent surveys that have attempted to assess, for instance, whether earthquake victims in California and in Japan display different degrees of fatalism, nineteenth-century felt reports open windows onto the cultural determinants of risk perception.[15] As the Swiss Earthquake Service put it in 1910, "To study the relationship of man to earthquakes was from the start a special goal of Swiss earthquake research."[16] In short, the information gleaned encompassed a disaster's physical impacts, human responses, and the conditions under which knowledge of either was possible. In today's terms, the perspective of nineteenth-century seismology would be described as "integrated." It addressed both the physical and human factors that define today's notion of "vulnerability"—including regional seismicity, building standards, and the social conditions that affect a society's ability to cope with disasters. In other words, this nineteenth-century research made apparent one of the core lessons of recent environmental history and disaster studies: that "natural disasters" as such do not exist, that catastrophic consequences are always the outcome of an unfortunate conjunction of geophysical circumstances and human choices.[17]

Today, seismologists are increasingly aware of the uncertainty of seismic risk assessments. The potential violence of even weak earthquakes has been magnified in an age of gravity-defying engineering and nuclear power. Assessing local risk means analyzing seismic processes over the course of centuries. Scientists therefore need to dig beyond the last few decades of seismographic records, deep into the human records of the past. The observations of ordinary nineteenth-century people have been dug out of newspapers and archives and mined once again for clues to long-term patterns of seismicity. Even today, the most sophisticated seismographs alone cannot reveal how the impacts of earthquakes vary at a local level; and there are many effects to which seismographs are blind, such as building damage and topographical changes. Moreover, it has become clear that earthquakes pose a threat in continental interiors, far from the edges of tectonic plates. The causes of such "intraplate" earthquakes remain a mystery: the experts themselves insist that there are no experts on this subject. As we will see, seismologists are once again enlisting the public as earthquake observers.[18]

Until recently, disasters have been conspicuously absent from historians' accounts of modern science. The current field of disaster studies typically locates its origins in the 1970s, with the rise of the risk-management paradigm. Before that, the assumption goes, disasters were regarded as "purely physical occurrences, requiring largely technological solutions."[19] This limited vision has been matched by a studied disregard for disasters within the field of history of science. In laying out the mission of the discipline in the 1920s, George Sarton demoted wars, pestilence, earthquakes, and the like from world-historical events to superficial contingencies. For Sarton, disasters were merely accidents that obscured the fundamental source of human progress: the work of science. Scientists "go on pursuing their life's work without seeming to be in the least concerned with the gigantic activities that surround them. Mere earthquakes or wars do not interrupt their work."[20] By contrast, recent research suggests that disasters have decisively shaped the historical trajectories of the modern sciences, creating a stage for the entrance of new classes of technical experts and new forms of expertise.[21]

Sarton represented one of two competing modernist tendencies, which have alternately pushed disasters toward visibility and invisibility. On one hand, the effort to orchestrate networks of citizen-observers to watch for earthquakes was a modernist project par excellence. It served the cause of popular enlightenment and informed plans for large-scale engineering. Still, the aim was not to offer false security. Scientists claimed to know little more about earthquakes than citizens themselves. They aimed not to suppress fear but to instrumentalize it. The goal was no less than to realign humanity's sense of its place in the cosmos. In the often quoted words of the geologist Eduard Suess, "the planet may well be measured by man, but not according to man."[22] Through their collaboration with the public, Suess and his colleagues intended, in short, to calibrate the human seismograph. This required nimble adjustments between geophysical gauges and lived experience. Only a science at once physical and human could ascertain how much fear a given shock induced and how much it actually warranted. What could be more characteristic of modernism's ambition of reconstructing society from the ground up than the impulse to rationalize fear itself?

Nonetheless, the hybrid science produced by this impulse offended modernist sensibilities. It undermined what Bruno Latour calls the modernist process of "purification," the separation of the analysis of nature from the analysis of society.[23] Making a science of disaster means constructing a basis for comparison, both geographic (how hard your town was hit versus mine) and historical (how much worse this one was than the one in your grandmother's stories). To this end, a science of disaster must constantly move back and forth between the natural and the social, the objective and the subjective, the global and the local. It must correlate geological formations and the built environment, instrumental data and human responses, planetary waves and local damage. Each informs the analysis of the other. Circa 1900, however, many scientists dreamed of a "modern" seismology, a "pure" science: one in which the objective, instrumental, mathematical, and global would no longer depend on the subjective, human, discursive, and local. For this reason, some of the same innovations that made possible a science of disaster—the seismograph, the observatory, the mathematical physics of seismic waves—simultaneously threatened to spawn a science that had next to nothing to do with earthquakes *as disasters*. In this sense, the history of the making of disaster as a scientific object has always also been the history of its unmaking.

Imagining Lisbon

Immanuel Kant was a thirty-one-year-old philosophy student, still a year shy of his doctorate, when Europe was rocked by the Lisbon earthquake of 1755. Three thousand kilometers away in Königsberg, Kant would not have read the first news reports of the disaster for several weeks. Then, for months on end, the German papers were packed with stories of tragedy, chaos, and ruin.[24] Attempts at natural scientific explanations vied for space with reflections on God's vengeance. Eyewitness reports abounded, and Kant collected them eagerly. With uncharacteristic impatience, he penned three essays for his local paper. He was taking on "the useful role of a scientific publicist," attempting to turn discussion from theological interpretation to naturalistic explanation.[25] His account was to be "not a history of the misfortunes that people suffered, not a catalog of the devastated cities and the residents buried under their rubble. . . . I will describe here only the work of nature, the remarkable natural conditions that accompanied the dreadful event and their causes."[26] What followed was a compilation of terrestrial and atmospheric phenomena observed across Europe in the days before and after the great earthquake. In this way, in the judgment of later commentators, Kant produced the first work of modern seismology. In the assessment of Georg Gerland, the founder of the International Seismological Association in 1901, the value of Kant's essay lay precisely in its exclusion of the plight of the victims: "in this omission of the doubtless exciting, but seismologically irrelevant trappings: in this broad view . . . he gives the first truly scientific treatment of an earthquake. . . . Kant was the first to give a scientific analysis that intends to depict only 'the work of nature' and the causes of the events, one that was exemplary well beyond his own day."[27] Likewise, for the historian of philosophy Kuno Fischer, the significance of Kant's essay lay in having "focused squarely on the lawful necessity of nature." The literary critic Walter Benjamin, not otherwise known as a connoisseur of natural science, judged that Kant's essay "probably represents the beginnings of scientific geography in German. And certainly the beginnings of seismology."[28] These verdicts have become part of the grand narrative of the Lisbon earthquake as a pivot of modern history. It figures as the origin of a secular, rational, statist modernity and a scientific approach to natural hazards. The conclusion appears inescapable: Kant brought Europe into the age of modern science by producing an account of disaster in which the human victims fell silent.

Hence the paradox that lurks within the history of the environmental sciences: Kant—the progenitor of a modern view of the knowing subject—founded a modern science of the earth precisely by eliminating the human subject from it. Kant's research on the Lisbon earthquake became part of the lectures that defined the field of "physical geography," to which the modern environmental sciences often trace their origin. Physical geography formed part of Kant's program of "pragmatic cosmopolitan knowledge," in which the human being figured strictly as "an object of experience in the world, and not as a speculative subject."[29] This demarcation kept natural knowledge at a safe remove from the implications of Kant's later critical philosophy, his inquiry into the conditions of mind and world that make human knowledge possible. As pragmatic rather than critical philosophy, Kant's physical geography treated nature—both the external world and human physiology—as passive raw material for human ambitions. "[Geography] teaches us to recognize the workshops of nature in which we find ourselves—nature's first laboratory and its tools and experiments."[30] The subsequent history of environmental thought is shot through with the tension between technical mastery and critical reflection that emerged with Kant's geography.[31]

Nonetheless, Kant's status as the founder of seismology is delectably ironic. For seismology soon departed radically from the course Kant had set. The human perspective, it turned out, could not be eliminated. The very empiricism that Kant urged in the study of earthquakes drove scientists of the late eighteenth to early twentieth centuries to collect the accounts of any and all witnesses. "Only through the cooperation of all," explained one researcher, "can a satisfying result by delivered."[32] In order to continue to work with eyewitness evidence, researchers had to investigate the capacities and limitations of their witnesses—as registers of seismic impact and as observers of nature in their own right. The eminently pragmatic science of earthquakes simultaneously cultivated the critical side of Kant's project: the investigation of the conditions of human knowledge.

Yet Kant implied that disaster itself could not be an object of scientific inquiry. To treat earthquakes as disasters, rather than as a strictly geophysical phenomenon, was to fall into what modern science characterized as the deceptions of anthropomorphism: "the fatal intellectual fallacy . . . precisely the antithesis and the nullification of science."[33] Gerland defined modern seismology as a Kantian quest for pure knowledge, dignified by "the grandeur and the novelty of the task, the vast insight."[34] This sublime perspective on earthquakes was supposedly the exclusive achievement of the modern scientist. Only the uninitiated would perceive earthquakes as disasters.[35]

Since 1935, scientists have described earthquakes in the briefest of terms: an epicenter and a Richter magnitude, a neat, quotable number that can be calculated from the readings of just three measuring instruments. Charles Richter developed his magnitude scale in Southern California in the 1930s, where he and his colleagues had been trying to enlist the local population as earthquake observers. By inventing a purely instrumental definition of earthquake strength, Richter hoped to be "freed from the uncertainties of personal estimates or the accidental circumstances of reported effects."[36] His idea caught on fast, and for much of the remainder of the twentieth century it seemed that Richter had indeed, once and for all, eliminated the need for human observers and the uncertainties they introduced. With the onset of the Cold War, seismology's status swelled, fed by defense funding for the detection of nuclear tests.[37] There was no time to think about the science that was being swept away: a field of knowledge that depended on the self-reported observations of ordinary people in extraordinary situations.

The Observers

Who were the earthquake observers? When the earth trembled, all had equal claim to this title. Nonexperts were often in a position to make the best observations. Scientists were often reduced to the status of experimental subjects, reporting on the states of their own bodies and minds. The line between expert and amateur was remarkably fluid in nineteenth-century seismology. Motivations were similarly varied. Some earthquake observers were romantics, zealously exposing themselves to extreme conditions in a quest to reach the nexus of mind and nature within their own bodies. Some saw themselves as heroic explorers on the model of Alexander von Humboldt, who judged his own body his most valuable instrument.[38] Some were simply eager to be of use to science. All were participating in a culture of scientific observation and self-observation that crossed the divide between expert and lay.[39]

This culture was rooted in the new public spaces of the eighteenth and nineteenth centuries and the questions they raised about bodily discipline.

In gardens and spas, instruments were often on hand to test one's sensitivity to sunlight, local winds, and barometric pressure. The *Wettersäule*, for instance—an elaborate tower displaying temperature and air pressure—became a fixture of European urban parks and spa towns. Readers of a popular science magazine in 1912 were urged to seek for themselves "the 'geopsychic' rule governing their individual psychic lives" and to devise their own remedies.[40] As historians have argued, this culture of self-monitoring reflected anxieties about the transition to urban modernity and about European colonization in the tropics.[41] But it was also a source of new knowledge. In the age before high-precision portable instruments, satellites, and an extensive network of observatories, the geosciences learned a great deal from monitoring physiological responses to environmental conditions. The idea was to begin with one's own reactions to telluric forces, and work outward toward an understanding of the operations of the cosmos. In the words of Walt Whitman, a Humboldtian was someone "who, out of the theory of the earth and of his or her body understands by subtle analogies all other theories."[42] In these ways, seismology participated in certain nineteenth-century efforts to expand natural knowledge by moving beyond a mind-body dualism, such as romanticism, *Naturphilosophie*, sensory physiology, and Darwinian ecology. To be sure, viewing human sensibilities as a register of nature's operations ran the risk of anthropocentrism. Yet Humboldtians never suggested that environmental phenomena could be measured *solely* in terms of their effects on man. What's more, these habits of self-observation had the potential to foster a new sensitivity to environmental change. As the political theorist Jane Bennett observes, anthropomorphic analogies can remind us of "the outside-that-is-inside-too"; "a chord is struck between person and thing, and I am no longer above or outside a nonhuman 'environment.'"[43]

Observing the mutual effects of mind, body, and nature was also a way to come to terms with the expanding horizons of the nineteenth century. By tracking weather patterns, seasonal changes, or seismic waves, a curious individual could situate herself with respect to continents and oceans. Registering the personal effects of geophysical processes opened the imagination to unfamiliar geographic scales. Still, how could one be sure that certain bumps and jerks were signs of planetary convulsions and not local (or psychic) artifacts? How could one judge whether one seismic event was causally linked to another, distant one? Nineteenth-century seismology tracks an emerging curiosity about interactions between local and global scales.[44]

Just as local stories shed light on the making of a global science, the lives of individuals can illuminate the organization of a collective effort like earthquake observing. Here too I am following the lead of the historical actors. As we will see, it was often unclear whether the object of investigation in earthquake research was geophysical or human. The results frequently said as much about the social psychology of the community of observers as about local geology. Moreover, macroseismological networks were experiments not only in the scientific observation of unanticipated, fleeting events; they were equally experiments in human relations. In explicit opposition to the modern trend of bureaucratization, the charisma of individual scientists mattered greatly to their success.[45]

In these ways, nineteenth-century seismology looks unabashedly anthropocentric: it studied earthquakes by means of their human impacts and with human interests at heart. In fact, its primary variable, seismic intensity, could not be determined at all in uninhabited areas. Intensity is a measure of shaking in terms of its effects on buildings and people. Yet seismology did not stop at the human perspective. Instead, it was a project of translation: among scientists, citizens, and instruments. It successfully mediated between the technicalities of physical science and the everyday experiences of people living with environmental risk. To twenty-first-century sensibilities, the discourse of nineteenth-century seismology is a paradox: a language simultaneously scientific and vernacular.

Scientists today often despair of communicating effectively with the public about environmental risks. How, for instance, should seismologists explain that—according to the latest research on intraplate earthquakes—the absence of nearby faults is no sure sign of seismic safety? Where might scientists and citizens begin a conversation about the risk of induced seismicity from hydraulic fracturing? As one seismologist recently put it, "How do we convey the results of our research when our most recent results tell us that we know less than we used to think we knew?"[46] This book investigates the historical conditions of possibility for such a dialogue.

It is customary today to think of science as "technical" knowledge. In Thomas Kuhn's words, science requires "translation for the layman." Kuhn even claimed in *The Structure of Scientific Revolutions* that scientists owe their efficiency as problem solvers to the "unparalleled insulation of mature scientific communities from the demands of the laity and of everyday life."[47] But scientists have not always been content to express themselves in jargon. A vernacular language for science was the goal of the eighteenth-century Swedish botanist Carl Linnaeus, who designed his taxonomic system in order to "make botany easy for people without schooling or wealth."[48] Nineteenth-century medical experts often eschewed Latinisms for the terms their patients used to describe their own experiences of illness. Nineteenth-century meteorologists formulated wind scales and cloud taxonomies on the basis of the lingoes of sailors and farmers. The brief heyday of earthquake-observing networks merits attention as a path not taken—as what Ted Porter has called a "living alternative" to the increasingly technical science of the twentieth century.[49]

Knowledge and Fear

To a world still reeling from the devastation and uncertainty unleashed by the 2011 earthquake and tsunami in Japan, "disaster science" may sound like a willful delusion. Indeed, the very idea is politically suspect. In the wake of natural catastrophes, rationalizing reforms have often been a mere pretext for the centralization of power.[50] In other cases, technical responses to catastrophe have provided false security, as in the tragic failure of Japan's sea walls to defend against the recent tsunami. In this context, disaster science seems like little more than a tool for the manipulation of popular fear. Jean Baudrillard has argued in this vein that any state capable of predicting and controlling natural catastrophes would be so coercive that its citizens would *prefer* a catastrophe.[51] Yet these twentieth-century perspectives on the relationship between science and catastrophe have obscured the history of quite a different and distinctively nineteenth-century project of disaster science.

As Lorraine Daston has pointed out, the makers of science policy today tend to frame their goal as the elimination of fear. Far better, she argues, to confront fear directly in a rational manner. Daston thus calls for a "debate about the philosophy of fear, traditionally the most unphilosophical of the passions."[52] The topic of fear is not new to political philosophers, who have long debated whether fear is more likely to cause action or paralysis, to inspire the excesses of revolution or the stranglehold of reaction.[53] Perhaps it is time for historians and philosophers of science to enter this discussion. The history of seismology suggests that fear plays a dual role in the sciences. It can motivate research that may ultimately provide a measure of control over the source of fear—in the form, for instance, of storm warnings, vaccines, or antidepressants. Even in the absence of practical interventions, however, scientific knowledge can respond productively to fear by helping us recognize, and come to terms with, the limits of our control.[54]

"Fear is implanted in us as a preservative from evil," wrote Samuel Johnson in 1751; "its duty, like that of the other passions, is not to overbear reason, but to assist it."[55] Similarly, David Hume posited that the mind produced hope and fear together in proportion to the probabilities of future joy or grief, like a prism decomposing a beam of light into two colors. In this sense, fear was itself a form of knowledge about the future. It was, moreover, a sentiment appropriate to the enlightened mind. One Hume scholar has described fear as "circumspect and open-minded. . . . It is the very opposite of that complacent reliance on acquired powers, on past prestige, on previous success, which is so detrimental to further advance and to open-minded recognition of new issues as they arise."[56] Hume noted, however, that fear was not always proportional to the objective likelihood of a grievous event; sometimes it reflected a subjective degree of uncertainty about the nature and existence of an evil.[57] Fear could thus signify a well-founded expectation of misfortune or a lack of information. In this way, the age of reason recognized that fear has epistemic value.

Like these eighteenth-century philosophers, those engaged in environmental politics today are often called on to evaluate the rationality of fear. Environmental conflicts often turn on the alleged tendencies of the public to over- or underestimate environmental risk: the problems of "anxiety," on one hand, and "apathy," on the other.[58] Experts frequently dismiss popular environmental concerns in gendered terms by contrasting their own cool-headedness with the public's "hysteria."[59] Unfortunately, we seem to have no language to describe a scale of more or less realistic forebodings. The Freudian notion of "anxiety" differs from "fear" in that it has no fixed object; it appears to be a state of mind incompatible with the exercise of rationality. Yet, as Iain Wilkinson notes, anxiety is not antagonistic to reason. Rather, "it is by so traumatising us with the knowledge of our own ignorance, that anxiety functions to alert us to, and prepare us for, the threat of danger."[60] Anxiety can be a spur to scientific inquiry. Nonetheless, the effect of greater knowledge is not necessarily a stronger sense of security. Despite the optimism of the Cold War sociology of disaster, one does not always gain comfort from "the healthy exercise of rationality involved in submitting the inconceivably terrible to scientific scrutiny."[61] On the contrary, sciences of natural disaster have taught us a great deal about the scope of our ignorance. In this vein, science can generate anxiety that successfully provokes public debate and political reform.[62] This book takes up the history of seismology with these concerns in mind. Following historians of the emotions, it considers how earthquake fears have been constructed and suppressed in different places at different times.[63] It recovers a science that had no intention of suppressing fear, but sought instead to learn from it.

Organization

In order to capture both the local and global dimensions of earthquake science, the chapters that follow alternate in scale. Half of them treat local experiments in planetary science. These four cases—Scotland, Switzerland, imperial Austria, and California—were, to my knowledge, the only places where networks of ordinary citizens contributed decisively to the emergence of modern seismology.[64] The scale of these public efforts was remarkable: the Swiss Earthquake Commission gathered approximately seven thousand reports between 1878 and 1910, while the Austrians had over 1,700 observers reporting from all sixteen crown lands.[65] Elsewhere, as in Japan and Italy, seismological observations were made primarily by men of science; occasionally by civil servants like stationmasters, telegraph operators, or postmasters; or primarily by instruments.[66] These four episodes span the "long nineteenth century," from the first glimpse of a technocratic regime in the Napoleonic Era through the triumph of technocracy in the wake of the First World War.[67] Alternating with these local experiments are chapters that follow the international circulation of the stories of earthquake witnesses as they were reconstituted as evidence for a global science of disaster. One might object that an earthquake in Japan is hardly the same scientific object as an earthquake in Switzerland: a potential catastrophe in one case, often a mere curiosity in the other. Yet the very possibility of such comparisons is a consequence of nineteenth-century seismology's expansive framework. To the extent that catastrophes are experienced as exceptions to a normal course of events, they resist comparison. Other catastrophes pass virtually unnoticed: they are experienced as "normal" hazards, as part of the "acceptable" risk of modern industrial life.[68] Nineteenth-century seismology resisted these extremes. By charting an entire spectrum of experiences of hazard, from the mundane to the overpowering, it cleared a space for sustainable adaptations.

ONE

The Human Seismograph

The word "seismology" was coined in the 1850s, not long after the word "scientist"—both harbingers of a new age of technical expertise.[1] Earthquakes, however, did not fit easily into the emerging rubric of professional science, not least because they forced scholars to rely on the testimony of common folk. Already in the sixteenth century, when stories of the New World were first circulating in Europe, Michel de Montaigne remarked that earthquakes compelled Europeans to trust the word of "barbarians." In a crucial twist, however, Montaigne suggested that the barbarian might prove the more able witness: "a simple, crude fellow—a character fit to bear true witness; for clever people observe more things and more curiously, but they interpret them; and to lend weight and conviction to their interpretation, they cannot help altering history a little."[2] Echoes of Montaigne's charitable perspective could be found in subsequent European studies of earthquakes, in the virtues sometimes attributed to untutored observers. Well after Montaigne's death, earthquakes were still widely discussed across divides of birth and education. Eighteenth-century sermons and news articles engaged the public in scientific and theological debates about earthquakes.[3] Accounts of the New Madrid earthquakes of 1811–12 became "a form of conversation," in which settlers modeled their descriptions of tremors on narratives of sickness and health.[4] In the early nineteenth century, however, this inclusive conversation was breaking down. Earthquakes figured counterintuitively in new geophysical theories as the effects of elusive electrical forces.[5] Seismology seemed ready to become the esoteric subject of expert knowledge.

The history of seismology since 1755 is traditionally seen as a progressive liberation of natural knowledge from the subjective impressions of earthquake victims. After the Staffordshire quake of 1777, for instance, Samuel Johnson warned that the event would "be much exaggerated in popular talk: for, in the first place, the common people do not accurately adapt their thoughts to the objects; nor, secondly, do they accurately adapt their words to their thoughts: they do not mean to lie; but, taking no pains to be exact, they give you very false accounts. A great part of their language is proverbial. If any thing rocks at all, they say it rocks like a cradle; and in this way they go on."[6] In this vein, the Calabrian earthquakes of 1783–84 typically figure as the first to have been described scientifically. Their importance—according to the most illustrious of nineteenth-century geologists, Charles Lyell—"arises from the circumstance, that Calabria is the only spot hitherto visited, both during and after the convulsions, by men possessing sufficient leisure, zeal, and scientific information, to enable them to collect and describe with accuracy the physical facts which throw light on geological questions."[7] Lyell made it seem self-evident that scientific descriptions of earthquakes could come only from men of science.

What Girls Will Tell

Others hoped to dispense with human witnesses entirely. Robert Mallet's study of the Neapolitan temblor of 1857, subtitled *First Principles of Observational Seismology*, is often cited as the founding work of empirical macroseismology.[8] Mallet, a British civil engineer, showed how cracks in masonry and overturned objects could be used to infer the direction from which seismic waves had propagated. He based his research almost exclusively on architectural damage, and found little use for eyewitness testimony. Indeed, Mallet complained to Lyell of the lazy Neapolitan savants; he saw no reason to believe their reports of nightly aftershocks, since he himself had slept just fine.[9] In Mallet's judgment, lay observers lacked the "observational tact and largeness [*sic*] of a disciplined imagination and eye that are amongst the accomplishments of the physical field-geologist."[10] He repeatedly insisted that the untrained eye simply failed to see. In the past, earthquakes had been studied in the absence of "any guiding hypothesis, of any distinct idea of what an earthquake really is, of any notion of what facts might have been of scientific importance to observe, and what were merely highly striking or alarming . . . —in the want of all these, as well as of any calmness or unexaggerative [*sic*] observation during such alarming visitations, few facts of the character and precision requisite to render them of value to science can be collected with certainty. The true observation of earthquake phenomena is yet to be commenced. . . . The staple of earthquake stories, in fact, consists of gossip made up of the most unusual, violent or odd accidents that befell men, animals, or structures, rather than of the phenomenon itself."[11]

Mallet's disdain for the "gossip" of human observers was echoed by many British, Italian, and Japanese contributors to seismology in the second half of the nineteenth century. When the Italian Earthquake Commission was established in 1883, it focused on constructing seismographic observatories, not a permanent network of observers.[12] This was in keeping with the move to exclude amateur observers from Italian astronomy in the aftermath of national unification.[13] The private network developed by Count Michele Stefano de Rossi in the 1870s consisted entirely of fellow gentlemen-naturalists, nearly half of whom owned seismoscopes or seismographs.[14] In Japan, where John Milne and Fusakichi Omori pioneered instrumental seismology in the 1880s and 1890s, lay observers played little role in earthquake investigations. Milne seems to have given up on the collection of felt reports after his investigation of the Yokohama earthquake of 1880, for which he sent out five hundred questionnaires and received only twenty-six responses.[15] A leading British seismologist speculated that such outreach efforts failed in Japan in part because Japanese earthquakes were so frequent. "The detailed inquiries which are possible in Great Britain can hardly be made in a country in which the recollection of one shock is soon after dimmed or erased by the occurrence of another or many more in rapid succession."[16] As Milne put it, the Japanese chatted about earthquakes the way the English chatted about the weather.[17] In any case, he had little patience for the analysis of felt reports: "Attempts to find out what sensations were experienced by the people at the time of the shock are unsatisfactory. People questioned will tell trivial circumstances—how they tumbled from the top to the bottom of the stairs whilst hurrying to get out of doors—girls tell how they began to cry, etc."[18] Likewise, many of Japan's historical sources on earthquakes were judged by Westerners to be "of a trivial character"—illustrated, for instance, by a 1707 account of a young man drinking with friends in a teahouse when an earthquake knocked him down a ladder into a barrel of pickled radishes.[19]

It was, therefore, against his better judgment that Mallet began to collect witnesses' descriptions of the sounds accompanying the Italian tremors. He warned his reader that reports of sounds, unlike the hard evidence of architectural damage, were compromised by numerous distortions: "Echoes, the disturbance of local noises at the moment, the uncertainty with which the ear judges of direction and sound, the evanescence of the phenomenon, and the difficulties inseparable from trusting to merely collected information of often incompetent observers, or unfaithful narrators, who observed under alarm, must ever deprive sound phenomena (except when heard by the physicist himself) of the unerring certainty of deduction, that belongs to the mechanical problems, presented by the phenomena left after the shock." With his insistence on the "unerring certainty of deduction" and the superior observational skill of "the physicist himself," Mallet verged on a caricature of the Victorian expert—the Sherlock Holmes of earthquakes.

And yet, to his own surprise, Mallet found that reports of sounds, despite their ambiguities, "are not without their seismic significance." Suddenly the voices of the earthquake victims themselves intruded into his treatise. On the periphery of the area in which sounds were audible, observers described a "low, grating, heavy, sighing rush," while those in the center reported a rumbling or (here Mallet felt compelled to quote the original Italian), "rombo, rumore di carozzo . . . fischio, sospiramente. . . . These descriptions, aided by the expressive gesticulation and imitative powers of the narrators, conveyed a far more exact notion of the sounds heard, and of the relative times in which they were heard, than I can hope to transmit in writing. . . . They were collected in my progress . . . without much idea of their leading to any very distinct or valuable conclusion. The result however now appears to support the conclusions arrived at, from the rigid methods of tracing the origin out from the wave-paths."[20] These rushes and rumblings, then, only found a place in Mallet's treatise because they corroborated his theory. Despite his "physicist's" instincts, Mallet reluctantly acknowledged the force of the testimony of earthquake witnesses, full of meanings that escaped translation.

A "Damned" Science

Despite the confidence of these new experts, the barriers to establishing seismology as a professional science remained steep throughout the nineteenth century. One problem was that there were no limits to the kinds of observations that might be relevant to earthquake research. According to theories widely accepted in the nineteenth century, earthquakes might be triggered by volcanoes, barometric fluctuations, atmospheric electricity, geomagnetism, humidity, or the positions of celestial bodies. Analysis of the course and impact of earthquakes required an even wider variety of data, from the geological to the zoological and psychological. Geographically as well, earthquakes presented no clear limits. Speculation was rife over the apparent coincidence of earthquakes across vast distances, and the subterranean channels that might account for such teleconnections. This uncertainty about what constituted seismology's evidence brought the field into precarious contact with such "pseudosciences" as astrology and spiritualism.

Earthquakes figured prominently, for example, in the manifestos of Charles Fort, one of the most celebrated skeptics of twentieth-century science. His followers in the International Fortean Organization continue to this day to hold annual conferences devoted to "anomalous phenomena." Fort's first major publication, *The Book of the Damned* (1919), was "a procession of data that Science has excluded." "Damned," in Fort's vocabulary, meant rejected by "Dogmatic Science." Fort wrote in an absurdist style that lurched between bombastic pronouncements and nuggets of humble common sense. Among the "damned" were correlations he uncovered between earthquakes and astronomical phenomena. To collect these "lost souls," he mined the reams of earthquake catalogs and observational reports published in scientific journals since the late eighteenth century. He was also a zealous collector of newspaper clippings—but then, so was any self-respecting seismologist circa 1900. The research process he described would have been familiar to any of them: "I have gone into the outer darkness of scientific transactions and proceedings, ultra-respectable, but covered with the dust of disregard. I have descended into journalism. I have come back with the quasi-souls of lost data."[21] Earthquakes suited Fort's hunt for "the damned" precisely because of the minor explosion of seismological observations in the nineteenth century.

The First Seismographs

Seismology's credibility hinged on the design of instruments to measure physical phenomena independently of their human impacts—to complement, not replace, the observations of human witnesses. While seismoscopes had been used since ancient times in China, modern European efforts to build them began in earnest in Italy in the late eighteenth century.[22] An eighteenth-century seismoscope consisted of a hanging pendulum attached to a stylus that made a mark when set in oscillation. Not until the 1840s were seismoscopes capable, in principle, of measuring the displacement of the ground during a tremor. In practice, they often did not work. The inverted-pendulum seismoscopes installed during an earthquake swarm in Comrie, Scotland, in the 1840s were state of the art. Yet residents of Comrie counted sixty shocks in one year, while the seismoscopes only recorded three. More useful instruments were developed by British engineers in Japan in the 1880s. These "seismographs" were able to trace the development of earthquake waves over time. They recorded pendulum movements on a revolving drum attached to a clockwork mechanism, producing a curve from which the waves' period and amplitude could be measured. Such traces became all the more interesting in 1889, when a German astronomer discovered by accident that his instruments in Potsdam had inadvertently recorded a strong earthquake near Tokyo: they had detected waves from a spot 5,500 miles away on the earth's surface. The realization that earthquakes could be mechanically recorded at such a distance inspired many new seismographic inventions at the end of the nineteenth century.

But were these ingenious mechanisms any more reliable than human witnesses? The first generation of seismographs was subject to small "self-oscillations" due to lack of damping, making it hard to record long-period motions. Few were able to record vertical motion. With further improvements, seismographs became more trustworthy; but humans did not necessarily become less so. At the turn of the twentieth century, scientists studying seismographic curves discovered two kinds of seismic wave traveling through the earth's core: a shorter period "primary" (compression) wave and a longer period "secondary" (transverse) wave. Later research also found several varieties of surface wave, which pass only through the crust. It soon became clear that many apparent contradictions in the felt reports of years past—conflicting accounts of jolts and rolling motion, "horizontal" and "vertical" shocks—were resolved if one admitted that people, too, could feel the differences between these types of wave.[23] Moreover, recent evidence has suggested that humans may be more acute observers than even the latest computerized seismographs. A 2008 study concludes that "events with a magnitude smaller than 1, and even negative magnitudes [meaning, on the logarithmic scale, fractions of a unit shock], can be felt, thus making the human being an instrument eventually much more sensitive than monitoring networks."[24]

Nonetheless, by the early twentieth century, the availability of seismographs allowed many earth scientists to dream of turning their discipline into a quantitative, objective science, modeled on physics. They transformed what counted as evidence of the earth's history. Out went data filtered by human bodies; in came the "hard" evidence of seismographs and accelerometers. This was the moment when scientists began to distinguish the "new seismology" from the "old." Among the achievements of the new seismology was the ability to use instrumental traces of the passage of seismic waves as clues to the internal structure of the earth. The seismograph became a telescope trained on the earth's hidden depths, where it revealed a core of iron buried under a mantle of rock.[25] The next "revolution" came with the acceptance of plate tectonic theory, the "new geology," in the 1960s.[26] Plate tectonics overthrew the belief that the positions of the continents and oceans were permanently fixed. Earthquakes took on new significance as evidence that the movement of the continental plates was still in progress. In principle, macroseismic observations should also have acquired new value as clues to the detailed distribution of faults and thus the contours of the plates.[27] On the global scale of the new plate tectonics, however, details of local seismicity tended to fade from scientists' attention. They took little interest in intraplate earthquakes, which remain poorly understood—though potentially destructive. At the same time, the globalization of news in the early twentieth century made reports of local tremors ever less likely to find their way into print. Working knowledge of the geography of seismic hazard was fast disappearing.

Local Earthquakes and Global Tectonics

Such knowledge had been built, inch by inch and tremor by tremor, by countless unrecognized observers. But the method behind it was associated above all with one man, the Austrian geologist Eduard Suess.[28] A veteran of the 1848 revolutions, Suess firmly believed that scientists had a duty to serve society. Geologists were responsible for keeping their communities safe and healthy from the ground up. He encapsulated his sense of the interdependence of environmental and social conditions in the subtitle of his 1867 study, *The Ground of the City of Vienna: According to Its Manner of Formation, Composition, and Its Relationship to Civic Life.* Suess turned projects of urban improvement into opportunities for geological research, and his conclusions bore directly on the lives of the city's residents. He was the first to discover, for instance, that a portion of the drinking-water supply in several neighborhoods came from the drainage of cemeteries.[29] Likewise, Suess's engagement with the regulation of the Danube stemmed from his experience of the flood of 1862, when he saw the poorest residents of his neighborhood fleeing their ground-floor apartments and witnessed the drowning of an entire herd of oxen.[30] As he argued in *The Ground of the City of Vienna,* engineering a healthy city was a municipal responsibility, not a private one. Even today, Suess's achievements are recalled in references to Vienna's "Sueß-Wasser" (freshwater). In 1873 Suess was elected as a liberal representative to the Austrian parliament.

Three days into that new year, just before seven in the evening, Suess was seated at his writing desk in Vienna when he felt a sudden jerk. Two hundred and eighty-three years earlier, an earthquake had famously damaged Saint Stephen's cathedral, in the center of Vienna. As Suess would soon learn, the heaviest damages in both cases occurred in precisely the same place. It could hardly be a coincidence.[31] First thing the next morning, Suess was on the road, stopping to make inquiries in villages across Lower Austria. "Along the entire line of this journey I received reports of the phenomenon."[32] Further reports arrived in response to a request he printed in the major Viennese papers, amounting to observations from a total of 203 localities. Suess also drew on meticulous firsthand observation of the geology of the region, gleaned from months of hiking and sketching (see figure 1.1), and he combed provincial archives for accounts of past quakes. Suess soon came to be seen as among the founders of a new school of seismological research, based on what was termed the "monographic method": "this investigates an earthquake in and for itself and is concerned to study its unique aspect and thence to draw conclusions about the various factors involved in this quake."[33] This style of research blended expert and popular knowledge: as the author of a study of the 1858 earthquake in Žilina (Slovakia; German: Sillein) had argued, in the study of earthquakes it was "very difficult . . . to specify the limit where someone starts or stops being an expert."[34]

Suess's contemporaries studied earthquakes as symptoms of volcanic forces, mine explosions, meteor strikes, the sinking deltas of lakes, or the collapse of subterranean hollow strata; they also probed the possibility of atmospheric or astronomical influences. For Suess, the significance of earthquakes was altogether different: he sought "the relationship between the structure of the earth's crust in a specific region and the direction and nature of the shocks."[35] The earthquake of 1873 would become an early and particularly convincing piece of evidence for Suess's epochal theory of global "tectonics"—the study of horizontal stresses in the earth's crust and the mountains and basins they produce.

As Suess saw it, the major stumbling block of nineteenth-century geology was the question of mountain formation. Charles Lyell believed he had rid geology once and for all of the biblical deluge and with it anything that smelled of "catastrophism." The new orthodoxy was that the earth's surface had been shaped over a previously inconceivable length of time by the slow, steady processes of sedimentation and erosion. The gradual rise and fall of the continents produced climatic changes, which in turn explained the appearance and disappearance of new species in the layers of the geological record. This vision formed the backdrop to Darwin's theory of evolution, and to most leading geologists it seemed impregnable. Yet Suess saw a contradiction. Geologists were equally convinced that the earth as a whole was contracting, as it cooled from an early molten state. On this view, mountains formed as the planet's crust wrinkled like the skin of desiccating apple.

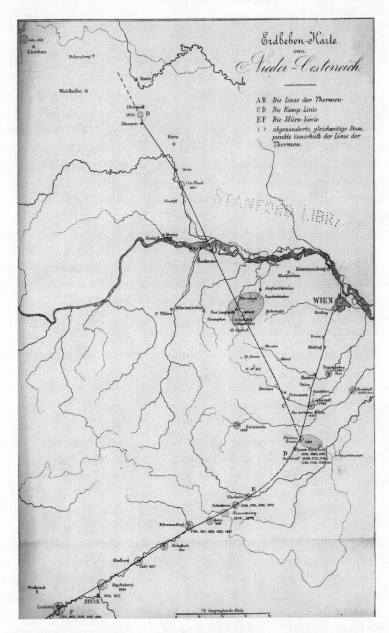

Fig. 1.1. Suess's seismic map of Lower Austria. Eduard Suess, *Die Erdbeben Nieder-Österreichs* (Vienna: k. k. Hof- und Staatsdruckerei, 1873), plate 1.

Mountains were thus the result of the tangential stresses of contraction, not sedimentation.[36]

Suess's study of Alpine earthquakes added another wrinkle to this situation. As he combined his data on the tremor of 1873 with historical records of the region's past earthquakes, Suess determined that seismic events in Lower Austria could be mapped along a single line. He concluded that these movements resulted not from forces of uplift but from "fractures or faults or some other discontinuity in the earth's crust."[37] Earthquakes in the Alps propagated perpendicularly to the chain of the mountains and were felt further away to the north than to the south. Other European mountain chains showed similar traits. Here was evidence that the Alps arose not simply through local compression but as the result of thrusting, as the entire European continent moved northward. According to his colleagues, however, continents did not move; they only rose and fell at a snail's pace. Viewed over the course of millennia, earthquakes could be but small fluctuations within this stable cycle. It would take another century for geologists to give up their belief in the fixity of the continents.[38] In the meantime, Suess advanced his notion of mobile continents as part of a bid to overcome, in his words, the "quietism" of the geology of his day, which lavished attention on "the little polyp building up the coral reef, and the raindrop hollowing out the stone." Suess thus accused his colleagues of confining their thinking to the puny human scale. Against this fondness for what Lyell called "trifling means," Suess mustered evidence for the formative role of catastrophic events in the earth's history. Much of this evidence, like the observations he gathered in 1873, consisted of the accounts of eyewitnesses, often written for purposes remote from natural science.[39] As a colleague put it in 1913, "since 1880 the predominant influence of M. Suess has again made the word 'cataclysm' acceptable in scientific circles."[40]

Suess's investigation of earthquakes in Lower Austria thus provided crucial early support for his tectonic theory, which continued to face resistance into the twentieth century. It also suggested a research program. In order to demonstrate that most earthquakes were likewise tectonic in origin, and to identify their associated faults, it was necessary to multiply such studies. Geologists would need to correlate the location and direction of as many earthquakes in as many regions as possible with features of local geology. Such observations held out the promise of identifying regions of elevated seismic hazard. Basic geology would go hand in hand with disaster mitigation. The tectonic hypothesis thus sprang from and supported a reliance on the earthquake observations of ordinary locals. Would the public comply?

TWO

The Planet in the Village:
Comrie, Scotland, 1788–1897

The village of Comrie lies just south of the Scottish Highlands, where three rivers meet in a green and gentle valley; the name Comrie supposedly derives from the Gaelic for "confluence." Its population peaked at three thousand in the 1790s. By then, according to contemporaries, the locals' "attachment to the soil" was already wearing thin: "the people are pouring down in numbers every season to the adjacent villages and towns in quest of labour and of bread."[1] The consolidation of small farms and the absence of a "staple trade" discouraged the young from staying; a nearby woolen mill, a prime employer, closed in 1866. By 1831 the population had fallen to 2,622, and by 1881 it was a mere 1,871. In these conditions, the locals had high hopes for tourism. The nearby spring had long drawn crowds for its healing powers, having been blessed by Saint Fillan in the ninth century. Sufferers of rheumatism were instructed to climb a hill and sit on a rock known as "the saint's chair," after which they were to be pulled down the hill by their legs. Now, in the nineteenth century, locals praised Comrie's scenery as "sublime," as a Scottish Switzerland. By the 1880s, there were two coach connections daily to the train station at Crieff, and a rail line finally reached Comrie in 1893.[2]

Comrie was, nonetheless, as quiet a village as one could find in central Scotland in the nineteenth century. Its tranquility seemed guaranteed when it became a center of the temperance movement. As one Temperance League lecturer noted in the 1880s, "No better spot could be selected for earnest thought and consecration . . . it sleeps quietly amongst an amphitheatre of hills. No railway train or noisy factory disturbs the stillness. Everything suggests peace and repose." A native son of the nineteenth century imagined the village back in the 1790s as "a most primitive place . . . unprofaned not only by railways but by stage-coaches; where the appearance of a gig or post-chaise must have been an event signaling a whole week; where two-thirds of the people could speak no English; and which has not as yet been tossed into importance by a single shock of an earthquake."[3]

The earthquake records begin in 1788. Initially, the tremors caused "great alarm, especially one which occurred on a Sabbath while the congregation was assembled."[4] Later researchers note two oddities of the Comrie earthquakes: the sheer number of slight tremors (an estimated eight to nine hundred within six months) and the small area in which they were sensible (no more than three miles away).[5] For the next thirteen years, these shocks were patiently recorded by two local clergymen, one of whom earned the moniker "Secretary to the Earthquakes." But they attracted little interest beyond the village. Then, for the first four decades of the nineteenth century, the earth lay quietly beneath Comrie. At last, on 23 October 1839, Comrie was the center of an earthquake felt across much of Scotland. "The consternation was such that the people ran out of their houses, and, late as was the hour, many assembled for prayer in the Secession Meeting-house, where religious exercises were continued until three in the morning"—during which two further shocks were felt. Robert Chambers's *Edinburgh Journal* reported that the shaking even burst a dam supplying water to factories in the neighboring county of Stirlingshire.[6] This event finally stirred Comrie to action. Two locals began independently to collect observations: Peter Macfarlane, the town's postmaster, and James Drummond, a cobbler. Macfarlane, also known as a temperance leader, drew up a crude yet innovative scale to weigh the strength of the shocks, setting the 23 October quake as a 10 and rating subsequent ones relative to it.[7] It is estimated that he recorded three hundred earthquakes in his thirty-six years as an observer.[8] Drummond, as we'll see, was by far the more ambitious of the two men. He was described by a neighbor after his death as a man with "much interest in antiquarian lore," whose "racy conversation" ranged across "Comrie earthquakes, Roman invasions, and opinions of political men and measures."[9]

Meanwhile, the Comrie earthquake of 1839 drew distinguished men of science to the humble village. The British Association for the Advancement of Science initiated its own investigation, beginning with the formation of an Earthquake Committee. In those years, the BAAS was still finding its feet. Founded in a reformist, public-minded spirit in 1831, the association was growing increasingly elitist. Its leaders were now at pains to police the boundary between expert and popular knowledge.[10] Yet the chair of the BAAS committee, David Milne (no relation to the later seismologist John Milne) was not a professional man of science; he was an advocate at the

Scottish bar.[11] Still, Milne's geological research had won him membership in the Geological Society of London and the esteem of the major British geologists of his generation, as his exchanges with the likes of Lyell and Darwin attest. And his wealth gave him the freedom to indulge his curiosity about Scottish earthquakes to his heart's content.

Milne was delighted at news of the 1839 shock. It was, he insisted, a "matter of regret and reproach" that British men of science had not investigated "volcanic action" on their home turf in the past. They were wrong to be deterred by the absence of volcanoes: "It is in foreign countries, that the British geologist has hitherto been in the practice of searching for and observing the indicia of volcanic action;—for it seems to have been thought that the phenomena were unsatisfactory or unworthy of attention, unless accompanied with eruption. But if, as is now generally admitted, active volcanoes serve the purpose of safety-valves, to give ready vent to the subterranean forces, the effect of these forces on the earth's surface ought to be greater where no volcanoes exist."[12] British colleagues concurred that the earthquake was a stroke of good fortune. Upon learning of Milne's growing collection of observations of Scottish earthquakes, Charles Darwin declared himself "astonished at their frequency in that quiet country, as anyone would have called it." The geologist Charles Daubeny was pleased to hear that "jets of smoke or steam" had been observed in connection with certain of these events: "I consider it a fortunate circumstance that a case of the kind should have occurred so near home; for the localities in which the evolution of flame from the earth is stated to occur"—such as the mountains of Albania and the Caspian Sea—"are for the most part inconveniently situated for exact experiments."[13]

Milne, the geologist-attorney, once described his investigative method to Darwin as "precognition"—the Scottish legal term for the collection of witness testimony in advance of a trial.[14] But there is little evidence of his activity in this respect. His lasting achievement was to publish for posterity the results obtained by Macfarlane and Drummond. These he saw fit to edit considerably, since "it has not been thought necessary to include a description of *all* the effects related of such shocks." Common observations such as damage to chimneys would merely "swell the register." "The object has been, rather to select and exhibit effects which seem calculated to throw light on the nature and causes of earthquake-shocks."[15] In this way, details crucial to the determination of intensity—and thus the mapping of hazard—were discarded, in favor of evidence that was clearly relevant to available theories. Yet the phenomena Milne chose to preserve may surprise modern readers: sounds (of two types, an explosion and a rushing or whizzing), unusual weather, seasickness, and frequent reports of two distinct sensations: the first tremulous, the second concussive (what "the country people" called "the 'thud'").

Meanwhile, the BAAS's Earthquake Committee busied itself with preparing an alternative to human reports. J. D. Forbes, professor of natural philosophy at Aberdeen, designed an instrument that would record earthquakes by means of an inverted pendulum; David Milne named it a "seismometer."[16] By 1841 the committee had supplied Comrie with ten such instruments, but it soon became clear that the machines were less sensitive than Comrie's residents themselves. In the first half of 1841, for instance, the seismometers registered only two tremors, while their human keepers felt twenty-seven.[17] Metropolitan science had yet to prove its value in Comrie.

Equally unclear was the significance of Comrie's earthquakes to the global science of geology. At the outset, Milne was eager to test his hypothesis that these tremors were a direct result of volcanic eruptions in Italy, via some subterranean connection. It was commonly assumed that earthquakes in Britain must be of foreign origin; newspapers reported them thus as early as 1824 and as late as 1957.[18] Fortunately, Milne soon had the benefit of Charles Darwin's experiences abroad: "There are many regions in which earthquakes take place every three & four days," Darwin wrote him, "& after the severer shocks, the ground trembles almost half-hourly for months.—If therefore you had a list of the earthquakes of two or three of these districts, it is almost certain some of them would coincide with those in Scotland, without any other connection than mere chance." Milne replied: "I am much obliged to you for warning me against too ready a belief in the connection between earthquakes felt on opposite sides of the equator . . . I am beginning to be sceptical as to any connection between our Scotch earthquakes and those in Italy."[19] On the other hand, Darwin lent support to Milne's suspicion that earthquakes bore some relationship to weather. The veteran of the *Beagle* voyage cited South American natives on this point. "Under the very peculiar climate of Northern Chile, the belief of the inhabitants in such connections can hardly, in my opinion, be founded in error."[20] Meanwhile, Milne was on to a new theory: that the earthquakes were caused by the conversion of underground water into steam, which might also generate electrical activity. In short, the proper geographic scale on which to study earthquakes was wholly undecided.

In fact, there was nothing resembling an expert consensus on what constituted the pertinent "details" of earthquake observations. Weather, sounds, the behavior of wells and springs, atmospheric electricity, steam and smoke, even seasickness—any of these and more were potentially evidence. And so the question arose: on what criteria did Milne continue to prune Macfarlane's and Drummond's observations in order to illuminate the quakes' "causes"?

Therein lay the seeds of controversy. Drummond seems to have been a pugnacious type, suspicious of the London experts and jealous of his local status as a scientific authority. He recognized that the BAAS committee had left itself open to a charge of bias in their selection of earthquake accounts for publication. He accused the experts of being wedded to a Plutonist conception of earthquakes as the expression of mountain uplift, "Observing that those who were taking scientific notes of the phenomenon ignored all facts which did not agree with their preconceived opinions on the subject, I resolved to discard all fancies and theories, and to try to ascertain from facts alone where the disturbance originated." Thus did Drummond make a virtue of his amateur status, casting himself as a plain-talking, unbiased observer. For his own theory of the earthquakes' origin, however, he offered little empirical evidence: "Suffice it to say, that no preconceptions of my own led me to that conclusion." Apparently, Drummond had initially decided to give his earthquake reports to a proper "man of science" and await the expert's conclusion. However, to his shock and dismay, the expert to whom he was referred "coolly offered me a bribe to let him have the credit of my labours. This I indignantly rejected." At that point, Drummond decided to publicize his own conclusions in a local newspaper, identifying the seat of the earthquakes and theorizing that they were galvanic in origin (he later specified that they were due to atmospheric electricity). "The offer of a bribe forced me to publish prematurely, and also to attempt to solve the problem of earthquakes myself." Drummond then tried to make a formal accusation of bribery against the unnamed BAAS expert, in the presence of David Milne. Before he could do so, both men walked out on him. The BAAS had likewise turned a blind eye to his observations and conclusions. "Considering the position in society Mr. Milne occupies, and his reputation as a scientific man, it was not likely that he would readily acknowledge himself outrun in a scientific race by one in the humble station of life in which I am placed. . . . Can it be wondered at, then, that the Committee paid no regard to my opinion,—indeed, completely ignored me."[21] Drummond did not stop at the charge of attempted bribery. He also claimed that the BAAS was neglecting to observe atmospheric phenomena in conjunction with the tremors. As "Plutonians," they "ignored all phases [of earthquakes] that are irreconcilable with their ideas."[22]

Having cast himself as a champion of common sense, honest empiricism, and local knowledge, Drummond turned to his neighbors to furnish evidence in support of his theory. In an article for the *Philosophical Magazine* published in 1842, he appealed to the superiority of local observers, who, "from greater habitude . . . are better able to distinguish the shocks."[23] He called on Comrie's residents to "pay particular attention" to the winds accompanying shocks and challenged them to produce a single observation that contradicted his theory: "As this was an indispensable condition of my theory, to guide the people of Comrie in their observations, I stated that if a great shock occurred when the lower regions of the air were clear, so as to permit distant objects to be clearly seen, then I would give up my theory as untenable. That was in 1842, and no year has passed, and no earthquake has taken place, without my reminding some one of the challenge. It is now more than thirty years since I gave the challenge, and during that long period my theory has triumphantly stood the test."[24] Drummond even extended this populist appeal to readers around the world. "I bid the reader pay no attention to any theory whatever, but simply observe facts for himself. . . . All I ask from the people of earthquake countries is justice. I bid him regard no theories, but simply observe facts for themselves." Drummond was willing to admit his amateur status, but he refused to defer to the "man of science." Instead he called on the expertise of the "practical electrician": "I have no doubt, when a practical electrician takes up the subject, he will give far more information than I am able to give."[25] This appeal was well calculated: electrical engineers at the time cast themselves as the protectors of British empiricism, against the "wild hypotheses" of mathematical physicists.[26] Ambitious engineers like Cromwell Varley were trying to parlay their knowledge of telegraph cables into expertise about geomagnetic, meteorological, and seismological phenomena.[27] Drummond likewise stressed the practical benefits his explanation offered. While it might not permit the prediction of earthquakes, proper observation of the atmosphere could make known "positively when all danger is past." Again, Drummond contrasted his "practical" motivation with the theoretical contortions of the scientific elite: "It was only a very strong sense of the great practical importance of the facts to be gathered from a careful study of the varied phenomena of earthquakes, that induced me to continue my labours in the midst of so much opposition."[28]

After 1844, the BAAS ignored earthquakes altogether. Comrie was undisturbed by seismic phenomena for many years, and its villagers fell back into their quiet way of life. In 1869, it looked as if a new swarm might be arriving. The BAAS revived the Earthquake Committee and began construction

Fig. 2.1. "Earthquake House," Comrie, 1874: "the first ever purpose-built seismic observatory" (http://www.panoramio.com/photo/40904671). It fell out of use between approximately 1900 and 1988, when it was restored and furnished with an up-to-date seismograph. Today, it's a tourist draw.

of what appears to have been the first purpose-built seismic observatory, completed in Comrie in 1874 (figure 2.1). Yet the expected earthquakes never came. In 1882, the local amateur historian Samuel Carment concluded that Comrie's earthquakes had been consigned to history: "they are now grown rather stiff in their joints by reason of age, and can only sit in their subterranean caverns and grin at the passers-by, and bite their long nails in impotent rage."[29] Aside from light shakings in 1894, 1895, and 1898, Comrie kept its peace.[30] Meanwhile, the seismic threat began to announce itself closer to the heart of the British Empire.

"British Earthquakes"

Comrie's quick shot to fame as an "earthquake center" was a function in part of the exotic quality of earthquakes in the British Isles. Where tremors were so rare, even weak ones drew numerous and elaborate reports. One witness even erected a stone monument to mark an 1840 earthquake, of which he had heard some rumblings but felt not even a bump.[31] To the British, the very word "earthquake" conjured a scent of Mediterranean dust and an echo of Catholic supplications: they appeared to be a scourge reserved for "tropical" lands. In the 1850s, the popular historian Henry Thomas Buckle argued that earthquake-prone lands were doomed to mental backwardness, for "there grow up among the people those feelings of awe, and of helplessness, on which all superstition is based."[32] Buckle's purpose was to prove that no earthly environment was more conducive to human progress than England's.

Certain of the terra firma of their island home, the British felt safe to indulge in seismic fantasies. In 1848, as revolutions erupted across the European continent, Londoners—spared the political turmoil—entertained themselves instead with a seismic upheaval. The Cyclorama staged a spectacular panorama of the great Lisbon earthquake of 1755, replete with moving scenery, atmospheric effects, and offstage screams. Such an amusement was amusing because the threat it represented seemed so foreign.

From time to time, however, the British were reminded that the stability of their isles was not guaranteed. No less an authority than Charles Dickens warned the public in 1860 that "we enjoy no immunity from the most sudden, the most irresistible, the most destructive of nature's powers. Another such shock as the Lisbon earthquake may happen this or next year. It may not happen in this country, but it may originate beneath our own metropolis." Still, the British remained confident in their belief that "earthquakes are meant for other countries." As the London *Times* put it in 1863, "Here, in these cooler climes, with more reasonable temperaments, and under a purer faith, it is hoped that we do not need this awful language. The ALMIGHTY footfall is soft here, even in the earthquake and the storm."[33] This time, however, these articles of faith were being repeated for ironic effect. As this editorial was printed, reports were pouring into the press office of a tremor felt across eighty-five thousand square miles of England.[34]

The *Times* printed a request for observations of the 1863 earthquake from the meteorologist E. J. Lowe, and two hundred and fifty-one individuals replied. Their observations formed the basis for a table Lowe published the following year: heavily discursive and laden with footnotes, it ran to thirty-four pages. Three months later, Lowe lectured on the history of "British earthquakes" at the Mansfield Mechanics Institute, where he urged his working-class audience to recognize the true instability of the ground beneath them. "No doubt more earthquakes occur than we are aware of, as unless the shocks happen to be tolerably severe, no notice is taken of them. . . . Over all Western Europe earthquakes are incessantly at work, the average in the zone being thirty-three shocks a year, or one in every eleven days" (a statistic he likely drew from the seismic cataloger Alexis Perrey). Lowe himself had conducted "experiments" that showed that faint sounds were more audible in the dark, and weak movements of the ground more sensible when lying down.[35]

Among the Englishmen eager to share their seismic observations in 1863 was Charles Dickens. Dickens had evidently conducted a fair share of research, for he was able to cite a report in the *Philosophical Transactions of the Royal Society* from 1683 on the behavior of a barometer during an earthquake. He cast his own observations with appropriate precision: "I was awakened by a violent swaying of my bedstead from side to side, accompanied by a singular heaving motion. It was exactly as if some great beast had been crouching asleep under the bedstead and were now shaking itself and trying to rise. The time by my watch was twenty minutes past three, and I suppose the shock to have lasted nearly a minute. The bedstead, a large iron one, standing nearly north and south, appeared to me to be the only piece of furniture in the room that was heavily shaken. Neither the doors nor the windows rattled, though they rattle enough in windy weather, this house standing alone, on high ground, in the neighbourhood of two great rivers. There was no noise. The air was very still, and much warmer than it had been in the earlier part of the night. Although the previous afternoon had been wet, the glass had not fallen."[36] Dickens's interest in earthquakes may seem surprising. His long ruminations on earthquake statistics in *All the Year Round*, in particular, seem out of keeping with his famous cynicism about statistical knowledge in *Hard Times*. What intrigued him about seismology was the suggestion of hidden connections between earthquakes and all sorts of other phenomena—electrical, galvanic, atmospheric, and astronomical. He was also drawn by the absence, as yet, of a satisfying scientific explanation. Dickens arrayed earthquakes alongside a host of newly discovered effects—geomagnetism, electromagnetism, photochemistry—that Victorian physicists hoped to unite in a single theory: "But even Newton's marvelous generalisations do but serve as the basis of still higher generalisations, arising from the rapid increase since his time in the number of facts accurately observed. Newton's so-called laws, once looked on as universal, are now becoming recognised as only subordinate to some other laws yet to be made out. All the recent facts about earth-magnetism are new; all the workings out of electricity in every department, are new; all we hear about certain rays of the sun not communicating light or heat, but having chemical effects, illustrated in what we call photography, is new; and what little is known about the interior of the earth has been learnt since Newton lived."[37] Dickens dreamed that Victorian science might ascend from the study of earthquakes to an all-encompassing cosmology.

Other Englishmen read more worldly lessons into the earthquake of 1863. The *Times* observed that "there are means, utterly beyond our ken and our computation, far below our feet, by which cities may be subverted, populations suddenly cut off, and empires ruined." Reflecting on the fate of the Ottomans, the editor noted, "We see, afar off, a great Empire, that had threatened to predominate over all mankind, suddenly broken up by moral agencies and shattered into no one knows how many fragments. We are safe from that fate, at least so we deem ourselves, for never were we so united. But there are other weapons of destruction in the arsenal of the OMNIPOTENT. Who can say what strange trial of shaking, or upheaving, sinking, dividing, or drying-up, may await us? We know by science these isles have gone through many a strange metamorphosis, and science cannot assure us that there are none more to come."[38] It was a remarkably quick leap from recognizing that England was not immune to earthquakes to sensing the fragility of the British Empire.

Nonetheless, after 1863 earthquakes once again vanished from the pages of British newspapers and the worries of the public. Until, that is, the morning of 22 April 1884, when the town of Colchester in Essex—never before the site of seismic activity—was hit by the most destructive earthquake ever recorded in Britain. The damage was significant enough, if geographically limited, that a national collection was undertaken. On the other hand, there were no fatalities, apart from one woman who was so disturbed by the event that she drowned herself.[39] As usual, the papers had to remind the British public that "such disturbances were by no means unknown in former times."[40] In the view of the seismologist Charles Davison, the 1884 quake was "the chief means of converting it [the present generation] from the belief in the absence or comparative harmlessness of British earthquakes."[41]

There was one crucial difference between 1863 and 1884, and that was the state of seismological science. In the intervening years, John Milne and Patrick Ewing had introduced seismometry in Japan and had begun to establish earthquakes as a proper object of quantitative physical science. Milne promoted his own work in the London *Times* in 1881, and that paper followed two years later with an article entitled "The Vigour of the Earth." It stressed the sophistication of Milne's new instruments: "A delicate seismometer is agitated when to human sensation not the smallest sign of earthquake is evident."[42] The following year, the Colchester quake offered Milne a prime opportunity to convince the British public of the value of his work. From Tokyo, Milne sent off a letter to the *Times* in which he offered a "few facts . . . based upon the observation of many hundreds of earthquakes with every variety of seismograph and seismoscope with which I am acquainted. My object in recording them is to give those who experienced the earthquake some idea as to the true nature of the phenomenon, the knowledge of which is unattainable without the use of instruments."[43] Milne's insistence on the inadequacy of human observations was repeated in subsequent reports on the temblor of 1884.

The 1884 earthquake also arrived at a moment when relations between metropolitan and provincial science were in flux. In 1883 the British Association had attempted to smooth its interactions with the many provincial scientific societies with which it cooperated. A committee chaired by the distinguished Francis Galton would "draw up suggestions upon methods of more systematic observation and plans of operation for Local Societies." The committee envisioned that the BAAS would become "an organizing centre of local scientific work," promoting "systematic local investigation."[44] The most thorough investigation of the Colchester earthquake was conducted by Raphael Meldola and William White, members of the Essex Field Club (founded in 1880), who framed their work as emblematic of this new era in the relationship between metropolitan and provincial science in Britain.[45] In the introduction to their report, Meldola and White presented their detailed empirical report as a model of "local investigation" in the BAAS's sense. "We are of opinion that such investigations should deal as exhaustively as possible with the *facts* relating to any particular subject." They expressed due deference to the experts of the BAAS, wishing "that this task had been taken up by some more qualified specialists." As locals, they did their best to "interpret the observations by the light of the results which have been achieved in Japan" by Milne and Ewing. Finally, they warned emphatically of the shortcomings of their data: "One point in connection with the report which will doubtless strike our readers is the general untrustworthiness of what may be called commonplace observations in any attempt to submit an earthquake to exact mathematical treatment. If our labours serve only to emphasize this inadequacy of non-instrumental methods of observation, we feel that our efforts will not have been altogether exerted in vain."[46] The new era in the centralization of British science thus coincided with a new era in seismology. The authority of metropolitan expertise over provincial experience rose along with the authority of instruments over human senses.

A Birmingham Schoolmaster

Comrie was a less-than-ideal site for earthquake research, too thinly populated and too close to mountains and sea. From a scientific perspective, the
Colchester earthquake was an improvement, but still not as propitious as
the earthquake of 1896. That year, a week before Christmas, a wide patch
of England between Cardiff and Birmingham awoke to an early morning
temblor centered in Hereford. This was a densely populated, urban region
"containing numerous capable and intelligent observers." While this shock
"was strong enough to be interesting, it was not too severe for accurate
study. Those who felt the shock were able to bestow their full attention
upon it."[47]

These were the judgments of Charles Davison, mathematics master at
King Edward's School in Birmingham. When he felt his bed tremble on the
morning of 17 December, Davison got to work. He had acquired an interest
in earthquakes by way of geology and the contraction theory of the earth's
development. As a graduate of the elite mathematics tripos at Cambridge
(thirteenth wrangler), Davison was interested in part in sophisticated analyses of the frequencies of earthquakes. Since 1889, however, he had grown
increasingly enthusiastic about the prospect of studying "British earthquakes" by surveying witnesses.[48] Like his Swiss and Austrian colleagues,
he was interested in mapping the areas affected by ground movement in
order to locate tectonic faults. Unlike many of the seismological pioneers
of the nineteenth century, Davison was not blessed with personal wealth
and leisure time. He wrote his numerous treatises on seismology, as well as
several mathematics textbooks, in his classroom during the lunch break.[49]
In the collection and analysis of evidence, he was equally indefatigable. For
the 1896 earthquake alone he processed 2,902 questionnaires and a stash of
newspaper reports, which, "if continuously extended, would form a column
200 yards in length."[50] By the 1920s, Davison's work had become definitive
of "British earthquakes," as it would remain until the 1980s.[51]

Davison was interested, in part, in what could be learned about the physical nature of earthquakes from human perceptions. But he was also keenly
interested in the human impact of earthquakes in its own right. Characteristic
was a letter he wrote to the *Times* on that classic tale of man's battle with nature, *Robinson Crusoe*. Davison wanted to draw attention to the verisimilitude
of Defoe's account of the earthquake that occurs six months after Crusoe's
shipwreck. In particular, he noted that Crusoe was not atypical to feel sick
to his stomach, "like one that was tossed at sea." He complained, however,
that Crusoe should have felt more terrified in the aftermath: "For hours after

a great earthquake like this the ground would not have ceased from trembling, slight as a rule, but broken every now and then by violent shocks that should have filled Robinson Crusoe with fresh terror. He certainly should not have felt composed for several days."[52] In "The Effects of Earthquakes on Human Beings" (1900), Davison reversed seismology's usual order of reasoning: rather than using human data to analyze physical phenomena, he used physical data to analyze human phenomena. He took the behaviorist position that self-descriptions of mental states during earthquakes were unreliable, and that "the resulting actions are less liable to error or exaggeration." He grouped these actions into four "rough" categories: "A) No persons leave their rooms. B) Some persons leave their houses. C) Most persons run into the streets, which are full of excited people. D) People rush wildly for open spaces, and remain all night out of doors." Mapping these reactions for the Charleston earthquake of 1886 generated a first insight. Initially, he had listed the "hasty dispersal of meetings" under category C, but his map indicated that this reaction was typical of weaker earthquakes: "A crowd in one room is more liable to excitement and fear than are persons in separate houses." Other conclusions followed from plotting the regions ABCD on a map of structural damage. "If the shock was strong enough to throw down chimneys or make cracks in the walls of buildings, then people thought it wiser to camp out for the night.. . . People rushed precipitately into the streets if the movement made chandeliers, pictures, etc. swing." More surprisingly, "If the shock was not even strong enough to cause doors and windows to rattle, some persons were so alarmed that they left their houses, and public meetings were dispersed. Whether these effects were due to the rarity of the phenomenon or to the highly-strung nerves of the American people, it may, I think, be inferred that in no other civilized country would such alarm be shown at a sudden and unexpected occurrence." Davison had turned the earthquake into an experiment in social psychology. Leaving aside the question of the fairness of the national stereotype invoked here, we can appreciate the significance of his conclusion that "the human effects of earthquakes" were culturally specific. Compare the perspective of his contemporary, the Hungarian seismologist R. de Kövesligethy. Kövesligethy hoped to measure human responses to earthquakes on a scale as simple and universal as the astronomer's new formula for defining the magnitude of a star in relation to its luminosity. From Davison's point of view, such a psychophysical law would be a hopeless simplification.[53]

Davison's work represents a lost opportunity. Late in life, he met and advised the young geologist A. T. J. Dollar. Dollar went on to found the British Earthquake Enquiry and, from 1935 to 1967, he continued Davison's surveys on a smaller scale, drawing on a network of approximately two hundred volunteer observers. This research received little support from the director of the Geological Survey at the time, Patrick Willmore, a seismometer designer who "had little time for Dollar." In 1974, Chris Browitt reinstated macroseismic surveys at the Geological Survey, "in the teeth of objections" from Willmore.[54]

Willmore reflected the direction seismology was taking in the twentieth century, in Britain and beyond. Davison's colleague John Milne returned to Britain in 1895, after two decades in Japan. Milne immediately began to campaign for government support for his project of a worldwide seismographic network. Milne promoted seismology as a science appropriate to the British Empire in the new century. For instance, he repeatedly stressed the need to safeguard Britain's global web of telegraph cables by monitoring submarine seisms. In the wake of the Hereford quake, Milne seized the opportunity to argue for the necessity of seismometric studies even for local research. As he put it in a letter to the *Times*, "At this particular time many may well ask whether in the Severn valley or in any other part of Britain, shiverings to which we are insensible are now in progress, whilst the thousands who are awakened every night may desire to know whether the causes have been fanciful or real. Until we establish seismographs, especially in suspected districts, the answers to these and other questions are in abeyance."[55] Next to Milne's high-tech planetary science, Davison's provincial inquiries appeared downright quaint.

The Seismic Goose and the Psychic Gander

In the last years of the nineteenth century, the human observation of earthquakes lost whatever dignity it had retained in Britain up to 1884. Seismology became known to the British public as a professional science that measured far-off vibrations with exquisitely sensitive instruments. "British earthquakes," meanwhile, became little more than a punch line. The village of Comrie was the butt of most of the jokes. An 1886 article in a Scottish paper reminded readers of the "extinct volcanoes that thirty years ago used to flourish in the Perthshire village of Comrie. The story went that they were got up by the local correspondent. As long as he was in settled employment, Comrie was quiet; but the moment the necessities of his life perceived shining merits in penny-a-lining, at once the earthquake appeared. It thus fell out that the newspaper reader was getting his shocks three or four times in the year in obedience to the fanciful wit and the domestic exigencies of the local correspondent. The story may itself have been fanciful; but the fact remains that there is not in Comrie one earthquake for ten that there used to be. In that scene the seismic forces have shaken themselves down into perfect accord."[56] In this way, the press poked fun at its own fondness for natural disasters. Such exposés undermined any credibility that observers in Comrie might still have been able to muster. In 1894 the same paper noted that a report of an earthquake at Comrie arrived "as usual, at the height of the season, just in time to attract the tourist who is fond of a little adventure without unnecessary danger. On one occasion some ill-conditioned person ventured to deny the annual summer earthquake in the interest, as he thought, of the prudent villagers, but what was his surprise when he found that a torrent of local contradiction was the result, backed up by a large display of cracked soup plates. It is dangerous, therefore, to interfere with the Comrie earthquakes. If not, strictly speaking, a source of danger in themselves, their denial is a source of danger to the scoffer."[57]

An 1898 report on a slight tremor at Comrie bore the subtitle "The Reputation of the Village at Stake." In the absence of earthquakes, the paper noted, "Comrie has been drifting in to the position of a mere commonplace summer resort." The recent disturbance had thus come "as a veritable godsend."[58] But what exactly had been observed? In the absence of instrumental evidence, the claims of the villagers no longer stood the test: "Sad to relate, despite the emphatic declarations of the villagers, there are sceptics who doubt whether there really has been an earthquake. A number of persons—visitors to Comrie—declare that at the time when the earthquake is said to have taken place nothing unusual occurred. There is a seismometer at Comrie, but owing to the earthquakes having failed to put in an appearance for several years, it has been neglected, and is not in a fit state either to support or contradict the statements as to the alleged 'disturbances;' so that, unfortunately, there is nothing beyond the bare assertion of the earthquake experts of Comrie with which to confound the sceptics, and the latter refuse to accept these assertions. Nay, more, some of them are unkind enough to recall the fact that the reputation of Comrie for earthquakes rests almost entirely on the assertions of its inhabitants. The seismometer house was erected over thirty years ago by the BAAS, but for years nobody has taken any interest in it. 'Cos why? 'It never recorded anything.' Scottish intense practicability again, you see."[59] Against the new authority of the seismometer, Comrie's inhabitants found it ever harder to establish themselves as "earthquake experts." It thus became a matter of course to attribute Comrie's former earthquakes to a certain "ingenious reporter who made the reputation of Comrie as a seismic centre."[60]

The last strike at the credibility of Comrie's seismic observers came in

1896, when the estate of Ballechin, about twenty miles away, became the site of an experiment by the London Society for Psychical Research. The society had been founded in 1882 to pursue studies of "mesmeric, psychical and 'spiritualistic'" phenomena "in the same spirit of exact and unimpassioned enquiry which has enabled Science to solve so many problems."[61] In 1896, several members of the SPR—including the physicist Oliver Lodge, Fellow of the Royal Society—occupied Ballechin for three months to test the claim that the house was haunted. A debate soon exploded in the letters column of the *Times*, in which the experiment was condemned by some of the very people who had participated in it. The initial attack, signed "A Late Guest at Ballechin," charged that "a so-called experiment had been carried on there for nearly three months at the time of my visit by the Psychical Research Society, without any attempt at either experimentation or research. Unscreened evidence of improbable phenomena had been collected in heaps, but the simplest and most obvious tests had not been applied. The residents and visitors, it seemed to me, had been sitting there all the time, agape for wonders, straining on the limits of audition, and fomenting one another's superstitions without taking any precautions to prevent deception or employing reagents to clear up turbid observations. . . . Practical joking, hallucination, and fraud will account for the bulk of the occult phenomena recorded at Ballechin during its occupation by the Psychical Research Society. What remains—if anything—may be explained by earth tremors (Ballechin is only 20 miles from Comrie, the chief centre of seismic disturbance in Scotland), by the creaking and reverberations of an old and somewhat curiously-constructed house, or by some other simple natural cause."[62] Thus were Comrie's extinct earthquakes revived at this late date. They had acquired plausibility only relative to the paranormal alternative. Apparently, no one bothered to consult Charles Davison, who had documented an "exceedingly slight" earthquake at Comrie in 1895, but none at all in 1896.[63]

Still, Comrie's earthquakes proved only marginally more robust than ghosts. The debate in the *Times* over the Ballechin experiment ran on for five months. It underlined the uncomfortable proximity of observational seismology to the "pseudoscience" of psychical research. Skeptics discredited the observations of the SPR researchers at Ballechin in much the same terms as the press mocked earthquake observations from Comrie. In seismology as in psychic research, the construction of evidence depended on cooperation between metropolitan experts and local residents. It hinged further on the trustworthiness of witnesses who testified to phenomena that were inherently "transient and elusive."[64] The same solution was urged in both

cases. No less an authority than John Milne suggested installing a seismometer at Ballechin, in the interest of determining the origin of the sounds
and serving "seismic research." He also made the potent point that the
former proprietor of Ballechin, Lady Moncrieff, had been cited as an earthquake witness by David Milne himself.[65] An earlier generation of psychic
researchers—including Cromwell Varley, the telegraphist and proponent
of an electrical theory of earthquakes—had applied telegraphic apparatus
to the detection of spirits.[66] Now, one defender of the SPR objected that it
would be "pedantic" to apply "elaborate tests," such as those for "seismic
disturbance," to the phenomena at Ballechin. On the other hand, the author of the 1897 *Book of Dreams and Ghosts* asked his reader to consider the
case of a house "where noises are actually caused by young earthquakes.
Would anybody say: 'There are no seismic disturbances near Blunderstone
House, for I passed a night there, and none occurred'? Why should a noisy
ghost (if there is such a thing) or a hallucinatory sound (if there is such a
thing), be expected to be more punctual and pertinacious than a seismic
disturbance?"[67] The same author suggested in the *Times* that the sounds at
Ballechin might be explained as "collective hallucinations."[68] The consensus, then, was that the same standards of evidence should be applied to
earthquakes as to ghosts.

Seismic testimony typically came from common folk who were, in the
eyes of the scientific elite, effectively anonymous. Spirit testimony often
came from witnesses who remained anonymous for other reasons. "It would
be hard to afford a more ludicrous example of the methods of the SPR than
is given by X. in her own letter. . . . Can anything be more useless or more
absurd than inquiry in which names of persons and places are omitted from
the depositions? What value can be assigned to evidence when the names of
the witnesses are kept secret? Such proceedings are the caricature of a legal
inquiry and the parody of a scientific investigation. The SPR cannot have
the remotest idea of the legal or scientific meaning and value of evidence."
Yet the authority of this anonymous witness, like that of seismic observers,
rested not on personal circumstances but on the local knowledge derived
from long-term residence.

In the end, the last word went to an anonymous reader who suggested
that the seismic and psychic explanations of the "disturbances" were equally
tenuous:

I learn with pleasure that the latest tenant of Ballechin has been absolutely
undisturbed by any unexplained noises. But what is the correct philosophical
inference? The noises of which earlier tenants complained were accounted

for, in a learned way, as the result of "seismic disturbances." Does the absence of annoyance in the late tenant's experience demonstrate that there are no such things as seismic disturbances, or, at all events, none in Athol? And, if this be not quite a logical inference, is it more logical to infer that the recent tenant's immunity from noises proves that there are no such things as bogies, in Athol at any rate? Or would it not be fair to say that what is sauce for the seismic goose is sauce for the psychic gander?

Obediently yours, An Earnest Inquirer[69]

In other words, the evidence for earthquakes and for ghosts would rise and fall together. Seismic and psychic research unfolded according to the same rules. Each depended on the negotiation of authority between metropolitan experts and provincial residents. Comrie's status as a center of either paranormal or seismic activity hinged on its ability to draw London's experts to the scene. Those experts expected to be able to issue the final interpretation of local shudders and creaks. However, from James Drummond to the proprietors of Ballechin, Comrie maintained a tradition of refusing to submit to metropolitan scientific authority.

"Who Killed the British Earthquake?"

In 1972 Queen Elizabeth II presided over the festive opening of a new exhibit hall at the London Geological Museum. It bore the romantic title "The Story of the Earth," inspired by Mahler's "Das Lied von der Erde."[70] According to the curator, Frederick Dunning, the hall was "designed expressly for people with no knowledge of geology whatever, but nevertheless with above-average intelligence: the level of IQ we had in mind was around 115." From the start, Dunning's vision for the exhibit had included a machine that would simulate the experience of an earthquake: "a mechanically vibrated platform on which a visitor can stand and at the press of a button experience a simulated earthquake shock accompanied by the familiar 'express-train' sounds."[71] The popularity of this "earthquake machine" recalled the Lisbon panorama to which Londoners flocked in 1848. Over a century later, the entertainment value of a simulated temblor was still a function of the exoticism of earthquakes to British sensibilities.

By 1972, however, the British could no longer afford to be so complacent. As the scale of human construction grew in the twentieth century, so did the vulnerability of edifices to weak seismicity. With the advent of nuclear power, even a small tremor could unleash catastrophe. As the Edinburgh seismologist Roger Musson recalls, "The first NPPs [nuclear power plants] were built in the 1960s and had no antiseismic design whatever. It was only in the late 1970s that it was realised that this was an oversight, and retrospective safety assessments were called for."[72] Scientists at the British Geological Survey in Edinburgh and at Imperial College London responded by reopening the question of "British earthquakes." They unearthed evidence collected by David Milne and Charles Davison and began to re-analyze it. They also prepared to monitor earthquakes as they occurred. For this research they had no trouble finding financial support from the nuclear industry and its government regulators.[73] What they needed, however, were earthquake observers. "In the UK, there are no suitable local government officials who can be charged with the task of reporting on local earthquake effects; therefore it is necessary to address questionnaires to the general public."[74] Following Davison, the BGS scientists took advantage of newspapers to circulate their questionnaires. Their early questionnaires, however, failed to describe common earthquake effects. In an attempt to extract descriptions of the shocks, the scientists asked, "What did you feel?" Many responded, "Alarmed!"[75] With time, British seismologists began to recognize that "questionnaire design is somewhat of an art."[76]

To the public, meanwhile, the very idea of British earthquakes remained comical. This was clear to Robert Muir-Wood, one of the first scientists to raise public alarm about the seismic risk to nuclear facilities in Britain. At the August 1983 meeting of the BAAS, Muir-Wood called for a national seismic monitoring network and decried the public's willful ignorance about the earthquake hazard. The British, he charged, "still believe earthquakes are about as English as pizza." The popular magazine *New Scientist* followed with a series of articles on the neglect of British earthquakes, including a cover showing a cup of tea thrown from its saucer and the title "Is Britain Prepared for Earthquakes?" Once again, it wasn't clear whether the topic of British earthquakes had earned attention as an acknowledged hazard or a gag. Under the headline "Shaken to the Core," the London *Times* editorialized: "A claim that parts of Kent and Canvey Island, with its vulnerable concentration of oil and gas installations, could be hit by a 'large earthquake' invites scepticism, like a report that the Four Horsemen of the Apocalypse had asked for clearance to land at Heathrow. There is something millenarian, almost Monty Pythonesque about it." In the end, the paper's attitude was dismissive: "But at least until really shaken, Dr. Wood's fellow-citizens will not be easily persuaded that they ought to be worrying about the movement of tectonic plates beneath the British crust. They have other things on their mind."[77]

Attempting to shake some sense into the public and fellow scientists, the British macroseismologists of the nuclear era seized on the example of Comrie. The Scottish village served, first, as evidence that earthquakes were not so exotic after all. Muir-Wood and a colleague opened a 1983 article in the popular magazine *New Scientist* with an account of a rural community terrorized by a shock that had thrown down paintings and cracked walls. "Italy, Greece, some Pacific state of South America?" they asked their readers. "No, just Britain in the 19th century."[78] Comrie recalled a time when, as the authors put it, "news that Britain had earthquakes was no news." In the nineteenth century, they argued, the public was well informed about past shocks, thanks to the local press and to oral tradition. Then the international media intervened: "Journalists had successfully convinced even rural communities that real history poured out of the radio and the television. The folk memory faded into collective amnesia." Finally, the new macroseismologists marshaled the case of Comrie against mainstream seismologists, who could not imagine the value of human earthquake observers in late twentieth-century Britain. Comrie demonstrated that noninstrumental research could generate fundamental scientific knowledge. As Musson argued in a 1993 article on Comrie's "place in the history of seismology," the Scottish village "indicates the usefulness of earthquake swarms as a testing ground for seismological investigation."[79]

Indeed, Comrie's tremors had prompted the development of one of the earliest scales of seismic intensity as well as the most sophisticated seismometer of the mid-nineteenth century; they had also inspired some of Davison's seminal contributions to the methods of macroseismic survey. The organization of Comrie's residents into a quake-observing network was a successful geophysical experiment; less obvious is its status as a human experiment. Comrie served as a laboratory for the study of the social relations of British science: for the apportioning of authority between local common sense and metropolitan expertise. For the British press, however, Comrie lives on principally as a punch line. In March of 2008 the *Times* reported that the Earthquake House had failed to record the strongest earthquake to hit Britain in twenty years, "because it had run out of paper."[80]

THREE

News of the Apocalypse

Anyone who studied earthquakes in the nineteenth century was used to having their fingertips covered in newsprint. The circulation of earthquake stories in the daily press was the lifeblood of nineteenth-century seismology. Scientific researchers culled the testimony of earthquake witnesses from local papers, posted advertisements to solicit observations, and used the major dailies as a forum for communicating their research to the public. Newspaper culture in turn shaped the ways in which seismologists linked local stories into planetary ones.

The term "journalism" was introduced into English from French in 1833, the same year that the word "scientist" was coined.[1] At a moment of fitful democratization in Britain, each of these neologisms labeled an unprecedented form of authority over the public, a new class of guardians of information. Newspapermen saw themselves as fulfilling a modern responsibility for the rapid and transparent exchange of information. By mid-century, this aspiration was realized via the telegraph: Reuter's sent its first commercial newswires in 1851, its first foreign dispatches in 1858.[2] By 1834, it was "a matter of universal observation" in Britain that "the great changes of recent times have been mainly owing to the influence of the press."[3] But what was the nature of its influence? In step with the expansion of the franchise throughout western and central Europe, concerns grew over the power of the press to sway public opinion. A French physician lamented in 1895 that the papers were no longer a voice of reason. Readers grew skeptical of "sensationalism," even as circulations rose. The papers themselves remarked frequently on the machinations of their competitors. According to historians, a widespread sense emerged that "truth" had to be defended against the distortions of wily publishers.[4]

No modern science was as intimately involved with these developments as seismology. To be sure, meteorologists contended in print with competing authorities, including farmers' almanacs and astrological predictions.[5] Unlike meteorologists, however, seismologists often drew their data directly from newspapers: they assigned values of intensity to newspaper accounts of ground movements. No meteorologist ever tried to take the temperature of newsprint. Indeed, nineteenth-century seismology was uniquely dependent on the proliferation of newspapers and the expansion of the reading public. As one seismologist put it, local papers published "numerous individual observations, general commentaries, and calls for participation in the observation of earthquakes." They were an "essential support of earthquake study."[6]

Still, like much of the reading public, seismological researchers often groaned about the shortcomings of newspapers as a source of information. Unlike accounts obtained directly from witnesses, newspaper reports often "prove to be . . . highly exaggerated or even pulled entirely out of thin air."[7] "Since the newspapers mainly cite the most striking events," complained one seismologist, "one can in general far sooner rely on reports that one draws from trustworthy individuals by means of questionnaires for a normal picture."[8] The press itself recognized this problem. In the wake of the devastating earthquake and tsunami at Messina in 1908, for example, the satirical German weekly *Simplicissimus* found comic relief in the behavior of journalists: "In the vicinity of Messina was overheard the conversation of two foreign special correspondents: 'You write of two hundred thousand dead, my dear colleague?'—'Why not!' answered the one addressed. 'I am paid by the corpse.'"[9] In the historical relationship between seismology and the press lie clues to the mutual constitution of two emerging forms of public authority—scientific expertise and mass journalism.

We can watch the emergence of a public dialogue about earthquakes in the late nineteenth century in Eduard Suess's hometown of Vienna. The 1873 temblor that launched Suess on his tectonic theory (estimated today as magnitude 4.4) at first went unreported in the major liberal daily, the *Neue Freie Presse*. Four days later, a few reports filled only seventeen lines. Subsequently, even minor seismic events triggered small landslides of telegrams to the *Neue Freie Presse*. In 1876, when a temblor was felt strongly in Vienna, the paper cited dozens of eyewitness reports; there was "not enough space for the innumerable smaller incidents." But even a less dramatic event like a magnitude 4.4 in Lower Austria in September 1885 drew numerous telegrams and received fifty-nine lines in the evening edition, with more observations reported the following day. Finally, in 1908, a magnitude 4.8

quake in Lower Austria took up two full pages for each of the next two days. Witness after witness was quoted as to the precise time of the tremor, its duration, direction of propagation, damage, weather conditions, and effects on humans and animals. The editors received too many reports to print them all, and those trying to reach the paper by phone were told all lines were occupied, "due to the earthquake."[10] For a brief moment, reporting on earthquakes approximated the ideal of "participatory journalism."[11]

With the decline of local interest reporting in the years leading up to World War One, citizens and scientists alike lost a key repository of knowledge about seismic hazard. As the California seismologist Harry Wood observed in the 1930s, "as country cross-roads in this western region have grown at an amazing pace into sizable cities, while villages have spread and coalesced to form great metropolitan communities, reports of small local earthquakes have been crowded out of the published news of the day (the smaller the shock the less its importance as an item of news) along with accounts of small fires, minor accidents and the like—matters which often find room in the local press of small places."[12]

"A Brief Sensation"

Let us begin with a single seismic event refracted by multiple passages through the press: a rare case of a widely felt earthquake in the eastern United States. It is a story that illustrates the role of the press in soliciting earthquake observations, communicating scientific conclusions to the public, and suggesting global connections. "On the night of Thursday, the 10th of December, 1874, there occurred, in and around the city of New York, a slight but very distinct earthquake shock, which caused considerable excitement at the time, and furnished material for a brief sensation in the papers of the metropolis." The New York Lyceum of Natural History (now the New York Academy of Sciences) formed an investigative committee chaired by the geologist Daniel S. Martin of Rutgers Female College, an active member of that society. Martin prepared a questionnaire, to be printed "in the principal daily papers of the city, with an added request that it be copied by local journals throughout the region affected." One thousand copies were sent out. "About a hundred responses were received, varying of course greatly in their character: in the main, however, they furnished useful and excellent data, which were afterward carefully tabulated. Though almost all written by unscientific observers, yet the incidents and impressions were related in most cases with a care and clearness beyond what was anticipated, while many of the letters were exceedingly vivid and detailed."[13] Martin used these reports to prepare a map on which each observation was marked by a red circle, with an arrow indicating the direction of motion, a system he compared to that of weather maps.[14] As Martin revealed to the Princeton geologist Charles Rockwood, "The labor involved in these preparations was, as may readily be seen, very great; and it was quite out of proportion to the results. No seismic center or central line could be detected." A majority reported that the shock traveled from southwest to northeast, and Martin decided "on other grounds that such was the true course,—upward along the line of strike of all the rocks of this region. The cracks produced in the ground at Closter, New Jersey had a course transverse to this, and confirmed my impression." Martin shared these conclusions with "a very intelligent and somewhat scientific reporter" at the *New York Tribune*, but the paper "mixed up my remarks as to the crystalline rocks greatly."[15] No wonder Martin's efforts proved frustrating: scientists have since found that earthquakes in the northeastern United States do not tend to occur along known faults. The geologic structure responsible for earthquakes in this region remains an "enigma."[16]

While Martin worked up these results, information reached him from farther afield. He received "two remarkable letters . . . detailing a marked disturbance of very similar character in eastern Massachusetts, on the same afternoon between 5:30 and 6." He also read a report in *Nature* of an earthquake atop the Pic du Midi in the Pyrenees at 4:45 a.m. on 11 December—nearly coinciding with the temblor in New York. According to *Nature*, the original source of that information was apparently a small-town French weekly with the romantic title *L'Echo des Vallées*. Then Martin read of volcanic eruptions in Iceland, beginning in December and continuing through January. The famed British explorer Richard Burton drew his account of the eruptions primarily from critical readings of Icelandic newspapers: "The local papers . . . give ample accounts of the late movements. . . . Much of the matter has been translated and published by our home press, but there are interesting details which have not been noticed. Generally—allow me to remark—the accounts, though utterly unscientific, bear an aspect of sobriety and truthfulness wholly wanting in the older Icelandic descriptions . . . , and they show that the spirit of enterprise has not wholly died out in Iceland."[17] Faced with these new clues, Martin wrote to Rockwood in mid-June that he would have to postpone further publication: "But now, we have these remarkable accounts of outbreaks, and disturbances in Iceland during the past winter; and the question at once occurs to me, when did these begin? . . . My impression from all our accounts, is very strong, that the shock in our region was *deep* and wide-spread. The two (distinct) afternoon shakings,

in North Andover and in Salem, Mass., and more curious still, the simulta-
neous shock on the Pic du Midi (Pyrenees), that was reported in 'Nature,'
point to some wide and deep movement, but lightly felt hereabouts at the
surface, and yet very likely to be stronger near some of the earth's great
safety-valves, like Iceland."[18] Did Rockwood know, Martin wondered,
whether any shaking had been felt at intermediary points like the Azores
and Newfoundland? Rockwood did not, but that did not stop Martin from
printing his transcontinental speculations in *Science*. As John Milne noted,
"The probability, however, is that these coincidences are accidental."[19]

The Earthquake Collectors

Few young followers of the trans-Atlantic stamp-collecting craze of the
1870s were as fortunate as the sons of the mathematicians Alexis Perrey and
Charles Rockwood, of Brittany and New Jersey, respectively. As the fathers
amassed reports on earthquakes from newspapers and independent observers around the world, the sons reaped the philatelic benefits of their correspondence.[20] Perhaps the fathers shared something of the sons' Victorian
passion for collecting, and something of their thrill at epistolary contact
with distant lands.

Charles Greene Rockwood came from a prominent New Jersey banking
family that had been in the United States since 1636. From 1877 to 1905
he was professor of mathematics at Princeton.[21] Rockwood bequeathed the
Princeton archives three fat volumes of newspaper clippings on earthquakes.
These had formed the basis for his catalog of "American Earthquakes,"
which covered North and South America and was published periodically in
the *American Journal of Science* from 1872 to 1886. When Rockwood took
up this work, there was no central source for data on earthquakes occurring in the United States. The Heidelberg geologist Carl Fuchs, who began
publishing annual earthquake lists for Europe in 1865, complained about
this lacuna in an 1880 letter to James Dana at Yale. "In my annual reports
the very small number of earthquakes in America is particularly striking. It is to
be presumed that this is the result of lack of information on earthquakes
and not of an extraordinary scarcity of such events in your country. . . . I am
unfortunately ignorant as to whether any one in America is engaged in collecting earthquake statistics nor do I know what are the sources from which
information can be obtained."[22] Dana forwarded the letter to Rockwood,
who sent copies of his "Notes" to Fuchs and brought Fuchs's publications
to the attention of American readers. Fuchs reciprocated with his list for
1877–78, but he noted its imperfections: "These can only rise to a degree

of completeness with respect to Europe. From Austria, Hungary, and a portion of Germany I receive reports from the meteorological stations. In other countries I have correspondents, who collect the earthquakes occurring in their countries."[23] Beyond Europe, then, most of the information Fuchs received came from newspapers.

Rockwood too drew heavily on the American press, and local editors were among his valued correspondents. A query to the *Boston Post*, for instance, received the reply that the tremor in question was not felt in Boston, but that Rockwood was "mistaken in supposing the center of New England is always undisturbed. We had a nice little earthquake here about six years ago, it seemed so pleasant that we have been longing for a repetition ever since."[24] A Connecticut newspaperman apologized to Rockwood for being unable to find an article he had written on an 1875 earthquake. "All I can say today is that it was about two o'clock in the morning. Were I really a *scientific personage*, I should have had the matter recorded. At the time, people here did not consider it an earthquake but merely the cracking of the ground or ice, as it was very cold then. . . . I am quite sorry I can give you no more information!"[25] Not all journalists proved reliable, however. Rockwood was dismayed to discover that an evocative article on an earthquake in Caracas, published in the *Atlantic Monthly* in 1883, could not be corroborated. A local naturalist replied to his inquiry: "The article in the Monthly Atlantic for Mar[ch] [18]83 must be altogether a forgery, for there was *never* in the month of *September* an earthquake of such an energy in Caracas. . . . I am really curious to see this piece of fiction, and will certainly do something to warn people against that HD Warner whoever he may be. . . . The thing is undoubtedly not pleasant for you; however it has some good: you shall have henceforward [illegible] scientific notes on our earthquakes, small or great, as I have established a Mallet seismometer which works allright. I never sent anything to Perry [*sic*], nor to Fuchs, but I published a note on the earthquake at Chio (1877) in *Nature*."[26] By then, Rockwood had decided that newspaper reports of earthquakes required corroboration. From 1878 he marked "uncertain" those events "based upon *single* newspaper reports and which could not be otherwise verified."[27]

How Rockwood organized these slips of newsprint was decisive. His system was chronological rather than geographical. Clippings about earthquakes in Japan, Syria, Australia, England, Oregon, Hawaii, and Nebraska wound up side by side, all pasted to a single page. Such juxtapositions were grist for speculation. What hidden mechanisms might link these far-flung tremors? Journalists asked themselves the same question. The London *Times* suggested in 1824 that an earthquake in Comrie, Scotland, was linked to one abroad, just as the 1812 earthquake in Caracas had reportedly been felt in a village near Comrie.[28] On the occasion of a stronger shock in Comrie in 1839, the *Times* noted a coincidence with earthquakes in Calabria and Savoy and concluded that the events must share an origin deep within the earth.[29] In the wake of the Messina disaster of 1908, one Swiss editor wondered, "Will the countries situated to the north of Italy also be threatened? We know that a strong shock of earthquake was felt on the 18th of December in Coutances. A few days later another was reported at Angers. Monday, one was indicated in the valleys of the district of Oloron. The seismograph of the faculty of sciences at Grenoble recorded a seismic shock Monday at four thirty-three in the morning. These two last phenomena coincided with the shocks felt in Malta at the same hour. Finally they are being reported in the United States as well; a telegram from Virginia City, in Montana, announces that violent seismic shocks were felt Monday in that city."[30] The papers also invited reports of seismic coincidences by printing earthquake forecasts that covered as wide an area as "Persia, Asia Minor, Greece, Northern Italy, or Carniola" and soliciting confirmations from around the world.[31]

In this apparent chaos, seismologists sought patterns. Perrey was the first to remark on the near simultaneity of earthquakes at distances as great as those between "Holland and Spain, Lisbon and Saxony, Calabria and Maurienne, Savoy and Scotland."[32] In 1911 John Milne pointed to seismographic evidence that one earthquake could trigger another remotely. He theorized that long-period waves from large earthquakes could "travel round the world, causing the crust of the same to rise and fall like a raft on an ocean swell."[33] Seismologists recognized the fascination of this small-world vision for newspaper readers. Albin Belar described planetary-scale standing waves in a Viennese paper in 1907: "After great catastrophes, the diametrically opposed points of the earth, in particular—that is, the antipodal points—are sympathetically affected, so that it is quite possible that one earthquake can trigger or at least prepare another. . . . One could say that the entire crust resounds, and where the frightful fundamental tone is struck, there death and desolation reign."[34] Belar asked his readers to imagine the entire earth ringing like a bell with a "world-shaking music," recorded by a world-spanning network of geophysical observatories pressed like gigantic ears to the ground (see figure 3.1). The British seismologist Charles Davison, whose first concern was always the human impact of earthquakes, once remarked somewhat bitterly that "it is the registration of a distant earthquake, not the havoc wrought by it, that appeals to the public with an unfailing interest."[35]

Rockwood's massive scrapbooks are an early case of a broader trend.

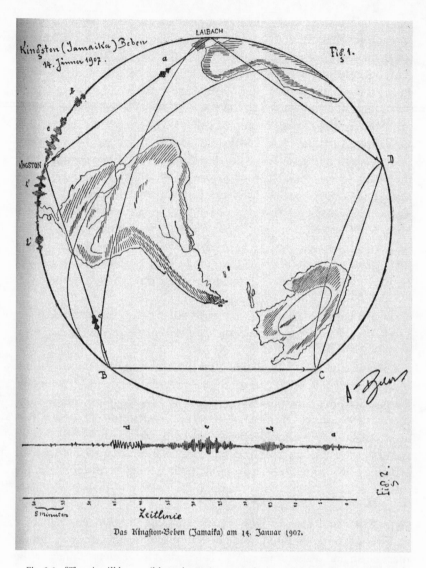

Fig. 3.1. "Then it will be possible to determine precisely the nodal points, antinodes, and superimpositions of the planetary and surface waves that shake the earth like a squall." In this image for the Vienna newspaper *Die Zeit*, Albin Belar illustrated the propagation of seismic waves through the earth from Kingston, Jamaica, to Ljubljana. Albin Belar, "Stürme und Erdbeben," *Die Erdbebenwarte* 8 (1909): 5–11, figure on 9, quotation on 10, originally published in *Die Zeit*.

By the early twentieth century, elaborate systems had emerged to gather, transmit, and archive newspaper clippings, driven by what one historian describes as a modernist "euphoria" for "completeness."[36] The Caltech Seismological Laboratory, for instance, maintained its subscription to the Allen Press Clipping Bureau into the 1930s.[37] Collectors of clippings "could not escape the medium" of newspapers, as Anke te Heesen points out. Try as they might, collectors never arrived at pure information. "Unintended and unavoidable juxtapositions generated their own meanings and fantasies, as a result of the cutting practices, rather than those intended by the collector."[38] Practices of newspaper reading, reporting, and collecting thus shaped seismology's epistemic ideal of completeness and its global vision. Rockwood's volumes of clippings juxtaposed distant seismic events, implicitly arguing for hidden global connections. At the same time, these collections literally severed tremors from their local context as indicators of seismic hazard.

The Earthquake Prophet

HUMAN SEISMOGRAPH FORETELLS DISASTER

—*New York Times* headline, 4 May 1909

Seismology's reliance on the press left it vulnerable to public challenges to its authority. Newspapers offered a platform to self-proclaimed earthquake prophets like Rudolf Falb, a former Catholic theologian from Styria who retooled as an astronomer and meteorologist. It was claimed that Falb had studied natural science in Prague and Vienna, but he held no scientific degree.[39] In fact, his outsider status was part of his persona as a genius unrecognized by the scientific establishment. In 1868 he founded the popular astronomical magazine *Sirius*, where he promulgated his theory that earthquakes were triggered by lunar and solar tidal pulls on the earth's fluid interior. (In France, Alexis Perrey advanced a similar theory at this time, but was not in the business of issuing predictions.) On this basis, Falb claimed to be able to calculate "critical days," on which the alignment of earth, moon, and sun would give rise to massive floods or earthquakes. In 1873, to widespread acclaim, Falb claimed to have predicted the earthquake that struck Belluno, Italy, and in 1880 he began publishing calendars of critical days.[40] Falb's theories and predictions were read and debated throughout Europe and North America in newspapers and scientific journals. The *Neue Freie Presse* attested that he had predicted a major subterranean explosion at a mine in

Moravia in 1885;[41] the *New York Times* reported on his apparent successes in 1887.[42] Scientists accused the press of feeding the frenzies surrounding his forecasts. Hoernes lambasted the *Neue Freie Presse* for sending Falb off to investigate the eruption of Mount Etna in 1874 with all the fanfare of another Stanley in search of Livingstone.[43] To be fair, the *Neue Freie Presse* also expressed skepticism. In 1887 it quoted the quip that, when it came to Falb's predictions, "a year has 365 'critical days' and a leap year 366."[44]

Clearly, Falb had struck a chord. This was an era when many newspaper readers were relinquishing religious interpretations of catastrophes and looking to science as an alternative source of certainty. In 1885, in Falb's home province of Styria, one village "spent a full night outdoors out of fear of a prophesied earthquake."[45] Meanwhile, the initial response to Falb from trained natural scientists was critical but not dismissive. The questions they raised concerned both theory and method. Falb attributed all earthquakes to volcanism, while central European geologists increasingly followed Suess in classifying most as tectonic.[46] Methodologically, Falb's predictions seemed so vague as to be unfalsifiable. Based on their own statistical investigations, other scholars acknowledged that a slight influence of celestial bodies on weather catastrophes and earthquakes had not been ruled out. However, they insisted that Falb was doing more harm than good with his predictions. The scientific questions required calm deliberation: one must do "one's statistics properly and not allow memory and judgment to be clouded by the sensational impression of isolated conspicuous coincidences, which it is well known is one of the most fundamental and most dangerous weaknesses of human judgment."[47]

The tone of Falb's critics grew more shrill in the wake of his claim to have predicted the severe earthquake on 9 November 1880 in Zagreb (German: Agram; chapter 7). It came several months after Falb's return from a research trip to South America, where a bout of rheumatism had left him with poor vision and swollen ankles. "Only with the Agram earthquake on November 9, 1880 did the old energy and motivation return fully," he later recalled. He hurried to the scene of the disaster and "had the satisfaction of seeing positively confirmed the secondary recurrence of the impact predicted according to the tide theory for the 16th of December."[48] Falb's claim to have predicted this catastrophe—and his prophecies of more to come—transformed the wary skepticism of some professional scientists into bitter opposition. But men of science learned a nuanced lesson from Falb. His demagoguery prodded them to cultivate their own mode of public outreach.

Rudolf Hoernes, one of the pioneers of the monographic method in Austria, was the first to draw this conclusion. "The popular form in which

Falb represented his theory in many writings, and perhaps even more the journalistic and oratorical activity with which he supported his work, are to thank for the wide recognition that his 'earthquake theory' found—if only, however, in circles that were not at first able to make a competent judgment on the question. The Falb theory was all the more able to succeed because opposing popular accounts were not available." Hoernes responded with a "popular account" of his own—a witty and accessible essay, which proved a source of "amusement" to him to compose.[49] Falb's true sin, as Hoernes saw it, lay in publicizing his speculative forecasts. Hoernes blamed the press for helping Falb do so. An anonymous writer in the popular scientific journal *Aus der Natur* likewise attacked the press for allowing Falb to terrorize vulnerable populations: "Via the ignorant daily press this putatively new perspective has been spread to wider circles, so that even in earthquake-prone regions it has found a fateful reception, especially for the reason that the author is a Catholic cleric. A panicked terror seized the inhabitants of the West Coast of South America, as Falb prophesied a repetition of the frightful earthquake of the previous year; everyone evacuated, and the prophecy was not fulfilled."[50] The Falb debacle made scientists worry "that public papers, especially those with a political leaning, are not the forum in which to air scientific questions."[51]

Scientists became anxious to gain some control over reports of earthquakes in the popular press. This question was on the agenda of a 1905 conference of central European seismic observatory directors. The participants were concerned that the reputation of the "young science" of seismology might suffer if observatories reported distant seisms and no one came forward to confirm a corresponding earthquake. More disturbingly, newspaper readers might misconstrue reports of distant seisms as earthquake *predictions*. In one case, the *Neue Freie Presse* had presented reports of an earthquake from abroad as a "remarkable confirmation" of a teleseismic observation from Belar's observatory.[52] Belar's point was that this confirmation was not remarkable in the least. One simply had to understand how seismographs picked up waves from afar. Belar and his colleagues voted against the obligatory reporting of seismographically recorded earthquakes to the newspapers. They decided that it was best, in some cases, to withhold information. As we will see, this was a dangerous precedent.

The Inverted Sublime

From the perspective of the nineteenth-century press, earthquakes were good for business. They could be exploited not only for pathos and melodrama, but also for their vast comic potential. Indeed, the nineteenth century saw the rise of a veritable genre of seismic humor. This is not as incongruous as it might sound. The romantic writer Jean Paul theorized that humor is an inversion of the Kantian sublime. Like the sublime, its effect derives from the tension between the finite and the infinite. The sublime comes from the attempt to grasp, via reason, something that exceeds the reach of imagination. Humor instead leads from awe of the infinite to a contrasting sense of the puniness of the finite; it "fastens us tightly to the sensuous detail."[53] Earthquake humor released the tension latent in the juxtaposition of the scales of seismic force and human life.

The jokes typically came at the expense of some blustering fellow who was reduced to whimpers, fled outside in his underwear, was somehow knocked by the shock into a compromising position, or hadn't got word that what had transpired was in fact an earthquake. In the last instance, the lesson was that those who failed to read the morning paper might become laughingstocks themselves. Predictably, racial minorities were often targeted for ridicule. After a shock in Brooklyn, the *New York Herald* described the "ludicrous sights" on the Lower East Side, where Jewish immigrants were throwing their belongings from tenement windows into the streets below.[54] Outside of big cities, the jokes strained with the desperation of small-town editors hungry for news. After a tremor in the Pacific Northwest in 1872, one paper promised "a great many serio-comic incidents, bordering on the ludicrous." A headline in an Olympia paper ran, "A Lively Shaking Up—Fright of Women and Children—Universal Consternation—Laughable and Other Incidents." Typical of these incidents was the plight of an initiate at a Masonic lodge, who mistook the earthquake for part of the rites, while the Masons fled. Newspapermen who delighted at the prospect of reporting quakes were another butt of seismic humor. "More earthquake," read one headline: "Several clocks were stopped . . . restless people waked up, timid people kept awake, everybody set to talking, some to trembling, others to laughing and editors to writing."[55] The jokes rose to about the same level in central Europe (see figure 3.2). Anecdotes of the Zagreb quake of 1880 described a young officer falling into the arms of a businessman's wife, a master cobbler grabbing his apprentice's hair to steady himself, and apartment dwellers running upstairs to demand that their neighbors quit moving the furniture around.[56] In one satirical sketch, a tremor makes a henpecked husband stumble, and his wife—oblivious to the earthquake—accuses him of being drunk. After seeing the news reports, she does her best to make it up to him: "Wife (crooning even more): '[Calls him by his nickname] Come on, be nice. I'll never do it again. As far as I'm concerned, every day now an

Wer hat das Erdbeben nicht gefpürt?

Fig. 3.2. "Who Did Not Feel the Earthquake?," *Die Bombe,* 23 July 1876. One figure is a
young woman engrossed in a romantic novel; the others are a kissing couple lost in
"seventh heaven." Seismic humor in the nineteenth-century press cemented the
earthquake's status as a journalistic "sensation."

earthquake can knock the washbasin out of your hand, and you can stumble as much as you like.'"[57]

The first master of earthquake humor was unquestionably Mark Twain. He honed his style writing short sketches for the California papers in the 1860s, before either the press or men of science took the earthquake threat seriously. His reporting on the severe shock of 8 October 1865 earned him a reputation as "that funny cuss, Mark Twain, who, when his last hour shall arrive, will probably laugh grim Death out of countenance."[58] As Twain later wrote, "The 'curiosities' of the earthquake were simply endless."[59] Certainly, he mined them endlessly. In his most extensive account of the earthquake, Twain parodied the style of a naturalist, an earthquake connoisseur: "I have tried a good many of them here, and of several varieties . . ." With scientific precision, he described a "specimen belong[ing] to a new, and, I hope, a very rare, breed of earthquakes." In what reads as a canny parody of later psychiatric studies of earthquake victims, Twain gave a second-by-second account of his thought process during the quake: "I will set it down here as a maxim that the operations of the human intellect are much accelerated by an earthquake. Usually I do not think rapidly—but I did upon this occasion. I thought rapidly, vividly, and distinctly. With the first shock of the five, I thought—'I recognize that motion—this is an earthquake.' With the second, I thought, 'What a luxury this will be for the morning papers.' With the third shock, I thought, 'Well, my boy, you had better be getting out of this.' Each of these thoughts was only the hundredth part of a second in passing through my mind. There is no incentive to rapid reasoning like an earthquake."[60] Twain even parodied the bombastic earthquake forecasting that infuriated nineteenth-century scientists:

> Oct. 22—Light winds, perhaps. If they blow, it will be from the "east'ard, or the west'ard, or the suth'ard," or from some general direction approximating more or less to these points of the compass or otherwise. Winds are uncertain—more especially when they blow from whence they cometh and whither they listeth. N.B.—Such is the nature of winds.
>
> Oct. 23.—Mild, balmy earthquakes.
>
> Oct. 24.—Shaky.
>
> Oct. 25.—Occasional shakes, followed by light showers of bricks and plastering. N.B.—Stand from under.
>
> Oct. 26.—Considerable phenomenal atmospheric foolishness. About this time expect more earthquakes, but do not look out for them, on account of the bricks.
>
> Oct. 27.—Universal despondency, indicative of approaching disaster. Abstain from smiling, or indulgence in humorous conversation, or exasperating jokes.[61]

Twain captured the pseudo-expert's blend of scientistic jargon, portentous prophecy, willful vagueness, and useless advice. His was perhaps the first satire of the mutual dependence of seismology and the press.

This comic genre largely died out circa World War One, as the weaker tremors that were its best source no longer found column space. A late instance from 1926 is telling for the sharper tone of its satire. The Munich observatory had registered a light tremor and printed a request in the papers for observations from the public. The German weekly *Simplicissimus* reported, "Incidentally, the entire yield of this appeal of science was a single report. It came from the foot surgeon Xaver Dorninger from the Au and said that the reporter was thrown from bed at about half past eight in the morning by the force of the earthquake and had received a painful bruise on the back of the head as well as several scratches. This bed-fall produced by the forces of nature of Xaver Dorninger was, for the material-poor newspapers of these sad sour-pickle times, a highly welcome sensation. The editors indulged in reflections of multiple columns garnished with cosmic prattle [*kosmischen Schmuss*], and demonstrated as clear as daylight why the shocks must have been felt most strongly precisely in the Au and nowhere else, and that the magnetic influence of the various machine factories in the Au evidently played a decisive role.—The Sunday columnist illuminated the topic exhaustively from the humorous side; and the professional meteorologists for their part happily took the opportunity to unload some of their excess expert knowledge into journalistic channels."[62] A week later, however, Munich's Central Institute for Meteorology apparently received the following letter, with spelling mimicking a thick local accent: "Dear sirs, it is not true that my husband Xaver was thrown from bed by an earthquake; it was by me because he was still lying around in bed at half past eight, the lazy good-for-nothing. Yours respectfully, Josepha Dorninger."

On one hand, this story contained familiar seeds—an earthquake suddenly turns a domestic spat into a public spectacle and exposes a browbeaten husband to ridicule. But the real butt of the satire was not the unfortunate couple, but rather scientists and journalists: namely, Munich's newspaper editors, with their "cosmic prattle" of pseudo-scientific explanations; the journalists who wrung humor from the situation; and the scientists who forced their expertise on the public. In this case, the light humor of earlier earthquake journalism had morphed into trenchant social critique. Very likely, *Simplicissimus* modeled this story on the satiric genius of one of its most controversial contributors, Karl Kraus.

"We Fled to Science"

The earth can take no more. It was only a nervous twitch,—and the chatter is endless. What then when it really loses its patience?

—Karl Kraus[63]

Karl Kraus has been called an "anti-journalist," a writer whose purpose was to expose the ways in which modernity disfigured language itself.[64] He attacked the sensationalistic, self-obsessed journalism of his day as a reflection of the hypocrisy, arrogance, and complacency of bourgeois culture. Walter Benjamin aptly likened Kraus to a premodern prophet of doom: "In old engravings, there is a messenger who rushes toward us crying aloud, his hair on end, brandishing a sheet of paper in his hands." Not by coincidence, some of Kraus's most acid critiques of fin-de-siècle journalism before 1914 were those dealing with the reporting of earthquakes and volcanic eruptions. Benjamin was one of the few critics who recognized that Kraus's vision centered on the revenge of the nonhuman world. Kraus was, in Benjamin's words, "Cosmic Man," who made the "insolently secularized thunder and lightning, storms, surf, and earthquakes" his "world-historical answer to the criminal existence of men."[65] Historians have noted the shift in Kraus's tone between roughly 1908 and 1911, from irony to "invective." But little attention has been paid to the rhetorical role of seismic disasters in this transition.[66]

The hubris of modern science and technology was a consistent theme in Kraus's early writings. A better known example of Kraus as scientific watchdog is his 1909 essay on the "discovery—or, as it has also been called, the conquest" of the North Pole. Kraus saw this as a feat of political chauvinism and technological arrogance. The dispute over the priority for the discovery was, to Kraus, a typical case of the machinations of science and the press. Each prolonged the dispute in order to bolster its own authority; each, while claiming to speak for "truth," ran roughshod over it. Science and the press seemed equally to blame for feeding the modern race to conquer nature. Nature, however, would have the last laugh:

> If nature knew that the news about the reaching of the North Pole has "heightened a feeling of superiority over nature" in all errand boys, it would split its sides laughing, and cities and states and department stores would then be thrown somewhat out of kilter. As it is, nature twitches a bit more frequently than is good for the superiority of its inhabitants. In a matter of weeks the elemental forces have so clearly evinced their readiness to meld into a realm of reason that even the masses must understand. They did so by destroying hundreds of thousands of lives and untold millions in property in America, Asia, and Australia by means of earthquakes, flash floods, typhoons, and torrential rains, leaving only European newspaper editors with the hope that "the human will" will shortly "move all levers of nature." Every parasite of the age is left with the pride of being a contemporary. They print the newspaper column "Conquest of the Air" and ignore the adjoining heading "Earthquakes"; and in the year of Messina and daily tremors of the earth man proved his superiority over nature and flew to Berlin.[67]

In this collage of headlines, Kraus set natural disasters as the counterpoint to technological "progress." Like recent theorists of the "risk society," Kraus warned against responding to the failures of modernization with redoubled modernization.[68]

Kraus's "apocalyptic satire" first exploded in a little-known essay entitled "The Earthquake," published in February 1908—a full ten months before the disaster in Messina and eight months before the Bosnian crisis that foreshadowed the Great War. The occasion was a minor earthquake centered in Lower Austria. Its timing—during the carnival season, in the year of the sixtieth jubilee of the reign of Emperor Franz Josef—was perfect for Kraus's purposes. The jubilee was, in his eyes, a shameful parade of hypocrisy, given the fractiousness of imperial politics. With such a rude interruption to their festivities, would the Viennese finally feel chastened? No, they seemed delighted with the earthquake. Suess's efforts at public outreach had paid off. Suddenly, all of Vienna seemed to be racing to report their experiences of the earthquake:

> Now, I thought to myself, at last there will be peace for a spell. We have been admonished. Our brains have been joggled into confusion; the Viennese will see that the patience of the earth cannot be counted on; he will learn humility and will be prepared to perish without causing a sensation. . . . Not a trace of it! Now all hell breaks loose for real. The oafs stumble into the street, snatch up "observations" where they can get hold of them, and run into the editorial offices to announce that they have felt a jerk. That they were there, too! Poles fall, windows clang, children prattle, mothers misspeak, and fathers write letters to the *Neue Freie Presse*.[69]

In this passage, Kraus presented typical earthquake observations as ready-made comedy. He went on to speculate that "the geologists of the *Neue Freie Presse*" were charging the public to print their observations as if they were advertisements. He even suggested that businesses were sending in earthquake reports as a form of publicity. What Kraus scoffed at was not simply the public's infantile need to be quoted about such banalities. More contemptible was their obsession with the least consequential effects of the earthquake, while its existential significance escaped them completely. "The Viennese greet the apocalypse with a Hallo! Hallo! . . . No, that was certainly not telluric, that was a cosmic earthquake. . . . And it was a test of how the Viennese will behave at the apocalypse, which will certainly take place this year."[70] These lines foreshadow Kraus's reaction six years later to the war fever of 1914.

Then Kraus had a flash of inspiration. He penned his own earthquake report for the *Neue Freie Presse*, signed by one "Civil Engineer J. Berdach" in Leopoldstadt, a heavily Jewish neighborhood:

> Just as I was reading your highly esteemed paper I felt a shaking in my hand. Since this phenomenon was only too familiar to me from my many years of residence in Bolivia, the well-known seismic center, I rushed right to the compass that I have in my house since those days. My intuition was

confirmed, but in a manner that diverged fundamentally from my observations of seismic data in Bolivia. Specifically, while in other cases I could
perceive a deviation of the needle towards west-southwest, in this case an un-
ambiguous tendency towards south-southeast was detectable. By all appear-
ances this is a case of a so-called telluric earthquake (in the narrower sense),
which is essentially different from the cosmic earthquake (in the wider sense).
The difference expresses itself even in the density of impressions. With this
form of earthquake it happens that someone in the adjoining room notices
nothing of what is revealed to us unmistakably. My children, who at that time
were not yet asleep, did not notice a thing, while my wife affirms that she felt
three shocks.

Respectfully,
Civil Engineer J. Berdach
Vienna, 2nd district, Glockengasse 17[71]

Needless to say, anyone versed in the seismology of the day would have
found elements of this letter sufficiently ridiculous to doubt its authenticity:
namely, the writer's distinction between telluric and cosmic earthquakes,
his report of a deviation of his compass, and his attribution of high seismicity to Bolivia. Nonetheless, Kraus was confident that the *Neue Freie Presse*
"will be pleased at last to allow an expert to speak in the midst of so many
laymen."[72] He was sure, too, that the editors would not cast doubt on a
correspondent who was so obviously Jewish.[73] He predicted rightly that the
letter would be printed. For ten years, he noted with satisfaction, the paper
had ignored his writing; now at last they had recognized him—but as an
engineer.[74]

The liberal press needed scientists, Kraus argued. How else could it sustain the "sensation" of a natural disaster without shaking faith in scientific
progress? To be sure, the papers craved catastrophe. As Kraus remarked after
a volcanic eruption on Martinique in 1902, the volcanoes "vomit and they
continue to vomit [*speien*], since our papers' insatiable greed for news is
busy churning up the catastrophe with hitherto unseen hideousness."[75] But
editors could not simply let the victims speak. As the seismologist Albin Belar explained, providing "reassurance and enlightenment to the frightened
inhabitants of the stricken region" was among the "practical" tasks of modern earthquake science.[76] In Kraus's words, each paper "managed to grub
for itself a man of science." Not that scientists had anything of substance
to add to eyewitness testimony in the immediate wake of disaster. Eduard
Suess could do no more than repeat "what could be gathered from previous

München, 18. Januar 1909   13. Jahrgang   No. 42

SIMPLICISSIMUS

Liebhaberausgabe   Herausgeber: Albert Langen   Abonnement halbjährlich 15 Mark

(Alle Rechte vorbehalten)

Messina

„Alles war so schön für einen Krieg hergerichtet. Geht das Rindvieh her und macht ein Erdbeben! Die ganze Menschheit verbrüdert sich wieder mal, und wir haben das Nachsehen."

Fig. 3.3. "Messina": "Everything was so beautifully prepared for a war. Then that meathead comes and makes an earthquake! The whole human race is fraternizing again, and we've lost our chance." *Simplicissimus* 13 (1909): cover.

reports." Kraus scorned scientists like Suess who made themselves "subservient to the press" (*Pressgehorsam*).[77] In the wake of the Messina earthquake, Kraus noted that the liberal papers barely disguised their contempt for the faithful, even as they clung to their own dogma—their unshakeable faith in science.[78]

"We fled to science," Kraus remarked after Messina (see figure 3.3). He quoted the *Neue Freie Presse* on the marvels of instruments capable of recording earthquakes thousands of miles away. "The greater the distance, the more reliably the instruments function. Only when they are at the site of the earthquake is there a danger that they will break." Technically, he was right, since strong motion seismographs had yet to be perfected. More profoundly, he was arguing that science refused to admit when it had been bested by nature. Kraus quoted further from the *Neue Freie Presse* as it invoked the authority of a famous representative of both science and liberalism: "What Eduard Suess, with such wit, has called the pulse of the planet, will be determined with scientific precision." Kraus immediately one-upped this bit of scientific poetizing: "That, however, will do nothing to slow the earth's pulse. And its *bon mots* are more astonishing."[79] As in his "North Pole," Kraus imagined that nature "spoke" in the language of disaster. He framed the earthquake as a witty retort to the inflated rhetoric of science and journalism. The *Neue Freie Presse* relied on Suess to prop up its liberal optimism at a moment when it should, by all rights, have acknowledged defeat: "The Sicilians will indeed allow themselves for once to be enlightened on the count that priests cannot protect them from earthquakes. The devout editors of the *Neue Freie Presse* will never renounce the belief that geologists are capable of it." The faith of the press in technical experts was matched, Kraus implied, by the experts' need for legitimation from the press. The real danger, in Kraus's view, was that science and the press were conspiring to lull the public into a false sense of security: "there comes science with its words of truth and comfort and bestows its benevolent solace on the frightened ones."[80] Kraus refused to let scientific rhetoric twist natural disasters into signs of progress.

The *Grubenhund*

Kraus attributed the power of modern science to the performative quality of its jargon, its rhetorical voice or "intonation" (*Tonfall*): "With the right intonation one can conquer the whole world," he remarked in 1910. "Scream murder and a murder is committed; whisper abracadabra and it is religion; write dynamo exhaust pipes and it is science."[81] The occasion for this reflec-

Der Grubenhund.

Fig. 3.4. *Grubenhund* caricature (1911). Arthur Schütz, *Der Grubenhund: Experimente mit der Wahrheit* (Munich: R. Fischer, 1996), 106

tion was once again an earthquake and the prank it inspired. Over lunch the day after the tremor, a group of engineers fell into conversation about "the unprecedented idiocy [*Schmockerei*] of the newspaper earthquake reports in general and those of the *Neue Freie Presse* in particular." Suddenly seized by "a wild desire," an engineer by the name of Arthur Schütz left the table; when he returned he read his friends the earthquake report of one "Herr Dr. Ing. Erich R. v. Winkler, Assistant at the Central Laboratory of the Ostrau-Karwin coal mines." This individual had supposedly observed the earth-quake's effects on a train's compressor—in excruciating detail and in the name of the "ceaseless efforts of our mining authorities for the protection of the lives of the miners." What would live in infamy, though, was the follow-ing sentence: "A wholly inexplicable occurrence is, however, that already half an hour before the start of the quake, my mining-dog [*Grubenhund*], asleep in my laboratory, gave conspicuous signs of the greatest disquiet." It helps to know that, to a mining engineer, a *Grubenhund* is a cart for car-rying coal (see figure 3.4). The letter reduced Schütz's friends to hysterics, but they refused to believe that a newspaper would print it. Schütz bet that the content of the report was irrelevant; only the tone mattered. Indeed, the letter appeared the following morning in the *Neue Freie Presse*. Kraus was naturally the prime suspect.

The incident gave birth to a new sociological concept. Schütz defined a *Grubenhund* as a false report that sneaks by editors but can be clearly recog-nized by readers. Thanks to his many imitators, the term came to be used more broadly to identify a tactic against the obfuscating jargon of technical experts. Schütz himself later cast his prank as an apolitical act of enlightenment:

"The *Grubenhund* is the symbol of the spoofing of pretended universal knowledge, the protest against the assumed authority of the printer's ink in everything, but especially in technical matters."[82] Kraus credited the *Grubenhund* with having "unmasked the scientific intonation [*Tonfall*]." It had exposed the complicity of science with the press: "For science is by nature so constructed that surprises are not excluded, and its credit rests on misappraisal. In duping journalism, it proved their equivalence and bedded down with it."[83] The laughter that greeted the *Grubenhund* had, according to Kraus, "a tragic feature: it comes from the heartlessness of a belief that has been disappointed."[84] In Kraus's view, the prank had accomplished what the earthquake itself could not: it had finally shaken the public's faith in scientific progress.

Conclusion

Before the advent of lay observing networks, the science of earthquakes relied heavily on the press for its data. This was a compromising position, as scientists well knew. Editors were in the business of selling papers, and a gripping account was far more valuable to them than one that was scientifically observed. Likewise, the prophecies of a Rudolf Falb might boost sales better than the more sober theories of a professional. Nonetheless, the liberal press also needed science. As Kraus demonstrated with his usual wit, disaster stories required the counterpoint of a few confident words from a scientific expert. Nature could not be given the last word. Hence the uneasy partnership that developed between seismology and journalism. It is easy to see why some seismologists would have been tempted to skirt the mediation of newspapers altogether—to build, in other words, their own direct avenue of communication with the public.

Kraus spied a profound irony in the fin-de-siècle culture of earthquake observing. It might seem that the public was more attentive than ever to nature's violence. In reality, though, the alliance of science and the press shrouded natural disasters in a veil of false certainty. Kraus charged that earthquake observations had become a genre unto themselves, a mere literary convention. By recasting seismological evidence as a modern style, a *Tonfall*, Kraus contributed to modernism's deconstruction of disasters as scientific objects.

Nonetheless, Kraus did not deny what was at stake in the practice of earthquake reporting. He agreed with scientists like Hoernes and Belar that cultivating the right attitude toward natural disasters was a matter of popular enlightenment. Kraus insisted, though, that enlightenment could not come from the mere posture of scientific observation. It could come only from a critique of that posture, an exposure of the limits of scientific authority. Yet Kraus's personal disdain for the masses often made his own notion of enlightenment seem chimerical. He sneered at the narrow-mindedness and self-regard of those who cared to report on the behavior of their pets or the state of their broken vases: "The idiocy that would never have thought of emerging from its life in private has discovered an opportunity for immortality; banality has been lured out of its hiding place; average humanity has been hauled out in triumph. A consuming greed to be named has taken hold of the Mr. Nobodies. Thousands besiege the press office, raise their hands to the miracle of the local department and call: Me too! Me too!"[85] Ultimately, Kraus was as uncertain as any scientist about what could be expected of the public. The difference was that seismologists were prepared to find out.

FOUR

The Tongues of Seismology: Switzerland, 1855–1912

The public is learning to observe earthquakes.[1]

Between 1878 and 1880, Switzerland, Italy, and Japan initiated the first national earthquake commissions, but only the Swiss made ordinary citizens a vital part of this undertaking. The nation was divided into seven regions, in each of which one scientist was responsible for soliciting and redacting observations of ground motion and related phenomena, to be penned by volunteers on questionnaires or postcards and mailed free of charge. It was a system that would be imitated around the world, echoed even today in the US Geological Survey's website Did You Feel It? The immediate aims of the Swiss observing network were to locate epicenters by mapping the felt intensity of tremors, determine their depth, and distinguish among different types of earthquake according to felt motion. Implicit was the goal of providing information on seismic hazards for the grand engineering projects for which Switzerland became famous in the nineteenth century—from mines to Alpine trains and tunnels to dams and canals along the Rhine and its subsidiaries.[2] Among the fruits of the Swiss Earthquake Commission's first two decades of work was the conclusion that tectonic quakes occurred principally in three "habitual regions of shock"—in the Rhine valley near Saint Gallen, in the corner between the Alps and the Jura near Lake Geneva, and around the three lakes of the Jura.[3] At the same time, geologists began to interpret tectonic earthquakes as symptoms of mountain formation. As Albert Heim, one of the commission's founders, explained to potential observers: "Most earthquakes are the expressions of the ongoing formation of mountains beneath our feet."[4]

Still, one might wonder, earthquakes? In Switzerland?

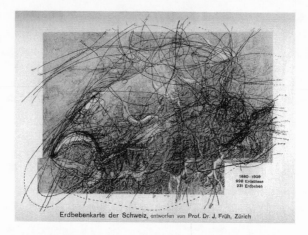

Erdbebenkarte der Schweiz, entworfen von Prof. Dr. J. Früh, Zürich

Fig. 4.1. Earthquake map of Switzerland, based on the first thirty years of the Earthquake Commission's activities, during which it collected seven thousand felt reports. Johannes Früh, "Ueber die 30-jährige Tätigkeit der Schweizerischen Erdbebenkommission," in 94. *Jahresversammlung der Schweizerischen Naturforschenden Gesellschaft*, vol. 1, *Vorträge und Sitzungsprotokolle* (1911).

Geologists today view earthquakes on Swiss territory as the continuing effect of the collision between the European and African plates that began the formation of the Alps approximately 80–90 million years ago. Strong earthquakes occur infrequently, but small tremors are common and often accompanied by distinctive sounds. In the Alps, earthquakes can trigger rock slides or avalanches or make rivers impassable. Today earthquakes are estimated to represent one-third of Switzerland's total disaster risk.[5]

According to the calculations of the Swiss Earthquake Commission, between 1880 and 1904 there was an average of six to seven "Swiss" earthquakes per year—meaning seismic events separated in time, observed by at least two people, and with epicenters inside the Swiss borders.[6] (See figure 4.1.) Before the commission began its work, earthquakes in Switzerland were often not recognized as such. Many residents mistook them for strong winds or thunder. It was not uncommon for people to accuse their upstairs neighbors of making noise that was actually seismic in origin. In some cases witnesses were aware only that the horse pulling their carriage had stopped dead in its tracks.[7] Some residents attributed the vibrations to an "invasion of a legion of cats, mice, or rats in the garrets."[8] These tremors rarely caused much damage: "In general buildings have very infrequently been damaged, thanks to their solid construction, in particular of the wooden buildings of the mountain region. In many cases cracks in walls and in the ground, avalanches, and rock slides have been reported, as well as the ringing of church bells and general panic of the inhabitants. Springs have been clouded, run dry, or appeared in new places. The local formation of waves on lakes in the absence of wind has repeatedly been observed."[9] In an age before reliable seismographs, such frequent sensible but not destructive seismic activity furnished ideal specimens for naked-eye research.[10]

1855: Volger's Democratic Example

On a wet and foggy day in late July 1855, the strongest Swiss earthquake of the past three hundred years struck the Visp Valley in the canton of Valais (Wallis). It was accompanied by "thunderous rumbling, stormy booming, heaving whooshing, cracking, crashing, crackling, braying, whistling and clanging . . . the mountains rocked up and down, countless rocks from all the peaks and cliffs slipped and rolled, the walls of the houses tilted and fell over in a heap of rubble, balconies broke, roofs slipped or collapsed."[11] The shock was felt through most of Switzerland and into neighboring parts of Germany, France, and Italy, and aftershocks were reported into November.[12]

At the time of the Visp earthquake, Otto Volger was a *Privatdozent* (unsalaried lecturer) in geology at the University of Zurich, having fled his native Saxony after participating in the failed revolution of 1848.[13] Apparently unaware of the magnitude of the project he was undertaking, Volger began collecting eyewitness reports of damage from the earthquake. In order to assess the quake's more remote effects, he put in place the first nodes of what grew into an extensive network of correspondents. Volger's evidence came from "people with a great variety of educational backgrounds" [*Personen von verschiedensten Bildung*].[14]

During the three years he spent writing his monograph on the 1855 earthquake, Volger held a part-time teaching position at a natural scientific society in Frankfurt. It was in Frankfurt that Volger became, according to Andreas Daum, one of the German-speaking world's first semiprofessional popularizers of science. Volger sought to popularize a stridently secular, cyclical view of geohistory, a challenge to biblical catastrophism. His earthquake research came to serve this goal. As he explained, "The rocking of the earth's surface does not announce the decline of the world, but rather the rule of the eternal order of nature, which in transformation furnishes the certain guarantee of its everlasting constancy."[15] Volger also became active as an organizer, working to counteract the growing specialization of the sciences. In 1859 he founded the Free German Foundation for Science, Art, and Public Education (*Freie Deutsche Hochstift für Wissenschaften, Künste und allgemeine Bildung*), a feat of networking among existing institutions. According to Daum, Volger was "driven by a democratic impulse. Through the FDH, [Volger] promised not only an improved exchange of information, but also the opening of research and teaching institutes to the public."[16] In keeping with his goal of making science accessible to the public, Volger studiously avoided foreign terms in his publications. Nonetheless, he styled himself as a researcher and scorned writers who viewed the world merely as "material for penny-a-line articles."[17] Indeed, in his own estimation, his monograph on the Visp earthquake offered no reward to its publisher beyond the satisfaction of seeing his name on the title page. (In the event, the publisher died before enjoying even that compensation.) While the book itself was no mere popularization, its methods and rhetoric conveyed Volger's democratic vision for the sciences.

Volger thought he knew what caused earthquakes, and his theory was not unrelated to his politics of knowledge. Earthquakes were caused by the collapse of underground caves due to the inflow of water. He branded those who insisted on volcanic causes with the eighteenth-century label "Plutonist." Referring to a "Plutonic hierarchy," Volger repeatedly characterized Plutonism as a kind of elitist superstition. For instance, correlations between earthquakes and volcanic eruptions amounted to "downright mischief with the temporal relations . . . to make 'the earthquake' look like a certain ghostly something which has its seat beneath the earth's surface and capriciously materializes first here, then there, in a way all too redolent of table-turning and spirit-rapping." He thus cast Plutonism as a highborn fallacy, and this rhetorical strategy shaped the way he reported eyewitness observations. For instance, he discounted reports that the quake had opened cracks in the earth or released thermal springs (potentially evidence for a Plutonist account of lava surging from below), by attacking the character of the reporters. These were either "superstitious" and "pseudo-scientific" "travelers" or sensationalistic journalists: "If the reports of travelers that have been spread in the newspapers spoke of ruptures in the mountains, of cracks in the ground, of the opening of the earth, of rivers of mud spewed from the earth, and similar scenes of horror, then what this depiction revealed was only the entirely amateur [*laienhafte*] viewpoint, defined by superstitious, pseudoscientific prejudices." Thus Volger embraced the testimony of eyewitnesses whose ability to "appreciate these phenomena" may have derived simply from their experience as locals, at the expense of the descriptions of "travelers" who refused to admit their "lay" status. By stigmatizing travelers' reports as both "superstitious" and "pseudo-scientific," both pretentious and "lay," Volger was challenging a class-based claim to adjudicate "science."[18]

At the same time, Volger's trust in the testimony of local residents suited his explanation of seismic activity. If water seepage could induce tremors by causing the collapse of subterranean caves, then rainfall and humidity were a proximate cause of earthquakes. Volger therefore needed reports on meteorological conditions not merely at the time of the quake, but also in the preceding weeks. He needed to trust the local residents, and he did. He spiritedly defended the experiential knowledge of locals against the exaggerations of well-to-do travelers. For instance, he rejected a "traveler's" account of a sulfurous smell (associated with "Plutonistic" explanations) on the ground that the witness lacked the experience to recognize the "half-musty, half-smoky scent of freshly overturned earth."[19]

Remarkably, what might appear to be the work of a pugnacious upstart turned out to be a pioneering contribution to seismology. Volger's research was original both in its conclusions and methods. His critique of Plutonism was apt, in so far as it cast earthquakes not as the product of eruptive forces but as a result of preexisting ground conditions, exacerbated by fleeting external factors.[20] Volger himself drew an analogy to the breakup of a layer of ice on a lake. His point of view stimulated further studies of the local geology of earthquake zones. It also inspired the compilation of historical earthquake catalogs on the model of Volger's own *Chronik der Erdbeben in der Schweiz*, the first volume of his publication on the Visp quake. Last but not least, Volger succeeded in developing one of the first quantitative scales for measuring an earthquake's intensity, as well as one of the first maps of the geographic distribution of intensity. His map introduced a standard convention of macroseismology today: isoseismals, or lines of equal seismic intensity. The final version used seven colors to plot the variation of intensity for the main shock; it also showed the limits of perception for three aftershocks. The map summarized Volger's research in a manner well suited to a popular audience. It displayed information about the distribution of seismic hazard at a glance and without technicalities. Indeed, the conventions of seismic intensity maps constitute a visual scientific vernacular, much like the weather maps that were becoming popular in the nineteenth century.[21]

There is little sign, however, that these innovations were appreciated by Volger's contemporaries. The immediate significance of Volger's study was instead its demonstration that a thorough survey of eyewitnesses and of historical sources could produce valuable evidence for the scientific analysis of an earthquake.[22] Equally influential was Volger's populist spirit: as one director of the Swiss Earthquake Commission later noted with approval, Volger "seemed not to have found the required tone of subservience toward the official custodians of scientific knowledge of the earth, but rather dared to forge his own paths."[23]

Earthquakes, Experts, and the Nation

Volger's study of the Visp earthquake was quickly overshadowed by the introduction of "objective" methods of earthquake research. In 1856 Luigi Palmieri put an influential new form of seismoscope to work on Mount Vesuvius, and in 1857 Robert Mallet conducted his widely imitated architectural survey of the Naples earthquake. As we have seen, Mallet and many of his colleagues in Britain, Italy, and Japan considered eyewitness testimony to be a poor substitute for architectural or instrumental evidence. Why then did the Swiss place so much trust in the observational abilities of ordinary people?

One answer is that the Swiss did *not* trust instruments, and with good reason. The pendulum seismometers of the 1870s were plagued by problems of self-resonance that were not addressed until the late 1890s. For assessing local earthquakes, the Swiss Earthquake Commission judged human observers more reliable than instruments: "the number of cases in which people took due note of tremors, while instruments in the same region remained indifferent, is in all countries too high. The horizontal pendulum and related complicated and expensive instruments adequately register distant earthquakes, elastically propagated through the earth, but not generally local earthquakes. . . . We still [1905] lack a simple instrument with which, as with meteorological instruments, any non-physicist can successfully make observations."[24] Observers thus recorded local earthquakes more reliably and less expensively than instruments.[25] (However, a full tally of the expense of using human observers would have to include the considerable cost of printing and mailing questionnaires and the time required for experts to evaluate them.) As Heim advised the Princeton mathematician and seismologist Charles Rockwood around 1885, "The simple reports of observers without instruments have consistently been more valuable to us than instruments. Even if they contain many errors, nonetheless one finds that if a really large number of reports are available from the same place, it is always easy enough to distinguish what is generally objective from an accidental error of a subjective kind [*das allgemeine objective von zufälliger Täuschung subjectiver Art zu unterscheiden*]."[26]

This judgment reflects the specific goals of the Swiss Earthquake Commission. While seismologists in Germany poured their resources into the study of distant earthquakes with the latest seismographs, the Swiss wanted to know about *Swiss* earthquakes and the seismic hazards of their own territory. But suppose that the Swiss could have found an inexpensive instrument that could reliably register local quakes? Would they have eliminated lay volunteers from their program? Note that the commission stipulated that such an instrument would need to be easy for the "nonphysicist" to use. The conclusion seems inescapable: Swiss scientists fully intended to remain dependent on nonexpert observers.

It is tempting to attribute this eagerness for popular participation to the Swiss political tradition, but this claim needs to be carefully qualified. An illuminating comparison can be made to the system of meteorological observation likewise set up by the Swiss Natural Scientific Society (*Schweizerische Naturforschende Gesellschaft*, or SNG) in the 1860s. The comparison is a natural one, since key members of the Swiss Earthquake Commission were meteorologists, and it was common at the time for seismic phenomena to be noted by weather observers. A national network of volunteer weather observers was inaugurated in Switzerland in 1864, less than two decades after Switzerland had become a federation with four official languages, significant cantonal autonomy, and direct democracy through referenda. Albert Mousson, an engineer-naturalist and the son of a chancellor of the Swiss Confederation of the Napoleonic Era, presented the establishment of this volunteer network as evidence of the success of Swiss federalism and liberalism: "Without enjoying the advantages of a strong centralized administration, as its strong monarchical neighbors are able to provide, Switzerland has in this respect, thanks to its freedom of movement and advanced learning, not entirely fallen behind; namely the Natural Scientific Society can pride itself on having inspired, initiated, and completed much on its own initiative, which elsewhere is achieved only on the basis of official organization, through learned academies and through numerous paid employees."[27] Switzerland thus managed to keep pace with the scientific progress of its monarchical neighbors "on its own initiative [*aus freien Stücken*]," thanks to liberal policies (*freie Bewegung*) and an educated public. Still, there was a high bar to becoming a meteorological observer. It required familiarity with instruments (if not their purchase, since instruments were supplied by the SNG), enough leisure time for regular daily observations, and a background in mathematics to perform data reduction. Indeed, more than half of the original weather observers belonged to the learned professions. In mountainous areas, the society was forced to rely on innkeepers, and this made

Mousson nervous: "there remains the question of whether it was appropriate to entrust such fine instruments to the hands of non-physicists. . . . The three-year experiment will at best teach us which observers can definitely be counted on and which would better be abandoned."[28] Looking back on the network's first year of operation, the best Mousson could say was that no instruments had been broken. Democratic rhetoric alone did not ensure that all citizens would participate in Swiss science on equal terms.

Earthquake observing proved a more genuinely populist enterprise than meteorology. This was evident in the Earthquake Commission's instructions to observers, which were distributed to "all telegraphic and meteorological stations, all natural scientific societies, to physicians, pastors and many teachers."[29] They were written explicitly "in order to stimulate interest and understanding for the observations in wider circles. . . . For the investigation of every earthquake numerous observations from as many different locations as possible are necessary. The scientist [Naturforscher] is dependent here on the help of numerous friends of science. He turns not only to his colleagues, but to everyone who takes an interest in the observation of nature." The author, Albert Heim, suggested that laypeople could contribute to scientific research in many ways. The questionnaires guided the observer through a set of questions that requested certain standardized responses, but also prompted attention to multiple "associated phenomena" and left ample blank space for elaboration. Moreover, observers were explicitly free to choose to dispense with the questionnaires and write free-form letters instead. Indeed, Heim instructed observers to report anything and everything: "The observations should not be limited by our questionnaires; rather every further observation, beyond those to which we have drawn attention, can be valuable."[30] This openness derived in part from the lingering uncertainty about what constituted seismology's data, which might range from the astronomical and meteorological to the electrical, magnetic, and zoological. As Heim put it, "Nature is far more complicated in its phenomena than we could guess."[31] In this sense, keeping an open mind was a sensible research strategy for seismology.

Heim made clear that the call for volunteer observers was motivated by more than a need for data. It was meant to condition a new set of responses to the earth's volatility. Citing Buckle, Heim acknowledged that the experience of an earthquake could impair reason, if unaided by scientific preparation: "When the most solid thing that we can perceive with our senses wavers, and we do not know whereby and how far, then the imagination is easily excited to a fever-pitch and clouds rational observation." Science had the potential to fortify human reason in the face of natural hazards: "Every new discovery about the relationship between cause and effect in nature has a great value for humanity, for it helps free the human spirit from the chains in which it was once imprisoned by the base fear of what is not understood and, therefore, seems arbitrary [*vor dem nicht zu Begreifenden und deshalb vielleicht Willkürlichen*]."[32] Heim echoed an ancient philosophical tradition, in which the highest aim of natural knowledge was to release man from fear. Consider, for instance, the famous response of the poet Seneca to an earthquake that struck the region of Campania in 62 CE. "What consolation—I do not say help—can you have when fear has lost its way of escape?" For the stoic, the proper response to such boundless terror was rational explanation. Seneca therefore invoked Greek theories of underground water and trapped air, ultimately explaining earthquakes as the result of strain on one part of the "body" of the cosmos.[33] In this vein, Heim suggested that seismology could serve a therapeutic purpose, consoling and fortifying the mind in the face of terror. Yet Heim's aspiration was unique to the nineteenth century in two respects. First, he reconceived seismology as a popular, empirical practice, and thus as a means of mass enlightenment. Second, the source of spiritual comfort for Heim did not lie in an all-encompassing explanation, which was forever beyond the reach of empirical science, but rather in the empirical attitude itself. A parallel was evident in Heim's response to the 1881 rock slide in Elm, in the Glarus Alps, which killed over a hundred people. He called for public participation in recognizing and reporting the warning signs of disaster: "Even with respect to rock slides, we want to enter the enlightened age [*Zeitalter der Erkenntnis*]."[34]

Scientific earthquake observation was thus fundamentally different from weather observation, and the difference derived not just from the absence of instruments. More fundamental was the human threat that earthquakes posed. Historians have argued that the Swiss nation was forged to a significant degree through common measures to protect against natural disasters. As Daniel Speich shows, Hans Conrad Escher's canalization of the Linth River at the turn of the nineteenth century aimed to improve not just the land but the moral fabric of the nation. Nature, Escher wrote, "forces us into a social condition, as we have to act united, if one man alone is without help." The swamplike conditions in the valley were considered morally debilitating to the local peasantry, whom Escher recruited to build the canal. To reverse these effects, Escher provided the workers with special schooling. Speich points out that such a combination of environmental engineering and public education was unlikely after the early nineteenth century, due to growing technical specialization.[35] In this light, the Swiss Earthquake Service is all the more surprising. It was a combination of environmental defense and popular enlightenment in the mold of Escher's canal. And it was an undertaking that was perhaps more urgent in Switzerland in the late nineteenth century than ever before, as the rise of ethnic nationalism in the rest of Europe threatened the civic nationalism of the federation. The Swiss cultivated an ethic of "civic exceptionalism," a "voluntary commitment to a set of values and institutions, which in turn secured Switzerland's existence as a polyethnic state." Swiss nationalism increasingly emphasized the role of the Alpine landscape in forging a supra-linguistic unity. In 1891, for instance, at a festival celebrating six hundred years of Swiss history, the character Helvetia, the embodiment of the Swiss nation, was said to be "shaped by mountain peaks and valleys alike."[36] The physical barrier of the Alps was likewise central to the emerging Swiss ethos of isolationism. In this context, earthquake observing was part of a lesson on civic identity. Not for nothing did the Swiss Alpine Club, promoter of the nation's identification with its landscape, contribute the cost of printing 1,450 German and 850 French copies of the commission's instructions to observers.[37] The Swiss were being taught to feel the forces exerted by the mountains that united them.

Switzerland's network of volunteer earthquake observers self-consciously harked back to a pretechnical era, in which science stood for enlightenment and civic conscience. It also looked ahead, to an age of rapid communication and precision timekeeping. It is a case that justifies Ted Porter's assertion that "the Enlightenment—if this term may be taken to designate a faith in progress through the popular diffusion of knowledge—took place in the *nineteenth* century."[38] What's more, enlightenment in this case was potentially a more radical project than the mere diffusion of knowledge. It aimed to enlist the public in the *coproduction* of knowledge.[39]

Albert Heim, Nature's Physician

We can better grasp the ambitions of the Earthquake Commission by considering the public personae of its directors. Many of its members were affiliated with the Swiss Federal Institute of Technology (ETH), founded in 1855 to serve the practical needs of a modernizing nation. Many also served as geological consultants on state-sponsored engineering projects. One such project was the Simplon Tunnel, a landmark feat that bored a road beneath a five-thousand-foot peak. Most of Switzerland's top geologists advised on the Simplon at some stage in its design and construction, which stretched from the 1870s into the 1920s. Earthquakes were not prominent among the hazards that imperiled its construction (though a quake did cause a fissure in the first section of the tunnel). The real dangers would turn out to come from flooding and unexpectedly high temperatures. Scientists faced accusations that they should have warned of the risks involved. Albert Heim, who consulted on the tunnel from the 1880s, gave a revealing response to these charges. Rather than shifting the blame, Heim retorted that a geologist's knowledge was inherently uncertain.[40] As he later put it, "It is self-evident that the viewpoint emerging from the geological investigation cannot alone be decisive. What strikes us though, is the uncertainty in the assessment of many natural circumstances."[41] Heim had cast his original assessment of the tunnel project in similar terms. It was, he wrote, "impossible for me to foresee" the subterranean dimensions of decomposing gneiss. Likewise, he and his colleagues could not say for sure how high the temperatures might climb during construction; they could only judge which of the proposed routes would be safest in this respect.[42] Repeatedly in his role as scientific consultant, Heim stressed the limits of his ability to predict the environmental consequences of engineering projects. In this way, Heim circumscribed his own authority as an adviser on matters of public interest. "How it would have pleased me to have been able to present a clear and completely satisfying solution to the question," he wrote in an expert report on the provision of drinking water to an alpine town. "Nature, however, which has so unequally distributed its gifts, cannot be made other than it is—we can only observe *how* it is, and use it according to its possibilities. Much of the *how* is hidden and only within certain limits to be discovered, so that much uncertainty still attends our judgment and our foresight."[43] With his unusual insistence that nature be used only "according to its possibilities" and with due regard to the scope of scientific uncertainty, Heim erected a wise counterweight to the ambitions of nineteenth-century engineering.[44]

In keeping with Heim's sense of caution, the Earthquake Commission helped remind the Swiss of the principle of unintended consequences. A remarkable example occurred in 1913, when moderate earthquakes shook the vicinity of Grenchen near Solothurn. Never before in the history of the commission had earthquakes originating in this region been reported. Two years earlier, construction had begun on a tunnel through the Grenchen Mountain. In early 1913, the engineers struck the local spring, flooding the tunnel and cutting off the village's water supply—the first of several such mishaps. Four months later the tremors began. The press began to raise questions. The director of the Earthquake Commission, Alfred de Quervain, studied the reports of earthquake observers alongside the engineering of the tunnel. This, he concluded, was the first clear evidence that the boring of tunnels could trigger earthquakes.[45] "Induced seismicity," as this hazard is known today, has recently been recognized as a major problem for twenty-first-century schemes of energy production and waste disposal.

Swiss seismology thus proceeded according to the principle that nature should be used only "according to its possibilities." This was the principle behind Switzerland's burgeoning nature protection movement, in which Heim and other members of the Earthquake Commission took leading roles.[46] The nature protection movement had its immediate origins in the effort to preserve a mammoth rock, fondly known as the "Pierre de Marmettes," from the hands of a granite speculator. This two-thousand-cubic-meter landmark was an "erratic block," one of the massive boulders poised at precarious angles on the outskirts of the Jura Mountains. These imposing objects had long been treasured as curiosities. In the late nineteenth century, they came to be recognized as relics of the last ice age, transported by glaciers to their unlikely perches. In 1906, an alliance of Swiss scientific societies paid over thirty thousand francs to rescue the Pierre de Marmettes. Thus was a movement born that stressed the aesthetic, scientific, and historic value of natural monuments, against their mere economic worth.[47]

The Pierre de Marmettes came to symbolize nature's need for protection against human self-interest. Tellingly, Heim likened himself to nature's physician. Writing of rock slides, he described the geologist as providing "diagnosis," "prognosis," and "therapy." Implicit was an analogy to the work of Heim's wife, Marie, Switzerland's first female doctor. Marie Heim herself saw the similarity: "My husband's occupation and mine are actually very alike and therefore make equal demands on us. As I get called to suffering people, he is suddenly called to avalanches, failures of tunnels and dams, and potential rock slides. Like the physician, the geologist is also very often exposed to dangers. Just as the physician first observes and judges and then must prescribe and treat, the main activity of the practical geologist is likewise observation, thorough investigation, and considered judgment."[48] Indeed, Albert Heim seems to have worked according to a kind of geological Hippocratic oath—a recognition of his limitations as planetary healer and a promise, at worst, to do no harm.

For Heim, enlisting lay observers was not a last resort but an end in itself. Heim went so far as to suggest that nonexpert observers might have an advantage over scientists themselves. In an essay on "Seeing and Drawing," Heim argued that modernity had dulled man's alertness to nature. Pointing to the verisimilitude of prehistoric cave drawings, Heim noted, "The savage sees much more consciously than civilized man; he devotes himself to sensory impressions with his full interest, without allowing his attention to be disturbed by other series of thoughts." From this perspective, the "technical exercises" of modern scientific drawing could do little to improve one's ability to represent nature. Improvement could come only from "intimacy with the things represented." Scientific travelers were often surprised to find that "savages" (*Wilde*) were often the better draftsmen. Heim suggested that man's observational capacity had been corroded by religion, which had "overstimulated the imagination."[49] In his instructions to earthquake observers, Heim urged an attentiveness that steered clear of the modern extreme of nervous sensibility: "The sense organs should be kept alert in the correct state of tension, without exciting them to a state that would exaggerate sensations."[50] Heim was hardly the first European to have credited "primitives" with exceptional observational powers. Charles Darwin had commented famously on the Tierra del Fuegians' "more practiced habits of perception and keener senses, common to all men in a savage state, as compared with those long civilized."[51] Darwin saw no reason to assume that a British scientist was a better observer of nature than a "savage"—nor even, he might have added, than an insect or reptile endowed with the capacity for mimicry. As we will see in the next chapter, this anthropological perspective informed nineteenth-century debates about the observation of earthquakes. Observers were often compared by race and even species—snakes and catfish were considered particularly keen.[52] If scientists might turn to "savages" for scientific drawings, why shouldn't geologists turn to Swiss citizens for earthquake reports?

Heim maintained this capacious idea of seismological observation even as other seismologists increasingly restricted their data to instrumental traces. He had little faith in the objectivity of self-registering instruments.[53] As he argued to the International Seismological Association in 1909, "Above all, a more precise observation of all earthquake phenomena is necessary, in order to learn to understand them. One cannot say from the start, *how* they are to be observed. One must experiment, experience; and theoretical perspectives and hypotheses can and will enrich observation."[54]

F. A. Forel: Earthquakes and Ecology

Heim was not the only scientist whose personality and politics shaped the Earthquake Commission's outreach efforts. The Neuchâtel-based naturalist F. A. Forel is better known today as a founder of an early strain of ecology: limnology, the study of the physical and organic conditions of mountain lakes. Limnology stood as a critique of the modern splintering of scientific expertise and the estrangement of scientists from those with practical interests in the natural world. As Forel wrote in his seminal treatise on *Lac Léman*, the study of limnology forced the scholar "to transform himself successively into a physicist, chemist, zoologist, botanist, archeologist, historian, economist."[55] On their own, none of these disciplines could come to know a lake in the holistic manner of limnology. Forel echoed the comprehensive vision of Alexander von Humboldt: "Geography has, in effect, a noble ambition and a magnificent program: propelled by its definition, it aspires to embrace in a vast generalization the ensemble of the sciences that deal with the earth and its inhabitants, that is, all the human sciences. . . . The description of the earth is not the enumeration and individual description of each of the categories of creatures and things that encounter each other on our planet; it is rather the picture of the whole furnished by the union of these diverse categories, by their relationships to each other, by the reactions that they have to the environment in which they are immersed and which they produce on this environment."[56] In its ambition to grasp a holistic vision of living things in relation to each other and to their physical environment, Forel's limnology can rightly be termed ecological.

Limnology was also a practical endeavor, and Forel addressed his study in part to a lay audience, a "diversity of readers," from ship captains and fishermen to those who made their homes on the lake. All such individuals would profit from knowledge of the lake's life cycles and the conditions that made it livable.[57] Forel wrote for "general readers," for whose sake he eschewed jargon: "Rarely can an idea or act not be expressed in familiar language. . . . If our beautiful science is not to become repulsive, we must avoid deforming it with too many foreign words. We must beware of making limnology incomprehensible to the many readers with an interest in limnology by the use of foreign words."[58] By the time he penned these lines, the ideal of communicating science in "familiar language" had also become central to Forel's work for the Earthquake Commission.

"Familiar Language"

Like Heim, Forel presented the commission's work as an opportunity to train citizens in a scientific mode of observation. Reflecting on the observations collected in the network's first year, Forel expressed guarded optimism: "The value of these documents is very unequal, as one would expect; some of them have all the characteristics of excellent scientific observations. But if there are some that are less precise and exact, there are very few from which there is absolutely nothing to be drawn. The general impression that results from the study I have made is that the method adopted by our commission, consisting of calling on the observations of the public, can, as it develops, give excellent results." More specifically, he deemed the majority of reports "very authentic," while only twelve were "imprecise, rather doubtful," marked by a lack of coincidence in the reported times. One of the "best" observations was that of a Monsieur Clement, a butcher in Yverdon. Indeed, as Forel's report for the following year would point out, observations came from "all the classes of society." To his satisfaction, moreover, the public was making progress: "A particularly interesting fact is that the relative value of these observations has notably risen, and that, after two years of collecting these documents, we recognize a very evident superiority in those that were sent to us most recently: the public is learning to observe earthquakes." Still, Forel urged a critical attitude: "Let us not however go too far, and let us not overestimate the precision of our documents."[59]

Forel did not place the burden of improvement solely on the public: scientists themselves needed to learn to interpret the observations of laypeople. Thus Forel judged "there are very few of them, even among the most simple and modest, from which an intelligent comparison and critique can not draw something useful." Not only would the public need to learn how to report to scientists; scientists would have to learn how to listen to the public. In this vein, Forel began, "I have, first of all, some definitions to make, which will spare us fruitless repetition and simplify description."[60] These definitions would lay the foundation for a dialogue between scientists and lay observers.

On one hand, Forel picked up on phrases that observers themselves were using. "The tremor [secousse] can be composed of several distinct movements that I will call, according to circumstances: *oscillations*, when there is a predominance of swaying movement; *vibrations*, when the movement has the character of a vibration; *shocks*, when there is a violent, brief, sudden impulse, etc." On the other hand, he offered new terms that members of the commission could use to compare the reports of different observers. For instance, descriptions of an event as "violent," "brief," or "sudden" could now be classed together and labeled as a single "shock." With the exception of the term *sismique*, which Forel derived from the Greek for earthquake, all the terms he used were drawn from "familiar language."[61]

Forel did not deny that the reports were colored by the subjectivity of the observers. For instance, reports of the wavelike movement of fixed objects were likely a "subjective illusion." Yet such reports were not insignificant. They could not be dismissed as the bias of observers familiar with theories of seismic waves. More likely, the slight movement of the observer's body during the earthquake produced the apparent wavelike motion of other objects. Subjective here implied a physiological rather than psychological bias. Forel further distinguished between "*objective* observations, based on the direction of swaying of a hanging object, or on the displaced of mobile objects, and *subjective* observations, based on the simple appreciation of the sensations of the observer." Here, "subjective" did not imply unreliable. It meant simply that the object under observation was the observer's own body.[62]

Critics of the use of lay observers often pointed to the difficulty of extracting precise reports of the time of an earthquake. The factor of time was of the essence, necessary for determining whether separate reports related to the same shock, calculating the velocity of the wave as it moved through the earth's crust, and localizing the epicenter. In the moment of alarm, however, people rarely bothered to look at a clock. And if they did, it was all too likely that the clock was wrong. With standardized time still a novelty, few people could be relied on to calibrate their house clock or pocket watch to the "telegraph" clock at the local railway station. At the start, Forel was disappointed to find that Switzerland's famed culture of precision timekeeping fell short of scientific standards: "the setting of the clocks of towns, railways, and telegraph stations leaves much to desire from the perspective of precision."[63] Even thirty years later, in the case of the strong 1910 tremor in the Berner Jura, Quervain was aghast to find that "among the many men of science who encountered the earthquake fully awake, none saw fit to look at his second hand and verify the setting of his watch."[64] Again, the commission saw an opportunity for public enlightenment. They "aspired to educate the public to be able to tell time with real precision." In 1907 Quervain published five hundred copies in German and French of an "Introduction to Precise Timekeeping for Observers."[65] The point is that in Switzerland it was not out of place for a scientist to berate the public to "to recognize the moral duty, during an earthquake, first thing to look at the second hand."[66] Already in 1881, "in the successive observations that are sent to us," Forel observed "a progress and an improvement in the determination of time, apparently the result of habituation and attention to the art of observing."[67] In 1910 the commission remarked, "Often it is complete amateurs who distinguish themselves by a pleasing sense for precision."[68]

In addition to codifying a language of observation and developing methods for interpreting observation reports, Forel presented a scheme for moving from discursive to quantitative description: a scale of seismic intensity, measured according to "the simple observation of the effects produced on man and his dwellings."[69] The ten degrees of Forel's scale ranged from barely perceptible to disastrous, making it far more widely applicable than the seven-degree scale that Volger had used to classify damage from the 1855 earthquake. Forel only intended his scale to be used "while awaiting more precise observation by standardized recorders," but it would form the basis of scales used in Switzerland and elsewhere into the 1960s.[70] Modified in collaboration with an Italian colleague, the "Rossi-Forel" scale was by far the most widely used circa 1900.[71] As part of a carefully constructed scientific vernacular, the Rossi-Forel scale bears comparison to the scales and idioms that emerged in the nineteenth century for the observation of winds and clouds.[72] Cloud taxonomies and the Beaufort wind scale were scientific tools based on the experiences of laypeople—artists, farmers, fishermen, and sailors. Unlike seismological intensity scales, however, they were not designed to register the subjective effects of geophysical phenomena on humans. This peculiarity of earthquake science was decisive.

"The Relationship of Man to Earthquakes"

The Rossi-Forel scale committed seismology to a culture-bound perspective on earthquakes. Intensity measures the *felt* effects of shaking and is assessed primarily with respect to humans and their habitations. It can be used both to compare seismic events and to identify regions of seismic hazard, but it is not possible to measure intensity in uninhabited areas.[73] Forel's scale confirmed seismology's dependence on lay participation. In fact, the scale operated in part by calibrating the reactions of the general public. For instance, degree 6, "Some frightened people leave their dwellings," versus degree 7, "general panic." In the Italian version of the scale, degree 3 tremors were "felt by people who are not concerned with seismology." By contrast, a degree 1 quake would be noted only by a "practiced observer" [*exercé, geubten*]. Word choice was telling: the relevant criterion for such sensitivity was not expertise but practice. The commission found that weak tremors were more frequently reported in the wake of a widely felt one, an effect that seemed too robust to attribute to suggestion. Indeed, the public *could* learn to observe earthquakes. And scientists could learn to work with their observations. Hence Forel's comment that "this scale will be able to be corrected once we have more experience in this line of research."[74]

Forel's scale reflected the ambiguous status of eyewitness reporters: they were both scientific observers and objects of observation. Certain constructions suggested that the witness's role was passive. Thus, for instance, the parallel constructions of "shocks registered by different seismographs" and "shocks reported by persons at rest" or "by active persons." Other phrases attributed a degree of discernment to the witness. Only a shock of degree

3, for instance, was "strong enough that the duration or direction could be appreciated." Only a shock of degree 6 would produce an "*apparent* shaking of trees and bushes." These phrases hinted at the mindfulness expected of witnesses. Inherent in the scale itself was thus the dual status of earthquake observers as experimental subjects and amateur naturalists.

Geography, according to Forel, was interested in "the reactions they [people] have to the environment in which they are immersed and which they produce on this environment." The same could be said for seismology. Details and idiosyncrasies of ordinary people's responses were recorded, remarked on, and preserved. An analogous approach could be found in the field of "cosmic physics" in late nineteenth-century central Europe. Phenomena like atmospheric electricity and the mountain wind called *föhn* were analyzed according to new theories of microphysics or hydrodynamics but also, in an ecological vein, as phenomena that shaped and were shaped by particular places. In all these cases, human reactions were part of a holistic understanding of the physical process. Thus *föhn* could trigger heart failure or even criminal behavior; atmospheric electricity could aggravate epidemics of cholera and influenza. A similar perspective underwrote the commission's commitment to using human observers alongside improved seismographs. As the commission's director put it in 1910, "To study the relationship of man to earthquakes was from the start a special goal of Swiss earthquake research."[75]

"Real Seismoscopes"

As the Swiss public was learning to observe earthquakes, an unusual picture was emerging of the scientific observer. By the late nineteenth century, scientific observation was typically associated with a self-effacing, sober, emotionally disengaged expert, by default male.[76] Some earthquake observers fit this description. For instance, a professor of physics in Mélan described how he had observed a tremor: "I was in the garden occupied with thermometric observations; I was watching a hair-hygrometer, and I saw with astonishment that its needle had convulsive movements, which I had never noticed to the same degree. Suddenly I heard an underground thunder, then I felt myself falter a little."[77] Certainly, there were observers who regularly engaged in other forms of scientific observation and happened to have a sensitive instrument at hand when an earthquake struck. In general, though, successful earthquake observers were a motley group. Interestingly, women were thought to excel. As the Italian seismologist Mercalli remarked when proposing his revisions to the Rossi-Forel scale in 1902, "very sensitive and nervous people, especially women, feel earthquakes much better than I do."[78] His degree 2 therefore specified "felt by many sensitive and nervous people." Looking back on thirty years of the Earthquake Commission's activity, one director noted, "Very interesting is the physiological fact that certain individuals of both sexes have an almost astonishing sensitivity for ground movements—[they] represent real seismoscopes."[79] A quiet, housebound lifestyle and close attention to the arrangement of domestic objects put many bourgeois women in an excellent position to detect tremors.

Already in the first years of the commission's work, it was becoming clear that the typical conditions of scholarly production were not the most favorable to earthquake observing. Attentive watching and listening seemed of secondary importance. More earthquakes tended to be reported at night than during the day.[80] Though some events were certainly missed by those asleep, people at rest—even if half-conscious—were more likely to feel the earth move than people in motion. Scholars who spent most of their days indoors were more likely to notice a slight tremor by means of an object's shift in position; but those who worked outdoors could more reliably report the direction of shaking, being free of the distorting effects of human constructions. In one vivid example of these paradoxes, a young mother sent in a report with observations made by herself and her two sons. It was the younger boy, playing on the floor with his toy soldiers when the earthquake struck, who "saw his little cardboard soldiers fall before him while saluting the Duke of Brunswick!" As a statue of the duke of Brunswick stood just to the south of the family's home in Geneva, the local commission member commented that the "fall of the little soldiers . . . seems to demonstrate this direction."[81] With no hint of amusement, the commission deemed the little boy's observation the best of the lot.

Building an Archive

Historians of science have just begun to explore historical developments in the methods of recording and storing information about natural environments in flux. This is, in part, a history of genres, including local natural histories and commonplace books in the seventeenth century; personal weather diaries in the eighteenth; the catalogs of botanical gardens, zoological museums, and synchronized global observing networks in the nineteenth; and the emergence of computerized databases in the late twentieth.[82] Switzerland's macroseismological network straddled two stages of this history, helping to transform methods of individual, local observation into geographically widespread, collective practices. Conscious of the challenge of organizing information on a new scale, the commission decided to maintain two record books, one in which each "Swiss" earthquake would be entered and numbered in order of occurrence, with a reference to the folders where the relevant observations were stored, and one in which records of past earthquakes would be collected "from the literature."[83] These books and folders would constitute the commission's archive, to be stored at the Bern observatory. "All these documents are preserved in the archives of the Commission," wrote Forel, "to be successively studied and compared."[84] In this way, the data would remain open to reprocessing and reinterpretation. Forel demonstrated this principle by publishing his own analysis of the 1879 earthquake, the first recorded by the commission, while Heim was still at work on his version. From the start, the commission recognized the need to preserve and periodically reexamine the original observations, a guiding principle of macroseismology today.

Somewhere along the way, however, the commission must have lost sight of this intention, for its archive is nowhere to be found. All that remains are two small sets of documents preserved in the personal papers of two commission members. From these incomplete sources, we can nonetheless reconstruct something of the texture of communication between scientists and lay observers. How did the network operate in practice? How were volunteers recruited and their observations systematized?

We can begin to answer these questions on the basis of the notes of one Christian Brügger of Chur (canton Graubünden), member of the commission since 1880. Brügger drew praise from the director Früh as "an avid collector of earthquake reports."[85] He was also one of the most active members of his cantonal natural scientific society, where he lectured on earthquakes among a host of other topics. According to his eulogist, Brügger was an old-fashioned naturalist, "who with wide-open eyes liked best to ramble in the open air and showed a lively interest for all the diverse manifestations of the natural and of the human world." He combed the hills and valleys of Graubünden for studies of zoology and meteorology alongside his specialty, botany. He also mined libraries to compose histories of spas, forestry, fishery, and dairying, and dug into local archives for a chronicle of natural events that reached back to the year 1043. He was, in short, a local notable and a fount of local knowledge—"the best connoisseur of our beautiful Graubünden."[86] His work for the commission was motivated not by an interest in seismology per se but by a commitment to local natural history.

Brügger was trained in plant geography, an important source of modern ecological thought. He was scathingly critical of the Stubenbotaniker, the armchair botanist, and of the futile quest to classify plants on the basis of dead specimens. "The plant species, like every organic creature, wants to be conceived and understood in its total appearance in the magnificent thousandfold interconnected organism of nature, in its total belonging to the infinitely diverse conditions that influence it powerfully and to the thereby determined relations to the external world, in short in its total *living* phenomenon as a *living*, dependent microcosm in the *living* kingdom of nature, in order to be judged rightly."[87] Brügger's view of nature evidently owed much to romantic *Naturphilosophie*, but it also resonated with Forel's more pragmatic search for an integrated view of life and its physical conditions.

As these lines also hint, Brügger was not an easygoing man. He was apparently fanatical about his ideas and work and intimidating to students. Nonetheless, he was a highly effective organizer. In the 1850s he organized a private network of meteorological observing stations, ninety in total, which became a crucial part of the Swiss national meteorological network in 1863. In conjunction with the network, he also published instructions for phenological observations. His seismological network presumably depended on the same circle of acquaintances.

Brügger's notes from the early 1880s shed light on the evolution of a system for processing seismological observations. For each seismic event, Brügger listed every locale to which he mailed a questionnaire and the person or institution to whom he addressed it (such as an observatory or cloister). With the exception of the first event (7 January 1880), he numbered these locations. He seems to have used the following annotations: "Ret" or a strike-through to indicate a response received; "not observed" for a negative response; in some cases, he also noted the direction of shaking. For instance, on 4 July 1880 an earthquake was felt throughout much of Switzerland, and on 6 July Brügger sent out forty-eight questionnaires (out of the two hundred given to each commission member). He seems to have received twenty-one responses, at least five of them negative. Some of his notes on this event are illegible, but they seem to include dates of subsequent communications with the observers, perhaps indicating that he followed up with questions about certain reports. Subsequently, his notation process became more streamlined, resulting in a single observer list for each event. In 1881, perhaps due to Forel's exhortations about precise timekeeping, Brügger began noting the reported time of observation to the minute. And in 1883 he began assigning intensities, in keeping with the commission's introduction that year of the Rossi-Forel scale. In 1883 he tallied how many locations experienced shaking of degree 3, 4, or 5. He then abandoned this practice, perhaps realizing that this information revealed little in the absence of a map. In later years, perhaps in growing recognition of the ambiguities of assigning intensity, he often assigned a range (e.g., 2–3) rather than a single degree. His final step at the end of each year appears to have been to review the list of events and number them. In this way he identified each separate and genuine "earthquake," according to Forel's definition. In June 1891, for instance, eight reports over a two-day period received one number (a single earthquake), while two reports from July of that year received no number—perhaps because they were not confirmed by other observers, or because they seemed to originate outside Graubünden and were therefore the responsibility of another commission member.[88] Here, then, is part of the process by which the observations of ordinary citizens were reinscribed as scientific evidence.

Coming to the Aid of Science

The most consistent and prolific lay earthquake observers in nineteenth-century central Europe tended to be individuals who were otherwise involved with *Landeskunde*, the term for regional history, both natural and human. As Lynn Nyhart argues, the suffix *kunde* in the nineteenth-century German-speaking world indicated a form of knowledge marked by three characteristics: it was comprehensive even at the expense of intellectual coherence; it derived from personal, sensory experience; and it was open to popular participation.[89] Likewise the adjective *lokalkundig* designated individuals endowed with scientifically relevant local knowledge, as in the geologist Melchior Neumayr's conviction that "a sharp and *localkundiger* observer who knows a volcano near his home in detail, will often foresee its eruption."[90] For proponents of the monographic method, the depth of their local knowledge was one marker of the distance between their *Erdbebenkunde* and the mere "theories" of upstarts like Rudolf Falb (chapter 3).

Habitual earthquake observers tended to be active in their local scientific society, in a mountaineering club like the Swiss Alpine Club, or, by the end of the century, in the emerging nature protection movement. For instance, the feat of observing over three hundred seismic shocks in the year 1885 was achieved by secondary school teacher David Gempeler in Zweisimmen (canton Bern), who won praise from the commission for his "conscientious, persistent observation and numerous, clear and precise reports." Gempeler was a member of the Alpine Club, a contributor to a regional botanical catalog, and the author of a *Heimatkunde* and a collection of local folklore, including tales reproduced in the local dialect. In 1885, the second most valued observer was a pastor in Mett (Bern) by the name of Ischer, brother of the geologist Gottfried Ischer. Pfarrer Ischer was a member of the Alpine

Fig. 4.2. The center of the town of Neuchâtel, circa 1871, not long after the start of major hydrological work in the Jura region to prevent flooding and improve irrigation. The view of the Alps at the top was added to the photograph in a nod to the local tradition of landscape painting. Source: Musée d'art et d'histoire, Neuchâtel (Switzerland).

Club and the Bern Naturforschende Gesellschaft and a collector of mineralogical specimens. His report on the events of 1885 was remarkable for its detailed exposition of the geography of the affected region. As the commission director Forster noted, "Herr Pfarrer Ischer was all the more suited to deliver a competent judgment on this situation, as he brings together precise knowledge of general geology with a special local knowledge [*Lokalkenntniss*] of the Obersimmenthal, where for years he had a parish."[91]

Still, devoted naturalists like Gempeler and Ischer could have contributed only a small fraction of the thousands of observations collected by the commission. The experiences of a wider assortment of observers can be glimpsed from the surviving original correspondence, which comes from the Jura region, with its capital in Neuchâtel. The city of Neuchâtel sits between the Neuchâtel lake and the Jura Mountains, and was home to a famed watchmaking microindustry in the nineteenth century (see figure 4.2). Socially, it was marked by a pronounced division between the hills and valley, between the "montagnons" peasants—celebrated by Rousseau as survivors of a golden age—and the patrician and bourgeois residents of the town below. Neuchâtel prided itself on several institutions of learning—it was an academic town, a *ville-école*. It was also home to a thriving community of naturalists, which included Louis Agassiz before his departure for Harvard. One might well expect the Neuchâtelois to go about observing earthquakes much as they collected minerals or mollusks. Indeed, many of those active in the Earthquake Commission also contributed as amateur naturalists to local scientific societies. In the eyes of one unforgiving observer, "In Neuchatel everyone is a boarding-house owner, the town being characterized by Protestant honesty, pedagogical pedantry, Prussian haughtiness, Dutch cleanliness, catacomb-like tranquility, and pastoral stupidity."[92]

Materials from the Earthquake Commission are preserved in the papers of the reporter for the region, the geologist Hans Schardt, and date from 1898 and 1907–10. Schardt was a specialist on the tectonics of the Alps and, like Heim, a frequent consultant on engineering projects like the Simplon. He was also active in local scientific societies, became president of the Neuchâtel branch of the Swiss League for the Protection of Nature in 1910, and compiled a glossary of the dialect of the Valais region.[93] His geology courses at the University of Neuchâtel "were attended not only by students, but also by adults eager to cultivate their minds, and by colleagues."[94]

In February 1909, Schardt received a nine-page letter from a boarding-house dweller on the outskirts of Neuchâtel, in which was described, among other things, sounds heard by one neighbor at half past two in the morning, another neighbor's nightmare of crossing a collapsing bridge, and a distant memory of a sudden storm of such violence that the writer had been thrown from her bed, only to be mocked the next morning when she recounted the experience to fellow lodgers. Two weeks later, the same writer reported to Schardt further shocks that admittedly might have been "an effect of the imagination."[95] Schardt might have been forgiven for dismissing these letters as the ranting of a lonely lunatic. Instead, he published them. In the report of the Swiss Earthquake Commission for the year 1909, much of the testimony of this lodger appears word for word, preserving even her metaphors and analogies.[96] What does this correspondence tell us about science circa 1900?

One common feature of the reports Schardt received was the desire for corroboration of an experience that many feared was a mere figment of the imagination. In several cases, observers had not mentioned their sensations to anyone until they saw Schardt's notice in the local paper. They "feared having perhaps had a bad dream"; "I don't know if other people will have noticed something." "I am sure that I then rose to go put on my hat and go outside, telling myself: One would say a light quake. But I changed my mind and sat down, thinking I was mistaken." "Even during the day I seemed to feel small tremors, but I didn't note them, fearing that they were an effect of the imagination"; "People made fun of my idea."[97] In an era when an individual known for his or her "sensitivity" might be diagnosed with neurasthenia or hysteria, earthquake observing offered affirmation to "sensitive," "nervous" individuals, especially women. "Being a person who is very sensitive to these phenomena, I have often felt them in this area without daring to speak of them."[98]

Schardt's correspondents also shared a respect for "science" and, more particularly, for Schardt as its representative. Most wrote of their hope that their reports would be of use to Schardt. Some expressed concern that their observations were not "scientific" enough. A witness identifying himself only as "un honnête citoyen" wished "to come to the aid of science, of which you are one of the representatives"; he closed: "hoping that my information will help you in your research."[99] Another observer begged Schardt three times to forgive him for information that was "imprecise."[100] On the other hand, the correspondents' respect for Schardt made them reluctant to send reports if they had *not* witnessed an earthquake. Scientists actively solicited negative reports, which were important for mapping the geographical limits of an earthquake's impact. Yet they were hard to come by. As one observer in Graubünden wrote to the canton's reporter Christian Brügger, "The earthquake was hardly felt here in Sils, it was so brief and weak; for that reason people naturally balk at being given such a long questionnaire, which they can fill out so very incompletely."[101]

Members of the Earthquake Commission "took pains to maintain and stimulate the interest of the public for our work, in part by communications in the press, in part by public lectures on the present state of the earthquake question."[102] Personal contact between scientists and volunteers had likewise been a cornerstone of the Swiss meteorological network since the 1860s. Members of the Meteorological Commission visited each station in their region regularly, not only to inspect instruments and practices but also because "personal acquaintance with the observers brought transparency and trust in the mutual relations and secured to no small degree the further course of the enterprise."[103] In the nineteenth-century American West, it was likewise personal contact that maintained local networks of scientific observation.[104] Still, Swiss seismology drew more specifically on the central European tradition of regional natural history societies, where bonds between scientists and amateurs were often forged through research outings to noteworthy local sites.[105]

Earthquake observing was also a social opportunity, particularly for those living in solitude. In the novels of Balzac and his contemporaries, boarding houses were a nineteenth-century symbol of the atomization of modern society. The injunction to collect testimony from other eyewitnesses offered a welcome excuse to penetrate the social walls that separated neighbors from each other. Observers reported conversations that transpired as individuals sought confirmation of their own experiences. They found themselves discussing their intimate moments alone in bed at night, even their dreams. "It is only in chatting with other people that I realized that it was indeed an earthquake."[106]

In one case, a Madame Bel-Perrin in Colombier responded to one of Schardt's requests for observations with a report that offered little in the way of seismology per se, but was highly suggestive in other ways. Bel-Perrin's letter is worth quoting in full. It offers a glimpse of the reverberations of the Messina disaster in the imagination of an individual with no scientific training, but with a thirst for scientific understanding.

Colombier, le 20 février 1909
Monsieur le Dr. Schaudt
Veytaux
Monsieur,

Last night I dreamed very clearly of an earthquake strong enough to set in motion the objects hanging on the walls of the room where I was staying. In my dream I immediately took note of all the particularities of the shock in order to communicate them to you. Unfortunately, it was only a dream, but it at least has the advantage of reminding me of the letter that I promised you and that I have long wanted to write you in order to excuse the slightly mischievous card that I wrote you in response to yours. [The card] contained a secret irritation at not having attended more precisely to the shocks that I certainly felt on one or another of the indicated dates. I have not yet forgiven myself this inattention. But between a dying father and a very ill sister, it was more or less possible, in the bustle of the year's end, to allow oneself to be distracted from the problems of science. Lowly and small-minded creatures that we are, it makes us uncomfortable to detach ourselves from our surroundings and to raise ourselves to those mysterious heights where one forgets everything that is not a mystery upon scrutiny. But there you have it, this has made me put my finger once again on my sordid ignorance, since, as I believe I have already told you, it may be that my sensibility is simply altogether bestial; I feel myself entirely ignorant of these great mysteries that govern worlds or disrupt them! I would like to know more about them. Is there any particular book from which I can extract some science? M. de Quervain recently sent me a report, I am very grateful to him, since I read it avidly. Would it interest you to read a letter related to the catastrophe of Messina and Reggio? It has nothing of a scientific or seismological point of view, but is simply an echo of the misery and human suffering. It was written in the first days, as 60,000 wounded were set ashore at Naples and cared for by the entire population. My sister who lives in Naples (an interesting region for the members of the seismological commission!) aided with these arrivals and still visits all the wretched and helpless. She gave a thrilling account, unfortunately all too true. It would be kind of you, Monsieur, after having read it, to return the letter to me (without haste). I would have liked to send it to you earlier; it has been passed around a large circle of acquaintances, and an indiscreet few kept it too long. Hence the long delay.

Present events will, I hope, no longer escape my seismological sixth sense; and, in order to be of service to you, I do not want my seventh like last night, but reality in full, *without* catastrophes and human sacrifices . . . even so! I have let you glimpse in these words a cruelty that I certainly do not encourage in myself. Rest assured. But I find that an earthquake should let no one escape when it begins its destruction. I pity those who survive, [whether] unharmed or maimed, since all have lost, a little or a lot. And would it not therefore be better that all perish without exception at the same moment? It is an end like another, perhaps even better than many others. I have heard—you too, I think—some very strange opinions raised by this recent catastrophe, I would even call them auguries! One must always know where they come from in order to understand and pardon them. The "good Christians" imagine that God permits and "wants" these things! They are to be pitied for having such a conception of their God. Above all that, there exists something, as beautiful as anything in our country, the absolute freedom to think whatever one wants.

Was there anything authentic in these reports issued from the seismo-logical observatory in Berlin concerning a violent earthquake perceived by instruments thousands of kilometers away? Do these observations support those of the recent catastrophes in Persia? As I know that there are no "liars" worse than journalists, even with quite precise information, I always remain skeptical about news they send from afar.

Monsieur, I am imposing on your time, which is precious and which deserves a more noble purpose than that of my lines. Perhaps you will say that a chatterbox of my stamp deserves to live in Messina or Reggio and be swallowed up there! And then, I do not know you and I find it very indiscreet of myself to act as if I have known you for a long time! But above the earth that trembles and thanks to it, there are affinities. . . . Perhaps it is in this zone, elevated and profound, that the gods decreed for us to meet by post.

With all my respect and my best wishes,
Marie Bel Perrin[107]

Over the following months, Bel-Perrin continued to supply Schardt with her observations and speculations. "If all the thoughts I have had for you were 'seismic sensations,'" she wrote, "I assure you that you would be satisfied and that science would perhaps benefit. One must believe that our thoughts encountered each other just the same, without, however, producing the expected tremor. Pardon my irreverence."[108] The following fall she was able to mail Schardt a completed questionnaire on a weak tremor. Characteristically, she managed to answer the most seemingly mundane questions with an air of cosmic mystery. Asked about sounds accompanying the shaking, she noted, "I *felt* more than heard the rumbling, however I discerned quite well the underground rumbling and the wind blowing outdoors." Asked for "external observations," she reported, "Every perception of shock produces inevitably (in me at least) a strange and real impression. It is an agitation that suspends all functions of the body, even respiration. It's very brief, but very clear. One remains breathless."[109] Would the breathless Madame Bel-Perrin have been pleased or disappointed to learn that her letters to Schardt were filed as research, rather than as personal correspondence?

Bel-Perrin's letters to Schardt illustrate how earthquake reporting could evolve into an intimate discourse of "sensations" and private thoughts. For the "sake of science," a presumably respectable woman could adopt a remarkably confidential tone with a distinguished man of science, to whom she was a complete stranger. This correspondence also shows how easily expert-lay communication about earthquakes could shift between a scientific and a moral register. Ultimately, Bel-Perrin subverted the Kantian opposition between the "scientific" and the "human" perspectives on earthquakes. She sought spiritual elevation in the union of science and humanitarianism. Together, naturalism and sympathy promised to lift Madame Bel-Perrin above the "lowly," "bestial" outlook of personal interest.

In these aspirations, Bel-Perrin was seeking a common point of view between her "ignorant" self and Schardt, the man of science. With her off-handed contempt for "good Christians" (presumably Catholics) and her complacence about Swiss liberty, she alluded to the cultural homogeneity that favored the spontaneous intimacy of her letters to Schardt. That homogeneity was undoubtedly a significant factor in the flourishing of an enterprise like the Earthquake Commission—even if Bel-Perrin preferred to imagine that she was linked to Schardt by cosmic "affinities."

Scientific Charisma

Even setting aside the case of Madame Bel-Perrin, it is clear that Schardt's correspondents were motivated not only by respect for "science" but, more particularly, for Schardt as its embodiment. They were "coming to the aid of science, of which you are one of the representatives." "Since I know that you like to be informed on the subject of earthquakes," one letter opened, as if the writer were doing Schardt a personal favor.[110] Some used their devotion to science to differentiate themselves from members of lower classes. Thus one writer noted that neighboring buildings were inhabited "by a population of workers who without doubt will not themselves relate observations of this sort." Most striking is their common desire to be of assistance to Schardt himself. It meant something to these writers to address a man of science, perhaps one they knew from his contributions to the local scientific society.

As macroseismological networks expanded toward the end of the nineteenth century, they became increasingly bureaucratic. Observers lost some of the motivation of personal contact with charismatic personalities like Heim, Forel, Brügger, and Schardt. In Austria, for instance, when the state Earthquake Service outgrew the oversight of the Vienna Academy of Sciences and was transferred to the Central Institute for Meteorology and Geophysics (ZAMG), scientists worried that the network would suffer. The ZAMG's director expressed his concern to the ministry of culture: "People always see a state institute only as a bureaucratic institution and regard the directorate as functionaries [Beamten]. Particularly for those circles that have enjoyed higher education, this is enough to make them stand aside or even to behave decidedly hostilely, because they are willing to subordinate themselves to the scientific authority of the imperial academy, but not to a functionary. This trait is certainly not justified, however it exists and must be reckoned with."[111] "In our experience," he added, "the best observers are those idealists, who make observations entirely voluntarily, out of the purest objective interest, and certainly for that reason demand a purely individual, personal, non-bureaucratic treatment."[112]

Telling Global Stories

How did scientists weave the idiosyncratic content of such correspondence into a publishable account of a geophysical phenomenon? In part, the challenge was to tell a convincing story: to stitch the partial reports of individual observers into a narrative of the earthquake. One approach was to reinterpret personal experiences as communal. Thus, Schardt once recounted that sleepers were jerked awake with the "impression of being instantaneously suspended before coming back into contact with the mattress. . . . There were many who went to find the cause in one or another floor of their house, entirely surprised to find there tenants busy with the same investigation."[113] Another tactic was to transfer the quality of seismic "sensitivity" from individuals to places. The church tower of Morges, for instance, was a "seismophone of the first rank; its bells ring every time that a seismic wave passes underneath." Various lakes and springs had also become known as natural earthquake detectors. One area of the Neuchâtel Lake was considered an "entirely peculiar point of pulsation" after school children spotted waves of fifty to sixty centimeters during the earthquake of 1898.[114] Forel explained this effect by analogy to the Chladni figures formed by vibrating membranes, thus certifying the lake's status as seismically sensitive.[115]

Attributing experiences to communities and places rather than individuals was one way to move between felt reports and earth physics. Another was to embed local phenomena in larger-scale narratives. As the commission explained in 1888, "The longer the observations persist, the more clearly certain lines of dislocation in Switzerland become noticeable, along which the tremors are far more numerous than in other places."[116] Three years before the founding of the Earthquake Commission, in 1876, a weak earthquake had piqued the interest of the Neuchâtelois. At the local scientific society, the event was described as an *Einsturzbeben*, caused by the collapse of underground caves as water seeped into them. A local geologist had deemed it an "essentially Jurassic phenomenon."[117] After three decades of nationwide earthquake observation, however, scientists recognized Neuchâtel to be part of a tectonically active area.[118] They reinterpreted "Jurassic" quakes as tectonic in origin, the consequence of a planetary process of contraction.

As they developed a picture of global tectonics, commission members looked to correlate Swiss earthquakes with reports from abroad. The push for global narratives was evident in the Swiss response to the Messina earthquake of 1908—the most deadly in European history, claiming over sixty thousand lives. Across central Europe, the disaster drew sympathy and aid;

it seemed that nature had undermined Europe's warmongers. Sensational-istic reports appeared in Swiss newspapers, describing crowds gone mad in southern Italy.[119] In the Jura, meanwhile, many observers reported tremors coinciding with the Italian catastrophe. Schardt framed these reports as evi-dence of "closely related movements" across the planet. "We are struck by the impression that we are in the presence of closely related movements, of which some are, by a kind of triggering effect, the consequences of others; as a strong barometric low can be followed by a light shaking of the ground. This is how we must consider the barely perceptible seismic phenomena that were observed in our region, while neighboring countries and those be-tween us and the great center of the quake were not perceptibly affected."[120] Nonetheless, the observers themselves seemed unsure about the bearing of these observations on events in Italy. None but Madame Bel-Perrin men-tioned the victims. One recalled thinking, "If Switzerland were in Italy, one would say that this is an earthquake."[121]

Even when scientists' narratives diverged from the perspectives of lay observers, they carried crucial traces of this dialogue. Consider the case of the 1885 earthquake in Bern, meticulously observed by the volunteers Gempeler and Ischer. Gempeler's experience of the shaking convinced him that this was an *Einsturzbeben:* "for the impression of this writer was none other than that under his feet a rockslide must be taking place." At first, commission director Forster agreed that this was indeed an *Einsturzbeben*. Then, like a detective inspecting his clues, he arranged and rearranged the felt reports. Epiphany struck in the "moment, in which the compilation of the times of impact . . . lay before me." It appeared that the earthquake had been felt nearly simultaneously throughout the affected region, with reports clustering along a long axis—a line that corresponded to the area struck by another quake four years earlier. Before drawing a conclusion, however, Forster introduced a skeptical interlocutor: "But the enormously strong ef-fects in Zweisimmen? and the hundreds of shocks that followed there daily for months on the smallest area? was this then not an Einsturzbeben?" This voice breaking into his train of thought was apparently none other than that of Gempeler and his neighbors, insisting, in Forster's imagination, on the legitimacy of their observations and even of their interpretations. Forster continued: *"Certainly!* but . . .." With that "but" he went on to reconcile the lay and expert points of view. Namely, a tectonic earthquake could have produced, as a secondary effect, an *Einsturzbeben* in the Bernese village of Zweisimmen. The horizon of the narrative had expanded, but the local story survived.[122]

Conclusion

By the second decade of the twentieth century, international seismologists were announcing the arrival of the "new" seismology, ushered in by the latest seismographs. The Swiss, by contrast, insisted that instruments could never furnish complete knowledge of an earthquake. As the director of the Swiss earthquake service argued in 1914, "In and of themselves instruments do not give the complete picture of earthquakes as a natural phenomenon." By then, the observational network, human and instrumental, had outgrown the capacities of the Swiss Natural Scientific Society. Between 1912 and 1914, it was transferred to the federal meteorological institute in Zurich, but every effort was made to maintain continuity—to ensure, in particular, "that also in the future full attention will be paid to the macroseismic data, for which the original instructions and questionnaires of the Earthquake Commission today still contain all essential points of view and guidelines." Scientists would still need to recognize and accommodate the uncertainty and cultural specificity of human observations: "It must be remembered that the earthquake map, which must be constructed essentially according to individual perceptions, will display an anthropogenic character—that is, it is in many respects a representation of the population density and of the culture of the inhabitants. . . . Therefore it is best in the descriptions to replace certainty and confidence with possibility and probability."[123] A decade later, the next director of the earthquake service was still arguing the same point: "One must still begin with the careful collection and analysis of the observations that have been made directly by man. . . . They will need to be tended to in the future as well."[124]

Far from disqualifying human observations, the development of seismography revealed new layers of meaning. It allowed Swiss researchers to trace "parallels between the results of the instruments and those of the observers," above all in the telltale sequence of longitudinal and transverse waves. Close comparison of observers' descriptions and seismographic curves showed "that in the subjective observations these facts have always found expression. It seems that relatively close to the area of greatest shaking in our land both waves are often felt as temporally separated and distinct shocks." If, as experience suggested, observers could indeed count the seconds between the arrival of the "vertical" and the "horizontal" shocks, this figure could be used to calculate the depth of the focus. "As evidence for the manner in which the different types of wave in the epicentral region are actually perceived by observers at rest, I quote deliberately a very old observation, that of Pastor Studer during the great Visp earthquake: he writes: 'I was clearly able to perceive that the first shocks were vertical, then however with an indescribable movement they became horizontal.'"[125]

The Swiss Commission thus set itself the task of bringing "the existing macroseismic and the new microseismic observations into the closest possible relationship, to exploit them as reciprocally as possible." Using a metaphor of "interpreting" or "translating" between instruments and humans, Swiss scientists pursued the question of "how in general objective and subjective conclusions can be translated [*umgedeutet*] into one another." The metaphor of translation is revealing for what it did *not* imply: it was not a matter of repackaging scientific conclusions for public consumption. Instead, in this nineteenth-century central European context, translation could refer to a writer's capacity to mediate the relationship of man to nature in an urbanizing world. As literary scholars note, this was the goal of the *Heimatkünstler*, the "homeland artist": to make a "native literature" accessible to a population removed from the land. As a form of *Landeskunde*, nineteenth-century seismology likewise confronted this need for mediation.[126] Earthquake observing provided one key for translating between urban experiences of the built environment and perceptions of elemental nature. What might sound like furniture being rearranged in an upstairs apartment could be recognized as a seismic tremor; a heavy wagon on pavement became seismic thunder. As Schardt's correspondence confirms, this particular skill of translation was highly valued by the late nineteenth-century inhabitants of a small European city like Neuchâtel. It offered them a sense of connection: to the surrounding countryside, to the earth and its raw elements, to the cosmos, and even to each other.

FIVE

Geographies of Hazard

Among the first desires which seize a visitor in an earthquake country, is to experience a shaking.

—John Milne[1]

Nineteenth-century Europeans had a tendency to think of earthquakes as a "tropical" phenomenon, characteristic of savage realms, where nature was said to operate on a grander scale.[2] When the "scientific" historian Henry Thomas Buckle set out to explain England's rise to the pinnacle of civilization, he divided the influences of nature into two classes: those conditions that appealed to rational understanding, and those that instead provoked the "imagination." Foremost among the latter were earthquakes and volcanoes, which thwarted every impulse to deduce the laws of nature:

> Of those physical events which increase the insecurity of Man, earthquakes are certainly among the most striking, in regard to the loss of life which they cause, as also in regard to their sudden and unexpected occurrence. . . . The terror which they inspire, excites the imagination even to a painful extent, and, overbalancing the judgment, predisposes men to superstitious fancies. And what is highly curious, *is, that repetition, so far from blunting such feelings, strengthens them.* . . . The mind is thus constantly thrown into a timid and anxious state; and men witnessing the most serious dangers, which they can neither avoid nor understand, become impressed with a conviction of their own inability, and of the poverty of their own resources. In exactly the same proportion, the imagination is aroused, and a belief in supernatural interference actively encouraged. Human power failing, superhuman power is called

in; the mysterious and the invisible are believed to be present; and there grow up among the people those feelings of awe, and of helplessness, on which all superstition is based, and without which no superstition can exist.[3]

Many of Buckle's contemporaries agreed that the longer one lived in a seismic zone, "the less one can withstand the agitation that takes hold of men and animals in an earthquake."[4] Buckle's theory divided the earth into nations destined for science and prosperity, and nations vulnerable to environmental hazard—a dichotomy still visible in the interventionist ethos of twentieth-century scientific internationalism.[5] By combining the standardized measurement of hazard with the comparative analysis of disaster response, nineteenth-century seismology had an unusual potential to complicate this deterministic geography of security and risk.

Seismic Tourism

In the course of the nineteenth century, a peculiar desire spread among Europeans: the wish to feel an earthquake. It was not unusual for a naturalist to refer to "a good, hearty shock of earthquake" or "a beautiful earthquake."[6] According to the American geologist Grove Karl Gilbert, "It is the natural and legitimate ambition of a properly constituted geologist to see a glacier, witness an eruption and feel an earthquake. The glacier is always ready, awaiting his visit; the eruption has a course to run, and alacrity only is needed to catch its more important phases; but the earthquake, unheralded and brief, may elude him through his entire lifetime."[7] The British naturalist Alfred Russell Wallace recalled his delight at first feeling an earthquake on Ternate: awakening in a shaking bed, "I said to myself, 'Why, it's an earthquake,' and lay still in the pleasing expectation of another shock."[8] And when William James left Boston for Stanford in 1905, a friend bade him farewell with the wish that he might experience "a touch of earthquake."[9]

Such seismic tourism was arguably part of what Aaron Sachs describes as a nineteenth-century quest for "extreme environments," from the Arctic to the Rockies, which attracted "travelers who embraced disorientation." Clarence King, torn between the life of a professional geologist and that of a romantic mountaineer, fell in love with the "unstable" landscape of the Sierra Nevada, "in every state of uncertain equilibrium." For King, vertigo was a state to be savored: "I found it extremest pleasure to lie there alone on the dizzy brink."[10] For those who dreamed of a life like King's, the *New York Times* reported in 1881 on the cost of a trip from San Francisco to the "Doomed town of Hilo, Hawaii," or "What it costs to see a volcano."[11] No traveler could have relished an earthquake more than John Muir, whose admiration for "wild" nature led him to found the Sierra Club in 1892. Muir's day came in the Yosemite Valley in 1872: "The shocks were so violent and varied, and succeeded one another so closely, one had to balance in walking as if on the deck of a ship among the waves, and it seemed impossible the high cliffs should escape being shattered." In the moment after the first tremor Muir's senses were preternaturally alert. He heard a "stupendous roaring rock-storm" followed by the "chafing, grating against one another, groaning, and whispering" of the settling rocks. He smelled "the odor of crushed Douglas Spruces, from a grove that had been mowed down and mashed like weeds." And he rode out the aftershocks, which "made the cliffs and domes tremble like jelly." When the ground stilled, Muir observed the reactions of the valley's inhabitants: "I found the Indians in the middle of the valley, terribly frightened, of course, fearing the angry spirits of the rocks were trying to kill them. The few whites wintering in the valley were assembled in front of the old Hutchings Hotel comparing notes and meditating flight to steadier ground, seemingly as sorely frightened as the Indians. It is always interesting to see people in dead earnest, from whatever cause, and earthquakes make everybody earnest." Muir observed the terror of his neighbors with detachment. He tried and failed to console two of his companions, who "fled to the lowlands." To Muir, they seemed to fail to grasp the true meaning of the experience: "Storms of every sort, torrents, earthquakes, cataclysms, 'convulsions of nature,' etc., however mysterious and lawless at first sight they may seem, are only harmonious notes in the song of creation, varied expressions of God's love."[12]

It was not only men who were drawn to seismic adventure. A *New York Times* headline of 1872 announced "A Lady's Earthquake Experience." What followed was a letter from an American woman traveling in Peru, whose brush with an earthquake had left her with "cold, clammy hands" and religious doubts; her husband "did not comprehend it." "I wish Mrs. L could experience one," the writer commented, "as she is so anxious to do so. I think one experience would satisfy her."[13] The novelist Gertrude Atherton offered similar advice in the wake of the San Francisco earthquake of 1906: "I can suggest no better 'cure' for those that live where nature has practically forgotten them and civilization has become as great a vice as too much virtue, in whom a narrow and prosperous life has bred pessimism and other forms of degeneracy, stunting the intelligence as well as atrophying the emotions, than to spend part of every year in earthquake country."[14] Anticipation ran so high that some travelers could not hide their disappointment when confronted with the real thing. Mrs. Campbell Dauncey, the wife of a British diplomat in the Philippines, felt her first "slight earthquake" in Manila in 1905: "It was my first experience of that form of excitement, and I am sure that I don't want another. . . . I don't think I like earthquakes." Still, she was "sorry now that I did not go on to the balcony" when her husband "invited" her to "see the street moving."[15]

Nineteenth-century travelers reported on earthquakes in all their facets. They interviewed native witnesses, described geological, atmospheric, architectural, and social conditions, and collected archival documents and oral traditions. To guide them, scientists furnished travelers with instructions for the observation of earthquakes. The first edition of the British Admiralty's *Manual of Scientific Enquiry* (1851) included an extensive chapter by Robert Mallet, still several years from his architectural method of earthquake survey. The manual was addressed to an educated elite, an audience that might be expected to follow Mallet's instructions for watching a second hand during an upheaval or building a seismoscope out of a barometer. Mallet instructed travelers to seek out and question local residents: "The opinions of old observers as to changes of climate or season; the occurrence of pestilences, failure of crops, &c., in relation to earthquakes, while they must be received with caution, should not be disregarded." Attention should likewise be paid to local "records or trustworthy traditions in volcanic countries or those neighbouring to them, as to the state of activity or repose of those vents for a long period prior to and during the quake."[16] A German counterpart to the Admiralty's manual was published in 1875 under the title *Introduction to Scientific Observation for Travelers*. The chapter on earthquakes was authored by Karl von Seebach, a Göttingen geologist who had studied the volcanoes of the Greek isles and Central America. Like Mallet, Seebach encouraged travelers to collect information on a wide variety of phenomena possibly associated with earthquakes, including weather events, geological changes, and animal behavior. He urged them to dig up records of past temblors in local archives, municipal and parish registers, and family chronicles.[17] In these ways, between the 1850s and 1880s, Europeans were charting worldwide seismic hazard as a Humboldtian project: eclectic, holistic, and cooperative.[18]

Humboldt at Cumana

The Victorian fascination with earthquakes reflected in part the place of the Lisbon tragedy of 1755 in nineteenth-century accounts of European intellectual history. Like many children of the Enlightenment, the poet and naturalist Johann Wolfgang von Goethe credited the Lisbon disaster with his boyhood loss of faith. Doubts about the veracity of Goethe's account do not detract from its influence.[19] In the nineteenth century, the Lisbon earthquake became a cultural shorthand for initiation into a skeptical, rational, and self-consciously modern search for natural causes. In an age fond of likening human history to a progression from infancy to maturity, Lisbon figured as the coming of age of the European mind.[20]

If earthquakes became virtually a rite of passage for Victorian naturalists, it was thanks largely to the seismic reflections of Alexander von Humboldt. Humboldt was born in 1769 in Berlin, the cosmopolitan capital of the rapidly expanding kingdom of Prussia. Humboldt matured in an age intent on exploiting the natural and human "resources" of the New World. Though he served European empires, he managed to retain a critical distance from their goals. Humboldt became the nineteenth century's most celebrated model of the scientific traveler. He embodied the romance, heroism, and freedom from social convention that seemed to await explorers in the name of science. He traveled in search of a "general physics of the earth." This meant tracing the most varied phenomena across the planet's surface—the characteristics of the atmosphere, the elevations of mountains, the distribution of plant species—in search of patterns and interconnections. Humboldt demonstrated how physical and organic nature could be studied together, region by region, landscape by landscape—comparing, say, deserts and grasslands, alpine lakes and tropical lakes. Against the nineteenth-century trend toward scientific specialization, this was an approach that united meteorological, geological, botanical, zoological, and even anthropological perspectives. Natural conditions posed the constraints within which cultures developed; cultural tendencies in turn shaped the uses and perceptions of nature.[21]

Humboldt felt his first earthquake in Cumana, Venezuela, in November 1799, two years after an earthquake there had killed approximately sixteen thousand people. His account appeared in his *Personal Narrative* of the South American voyage, which was required reading for aspiring nineteenth-century naturalists. (Charles Darwin, for instance, copied out long passages from it to carry with him on the *Beagle*.) It begins with Humboldt and his companion Aimé Bonpland crossing a desolate beach one evening, when out of nowhere appeared "a tall man, the colour of a mulatto, and naked to the waist," brandishing a large stick. Humboldt ducked his blow, but Bonpland was struck on the head, hard enough to leave him with a fever and dizzy spells. The man then pulled a knife, and the travelers "would surely have been wounded if some Basque merchants taking the fresh air on the beach had not come to our aid." Despite their "accident," Humboldt rose at five the next morning to observe a partial solar eclipse, congratulating himself on the clear skies. In the days before and after the eclipse, Humboldt noted "strange atmospheric phenomena." The night sky was thickly veiled in a "reddish mist," "at night the heat was stifling yet the thermometer did not rise beyond 26°C." "The air was sweltering hot, and the dusty, dry ground started cracking everywhere." Then, one week after the eclipse, "around two in the afternoon, extraordinarily thick black clouds covered the tall Brigantin and Tataraual mountains, and then reached the zenith. At about four it began to thunder way above us without rumbling, making a cracking noise, which often suddenly stopped. At the moment that the greatest electrical discharge was produced, twelve minutes past four, we felt two successive seismic shocks, fifteen seconds from each other. Everybody ran out into the street screaming." Humboldt's account is cool and precise. "Bonpland, who was examining some plants, leaning over a table, was almost thrown to the floor, and I felt the shock very clearly in spite of being in my hammock. The direction of the earthquake was from north to south, rare in Cumana." He must have gathered reports from the locals, for he added that "some slaves drawing water from a well, some 18 to 20 feet deep next to the Manzanares river, heard a noise comparable to artillery fire, which seemed to rise up out of the well." He continued with a vivid description of the sunset following the quake, the sun's disk looking "incredibly swollen," spreading "clusters of rays coloured like the rainbow."[22]

Then Humboldt's register shifted, as he looked inward to assess the psychological impact of the Cumana earthquake. It "made a great impression on me," he explained: "It is not so much a fear of danger as of the novelty of the sensation that strikes one so vividly when an earthquake is felt for the first time. When shocks from an earthquake are felt, and the earth we think of as so stable shakes on its foundations, one second is long enough to destroy long-held illusions. It is like waking painfully from a dream. We think we have been tricked by nature's seeming stability; we listen out for the smallest noise; for the first time we mistrust the very ground we walk on."[23] These often quoted lines seem to allude to the place of the Lisbon earthquake in European intellectual history. If earthquakes have the power to destroy "long-held illusions," Humboldt suggests, this is in part because they produce a visceral transformation. It is not merely religion that comes to seem suspect in their wake, but "the very ground we walk on." The immediate result is a heightened attention to the natural world, to the "smallest noise." This scientific mindset is not a consequence of deliberate reflection on the catastrophe; it is an immediate reorientation of the body's relationship to its environment. In this new state of alertness, Humboldt and his companions observed a meteor shower the following night, a phenomenon that he and the inhabitants of Cumana speculated might be related to the temblor. The earthquake thus primed Humboldt for empirical science—not through conscious reflection on its causes and effects, but by plunging him quite literally into a world in which nothing could be taken for granted.

The allegorical quality of Humboldt's account further depended on latent metaphors. Believing that earthquakes were the effect of subterranean uplift, Humboldt worked with an implicit analogy between the earthquake as a force of planetary formation (*Erdbildung*) and as a force shaping the human mind (the humanist concept of *Bildung*). In this respect, Humboldt's earthquake narrative reimagined a lesson of classical philosophy. Plato had conceived of vertigo as a formative experience for philosophers. In the "dizzy" confusion over the significance of perceptual judgments lay the state of mind that was "the beginning of philosophy." As recent scholars have noted, echoes of this position can be found from Descartes to the romantics.[24] In the nineteenth century, though, the earthquake became more than a metaphor for knowledge making; it became a natural laboratory.

Humboldt's note therefore came with an important coda: "if these shocks are repeated frequently over successive days, then fear quickly disappears. On the Peruvian coasts we got as used to the earth tremors as sailors do to rough waves."[25] Humboldt made this point twice for emphasis: "I would never have thought then that, after a long stay in Quito and on the Peruvian coast, I would get as used to these often violent ground movements as in Europe we get used to thunder. . . . The casualness of the inhabitants, who know that their city has not been destroyed in three centuries, easily communicates itself to the most frightened traveler."[26] In other words, earthquake fears were by no means crippling. Like the natives, a visitor could adjust to the unstable terrain. In this sense, Humboldt's story of the Cumana earthquake and its aftermath encapsulated his larger ambitions for his scientific voyage to the New World. As Michael Dettelbach has argued, Humboldt styled himself as a radical empiricist. He cast himself as an exquisitely sensitive measuring instrument for the registration of a dynamic interplay of natural forces. Politically as well as philosophically, Humboldt's empiricism had radical implications. He aimed at the "cultivation of free individuals, a Reason free from the prejudices of theory, theology, or self-interest."[27] Since 1789, earthquakes and other natural cataclysms had become common metaphors for political revolutions, and natural cataclysms were increasingly perceived as threats to public order. Humboldt's tribute to the earthquake of 1799 thus had revolutionary potential. No single moment of his controversial South American journey better expressed his liberatory vision than his embrace—in mind and body—of the shaking ground of Cumana.[28]

Humboldt would return to this theme late in life in his magnum opus, *Cosmos*. In keeping with what one commentator calls his "subversive vision of science for the people," *Cosmos* was meant for the widest possible audience.[29] Humboldt urged his readers to recognize that their own existence was implicated in the forces shaping the earth, to "perceive that the destiny of mankind is in part dependent on the formation of the external surface of the earth." This was not only a work of popular enlightenment, but also an invitation. Humboldt sought contributors to his physical geography far beyond Europe's cultured urban centers. *Cosmos* was a bid (if perhaps a misdirected one) to enlist the "men who live in the fields, the forests and the mountains" in his globe-spanning program of precision measurement. "The great problem of life is to produce a lot in a little time, and it is certain that if methods of measuring with very simple instruments were more widespread among the public, if the attention of men who live in the fields, the forests and the mountains were directed more towards the magnitude and distances of objects, after so many voyages and investigations pursued in the two hemispheres, our geological ideas (the most beautiful, most interesting part of human knowledge) would be advanced threefold."[30] In Humboldt's vision, natives of every continent would become scientific observers and contribute to a truly global science.

In *Cosmos* Humboldt offered his most extensive and dramatic discussion of the mental effects of earthquakes, of "the deep and peculiar impression left on the mind by the first earthquake which we experience, even where it is not attended by any subterranean noise." Here he raised the vexed question of whether earthquakes affected Europeans differently than inhabitants of the New World. He cited the observations of the younger Swiss naturalist Johann Jakob Tschudi:

The inhabitant of Lima, who from childhood has frequently witnessed these convulsions of nature, is roused from his sleep by the shock, and rushes from his apartment with the cry of "Misericordia!" The foreigner from the north of Europe, who knows nothing of earthquakes but by description, waits with impatience to feel the movement of the earth, and longs to hear with his own ear the subterranean sounds which he has hitherto considered fabulous. With levity he treats the apprehension of a coming convulsion, and laughs at the fears of the natives: but, as soon as his wish is gratified, he is terror-stricken, and is involuntarily prompted to seek safety in flight.

Tschudi's description exposed European hubris in a manner perhaps inspired by Humboldt himself. But Humboldt could not agree with it completely:

This impression is not, in my opinion, the result of a recollection of those fearful pictures of devastation presented to our imaginations by the historical narratives of the past, but is rather due to *the sudden revelation of the delusive nature of the inherent faith by which we had clung to a belief in the immobility of the solid parts of the earth.* We are accustomed from early childhood to draw a contrast between the mobility of water and the immobility of the soil on which we tread; and this feeling is confirmed by the evidence of our senses. When, therefore, we suddenly feel the ground move beneath us, a mysterious and natural force, with which we are previously unacquainted, is revealed to us as an active disturbance of stability. A moment destroys the illusion of a whole life; our deceptive faith in the repose of nature vanishes, and we feel transported, as it were, into a realm of unknown destructive forces. Every sound—the faintest motion in the air—arrests our attention, and we no longer trust the ground on which we stand. Animals, especially dogs and swine, participate in the same anxious disquietude; and even the crocodiles of the Orinoco, which are at other times as dumb as our little lizards, leave the trembling bed of the river, and run with loud cries into the adjacent forests.

To man the earthquake conveys an idea of some universal and unlimited danger. We may flee from the crater of a volcano in active eruption, or from the dwelling whose destruction is threatened by the approach of the lava stream; but in an earthquake, direct our flight whithersoever we will, we still feel as if we trod upon the very focus of destruction. This condition of the mind is not of long duration, although it takes its origin in the deepest recesses of our nature; and when a series of faint shocks succeed one another, *the inhabitants of the country soon lose every trace of fear.*[31]

Humboldt himself had experienced this transformation, becoming, by the end of his five-year journey through the Americas, a veritable earthquake aficionado. With greater familiarity came greater admiration for the resilience of the native South Americans in the face of the "incredible instability of nature."[32] In his *Personal Narrative,* he recalled being awed by the "extraordinary" "presence of mind" of a native woman during an earthquake. When asked how she remained so calm, she "answered with great simplicity": "I had been told in my infancy, if the earthquake surprise you in a house, place yourself under a doorway that communicates from one apartment to another; if you be in the open air, and feel the ground opening beneath you, extend both your arms, and try to support yourself on the edge of the crevice." Humboldt reflected: "Thus in savage regions, or in countries exposed to frequent convulsions, man is prepared to struggle with the beasts of the forest, to deliver himself from the jaws of the crocodile, and to escape from the conflict of the elements."[33] Here was evidence of a vast human capacity for environmental adaptation rather than control. As George Eliot's narrator puts it in *Middlemarch*, inhabitants of earthquake countries seem to possess the wisdom that comes of long experience of crisis: "We are told that the oldest inhabitants in Peru do not cease to be agitated by the earthquakes, but they probably see beyond each shock, and reflect that there are plenty more to come."[34]

Humboldt's Reverberations

Humboldt helped establish earthquakes as a phenomenon inviting a synthetic study of the physical, zoological, and human aspects of a region. As one anthropologist would remark in 1909, earthquakes could demonstrate "how closely related geography, ethnography, and linguistics are."[35] Twenty years later, a geographer would lament the more recent tendency to consider the physical and human aspects of earthquakes separately. He reminded his readers that earthquakes could start avalanches and forest fires, create the conditions for epidemics, and influence every aspect of a society from its religious views to its economy.[36] Humboldt also called attention to the dignity and humility with which some non-Europeans faced seismic disasters. In this spirit, Edward Tylor—perhaps the most influential British anthropologist of the nineteenth century—wrote admiringly of the earthquake myths of "primitive" peoples. Indeed, Tylor's belief in an evolutionary hierarchy of cultures precluded any clear divide between scientific and mythical explanations. In his view, anthropomorphic myths were a form of poetry that served to explain the natural world to those who could not understand the "technical language" of science. He lamented that "the growth of myth has been checked by science, it is dying of weights and measures, of proportions and specimens." It was the poet's task to relate "the being and movement of the world to such personal life as his hearers feel within themselves."[37] Inspired by Tylor's work, the Austrian anthropologist Richard Lasch took up a comparative study of practices meant to ward off seismic disturbances.

Many cultures, it seems, responded to earthquakes by crying out a variant of "We're still here!" They intended to "alert the beings who have produced the earthquake . . . and who are not by nature hostile to men, that men are still on the earth, so that they will put an end to their earth-shaking movements." They were calling out not in anger, but as a plea for mercy. Like Tylor, Lasch observed that earthquake myths were a form of causal reasoning that could not be neatly distinguished from science. In the Niassa province of southeastern Africa, for instance, "wise men" attributed earthquakes to the reverberations caused by a star crashing into the ocean. Friedrich Ratzel had deemed this a sign of "remarkable progress towards a rational perspective."[38] As Lasch put it, earthquakes made it difficult for anthropologists to "differentiate between scientific theory and myth."[39]

Humboldt thus provided a template for the earthquake narratives of later scientific travelers. When Charles Darwin felt his first temblor at Concepción, Chile, in 1835, he had been primed by his readings of Humboldt and Tschudi. Paul White has shown that Darwin successively revised his impressions of the earthquake in order to transform the emotion of fear into an acceptable and productive scientific attitude. By viewing the earthquake through the romantic lens of the sublime, he could transform terror and sympathy into intellectual pleasure and detached scientific interest.[40] Darwin's published account of his experience at Concepción echoed Humboldt: "A bad earthquake at once destroys the oldest associations: the world, the very emblem of all that is solid, has moved beneath our feet like a crust over a fluid; one second of time has conveyed to the mind a strange idea of insecurity, which hours of reflection would never have created."[41] Like Humboldt, Darwin framed the earthquake survivor's doubts as a visceral reaction, rather than the result of intellectual "reflection." Humboldt's model would also have led Darwin to expect a heightening of perception in the earthquake's aftermath—and that is indeed what he experienced. Surveying the Chilean coast, Darwin noted that the shaking had apparently produced a permanent elevation of the land. Here was evidence that, in his own day, the earth's surface was subject to lasting changes, not just temporary fluctuations. Those dry inches of coastline became another piece in the great puzzle of evolution.[42]

Humboldt's earthquake narrative was philosophically and politically subversive. By contrasting the undue panic of Europeans with the sobriety of New World natives, it inverted the hierarchies of colonizer and colonized, scientific observer and savage. It further celebrated nature's own "revolutionary" force, prescribing a dose of earthquake to free Europeans from the dead weight of tradition and initiate them into the radical empiricism of the Humboldtian traveler.

The Uneasy Earth

Until the Victorian Era, earthquakes had only been cataloged on a regional basis. These were nonetheless painstaking feats, and one cataloger even went blind in the process.[43] The work of soliciting observations required the leisure to maintain a wide correspondence network and tended to be a gentleman's pursuit. The first global catalog of earthquakes was assembled by Karl von Hoff, a high-ranking official of the Duchy of Gotha in the Napoleonic Era, and published in 1840, three years after his death. It included 2,225 earthquakes, from 1606 BCE (the eruption of Mount Sinai) to 1805.[44] Yet the most famous earthquake catalog of the nineteenth century was the work of the French mathematician Alexis Perrey, who was denied the advantages of wealth and birth that would have eased his labors. The son of a forest ranger, Perrey gave up plans for the priesthood in favor of a series of modest academic posts, but he maintained monastic work habits. He began to compile annual lists of global earthquakes in 1843; by 1850 he was recording more than five hundred events per year. As Charles Davison noted in his 1927 history of seismology, "Perrey's devotion to science was so intense that his health began to suffer from the strain."[45] Perrey retired to Brittany in 1867 in order to recover, but he maintained his network and continued to publish his lists each year.

Dismissing these earlier efforts, the British engineer Robert Mallet claimed for himself the distinction of having made "the first attempt to complete a catalogue that shall embrace all recorded earthquakes." Noting the sharp increase in the number of earthquakes recorded since 1700, Mallet commended "the advance of human enterprise, travel, and observation."[46] Indeed, Mallet was quick to assume that his work represented another victory in the British Empire's scientific conquest of the globe—"of the rise, progress, and extension of human knowledge and observational energy, and also of the multiplication and migrations of the human family and its progress in maritime power."[47] What was indeed unprecedented was the scale of his research into original historical sources. In all, he listed 6,831 earthquakes between 1606 BCE and 1842 (up to the beginning of Perrey's annual lists). His 1858 seismic map of the world was "more than a picture"; it was a faithful depiction of the "most formidable" seismic regions of the earth.[48] Mallet began by grouping all known earthquakes into three classes, "great," "mean," or "minor," which he matched to three different shades of watercolor. In principle, regions would vary in shade according to both the relative frequency and intensity of past earthquakes. In very few cases was Mallet able to determine the area of shaking according to historical accounts; otherwise, he made his best guess, based on "the physical, geological or other conditions of each area, known to modify the distant propagation of shock."[49] Sparsely populated areas, even regions of Europe like the Carpathian Mountains, were likely to be more seismically active than the map could indicate. As Josiah Whitney soon pointed out, Mallet improbably depicted the eastern half of the United States as more seismically active than its western half! Mallet's confidence reflected the distortions of an imperial lens, but his catalog served seismology well. Mallet judged that he had completed once and for all the task of a global historical earthquake catalog, and in 1927 Charles Davison agreed with him: "Mallet may have erred in regarding his catalogue as having no forerunners. There can be little doubt, however, that it is the last that will ever be published on so extensive a scale."[50] In fact, the catalogs of Mallet and Perrey remain a standard reference for historical seismologists today.

Mapping the Primitive Mind

Who knows how much that is strange and bizarre in Japanese art and life may be due to the constant appalling effect on the mind of these mysterious phenomena of nature?[51]

—*Times* (London), 1885

Victorian assumptions about the environmental determinants of civilization and barbarism sat uncomfortably with scientists' experiences on the ground in the colonial world. Seismology's engagement with native testimony and folklore in Asia and Latin America raised provocative questions about the relationship between science and myth. Among the products of the long seismological career of Count Fernand Jean Baptiste Marie Bernard de Montessus de Ballore (1851–1923) was the posthumously published *Seismic and Volcanic Ethnography*. Montessus was a captain in the French army, trained at the elite École Polytechnique, and a member of the missionary Society of Saint Vincent de Paul. In 1881 he led a mission to El Salvador, which looked to France as a model for its military. Montessus was to provide artillery training to native troops; he played an active role four years later in El Salvador's war with Guatemala. In the meantime, he developed a fascination with the country's violent earthquakes and volcanoes, which he pursued both in libraries and across the Central American landscape from Tehuantepec to Panama.[52] In the course of his research in Latin America and his correspondence with naturalists around the world, Montessus gathered a vast collection of popular ideas about earthquakes—the basis of the ethnography published, with his brother's oversight, after his death. At face value, Montessus's collection of seismic myths was an attempt to discredit the *sciences travesties* (pseudosciences) and thus "extirpate" the residue of the savage mind from modern science. Theories linking earthquakes to weather, for example, were said to be the result of "folklore indefinitely surviving in science." "There exists in seismology a series of opinions, or even of theories, that the efforts of the modern science of earthquakes has not yet succeeded in extirpating from the preoccupations of scholars. . . . Here we collide with the extreme difficulty of logically classifying this crowd of incoherent beliefs that originate in the depths of the thought of primitive man or of the savage, and which, for more highly evolved man, emerge from the mythology and religion of each people."[53] Indeed, the value of his anthology was said to lie in its demonstration of the enduring "unity of the human mind." Moreover, Montessus refused to include one source of legends in his ethnography: the Bible. He insisted that its stories did not admit of naturalistic explanation. To interpret biblical stories alongside pagan myths (as Eduard Suess did in *The Face of the Earth*) was to fall into an error that derived "either from the non-observance of [science's] own rules, or from the hubris of man to want to explain what is not capable of being explained."[54] Despite himself, Montessus blurred the lines between European science and religion, on one hand, and "savage" views of earthquakes, on the other.

Meanwhile, in Japan, the seismologist John Milne was "compelled to wade [through]" traditional Japanese writings on earthquakes (in translation), as he searched for "facts of scientific importance."[55] Milne, a mining engineer, arrived at the Imperial College of Engineering in Tokyo in 1876 and took up research on earthquakes soon after. He expected to return home to Scotland before long, assuming that the Japanese would take over his research and seismology's "foreign element" would "die out."[56] Like many Europeans, Milne was uncertain about Japan's status as a "civilized" nation. As Gregory Clancey has shown, seismology and antiseismic engineering offered the Japanese opportunities to surpass European science, and to do so on their own terms. Many of the characteristics of nineteenth-century seismology—its emphasis on empirical research over theorizing, its reliance on untrained observers, its lack of professionalization—made it accessible to East Asians at a time when Asia was still perceived as backward in Western eyes.[57] Seismology's ill-defined boundaries allowed both Japanese and

Chinese scientists an unusual latitude to define for themselves the meanings and values attached to "traditional" and "modern," "indigenous" and "foreign."[58] In both countries, seismologists melded aspects of folk knowledge (such as the predictive use of animals) with approaches legitimated by the international scientific community.

In one of Milne's earliest studies of stone-age tools, he proposed an unusual marker of civilization: "It would seem that the number of relics of a barbarous age in any civilized country, will, amongst other conditions, very largely depend upon the number of years which separate that age from its present civilized condition."[59] Anthropology was thus the counterpoint to Milne's seismology: it allowed him to measure Japan's proximity to the "barbarous," even as he used seismology to secure the nation's status as "civilized." Milne began to speculate on the cultural effects of earthquakes in 1879, in an anonymous article for the English-language *Japan Gazette:* "Why should we not study the connection between earthquakes and the human species? Why do earthquakes produce feelings of nausea and sickness. . . . Is it an effect upon the nerves or what is it,—nature? Further, might we not enter upon a broader question and consider what general effects have been produced upon the inhabitants of an earthquake country. What was the action of the last earthquake upon us? . . . Living under such conditions we might perhaps grow reckless; and drinking, gambling and other vices might consequently be a characteristic of the residents in Japan. Perhaps imbecility might become prevalent." At this time, the only authority to cite on this topic was Buckle. Yet Milne's anonymous speculations differed from Buckle's in one key way: he used the collective first person. It was unclear whether his "we" was limited to the English-speaking readers of the *Gazette,* or whether it included the native Japanese.[60]

Many of the anecdotes Milne gathered from Japanese historical sources found their way into his 1887 address to the Seismological Society of Japan on "Earthquake Effects, Emotional and Moral." Like Buckle, Milne argued that capitalism required a natural environment predictable enough to make investment in the future a rational act. Seismicity instead raised the specter of a "continually approaching and receding death." In earthquake-prone countries, people became "careless of the morrow," "passions have been unbridled and refuge has been sought in mirth and gaiety"; in the place of commerce and science, people turned to gambling and to "arts conducive to pleasure." Milne expressed confidence that "the successful or serious nations of the present day, characterized by their enterprise and commerce, are not those whose misfortune it has been to fight against unintelligible terrorisms of nature. Not only may seismic forces have stimulated the imagination to the detriment of reason, but amongst the weaker members of a community, by the creation of feelings of timidity resulting perhaps in mental aberrations like madness and imbecility, the seeds have been sown for a process of selection, by which the weaker members in the ordinary course of racial competition must succumb."[61] Milne went so far as to invert Humboldt's scheme, claiming that "the oftener you feel earthquakes the less they are to be disregarded" and the stronger the feeling of "timidity" they produce. This "complete demoralization" affected even the victims' descendants. "For years after such a catastrophe every tremble in the earth will produce a panic. The experience and fears of fathers are handed down to their children and before these terrors have become things of the past, a fresh disaster adds fuel to the fire consuming the moral constitution." As evidence of this lasting "demoralization," Milne cited superstitions and religious interpretations of disaster, Catholic devotions, and commemorative feasts. Like Buckle, Milne made little effort to hide his contempt for Catholicism. Predictably, he closed by citing Darwin on the fate of England were it suddenly to fall prey to earthquakes, thus attaching Darwin's authority to a notion of environmental determinism that Darwin never espoused.

Milne's paper was the launching pad for a wide-ranging discussion at the turn of the century on the human effects of earthquakes, as we will see in chapter 6. Lost in the subsequent debate, however, was the objection raised by a Japanese seismologist in the audience, Seikei Sekiya. Sekiya had just taken up a post as the first professor of seismology at the Imperial University in Tokyo. The Seismological Society's journal (edited by Milne) reproduced Sekiya's comment as follows:

Prof. Sekiya said that Prof. Milne had at one time written monographs on special earthquakes, next he wrote up on experiments he had made upon artificial disturbances, then he described instruments for measuring earthquakes, and now he takes up the literary portion of the subject; in short he threatens to exhaust all that there is for workers in seismology to investigate. Whenever large earthquakes have occurred in Japan the mental excitement has been great. Relating to the Ansei earthquake of 1855 there are 80 different works, some giving observations of scientific value, while others only tell us about the terror and misery of the people. One work called Ansei Kenbun Roku tells us that at the time of great calamities like the earthquake it describes, the hearts of mankind are shown in their true light. It substantiates this by describing the vices and crimes which were committed at the time of the Ansei earthquake as well as the acts of charity and assistance which were offered to the sufferers.[62]

On the surface, Sekiya seemed to be praising the breadth of Milne's research, and he seemed to agree that earthquakes produced mental "excitement" in Japan. Between the lines, however, Sekiya was challenging Milne's imperializing approach to both the study of seismology and the Japanese people. Sekiya nimbly expanded the meaning of "mental excitement" to include "observations of scientific value." Moreover, he cast Japanese reactions to the earthquake in universalist terms—they revealed not *Japanese* hearts but the *hearts of mankind*. Finally, Sekiya's conclusion was refreshingly nuanced compared to Milne's: disaster did not produce an inevitable "demoralization" and social breakdown, but rather a variety of individual responses.[63] In reply to Sekiya, Milne grew more equivocal: "All great calamities produced mental effects, and with savage nations these were more permanent than with civilized nations. With civilized nations the effect of natural terrorisms die out more rapidly than they do among the uncivilized. Many of our present mental peculiarities are undoubtedly the result of a complexity of causes, and with the exception of those countries where large earthquakes are frequent it is difficult to indicate the results due to earthquakes as distinguished from those due to other phenomena." Milne's own research into Japanese vernacular sources had opened a door to Sekiya's skepticism about the typology of "civilized" and "savage."

The Decline of Seismic Tourism

European men of science of the mid-nineteenth century attempted to channel the earthquake enthusiasm of European travelers into a global Humboldtian science of disasters. By the 1880s, many spied an alternative approach to the globalization of seismology. Already in his 1851 guide to seismological observation for travelers, Mallet cautioned against relying on human observations. Evidence should, "as far as possible, be circumstantial. Nature rightly questioned never lies; men are prone to exaggerate, at the least, where novel and startling events are in question."[64] Mallet's agenda was already clear: he believed that architectural evidence would bear out his hypothesis that earthquakes consisted of elastic compression waves.

By the time of the *Manual's* fourth edition in 1871, Mallet was prepared to dismiss all observations other than those made explicitly in light of his theory and on the basis of architectural damage. "Observations undertaken without such preliminary knowledge will for the most part be valueless, and uninteresting even to the observer."[65] Similarly, the German *Instructions for Scientific Travelers* of 1875 cautioned: "It lies in the nature of the matter that the observations of such an unexpectedly occurring phenomenon cannot all be of equal value and equal credibility, since illusions and preconceived ideas enter very easily, apart from unintentional inaccuracies, which likewise slip in and the source of which can sometimes lie in careless observation and sometimes in inadequate description."[66] The traveler should therefore assume that earthquake observers would offer "frivolous opinions" and "incomplete descriptions." The residents of earthquake-prone lands were especially suspect, since (contra Humboldt) "fear of earthquakes tends to grow with the number of shocks experienced." When the second edition of the German handbook appeared a decade later, the situation of seismology and of German travelers abroad had changed again. The 1875 edition of the German manual had been inspired by the far-flung expeditions to observe the transit of Venus of 1874. The edition of 1888 was devoted instead to the "colonizing ambitions of Germany." In the interim, seismology had become—according to the editor—an autonomous discipline, relying heavily on instruments and on expert knowledge of local geology. The traveler could contribute little to such studies. Only in "populous and civilized lands" could seismological research be undertaken. The editor judged that in the first edition, the chapter on earthquakes had "far exceeded the limits of an 'introduction for travelers,'" and in the new version earthquakes were relegated to a small section of the chapter on geology.[67] Similarly, the 1886 edition of Richthofen's *Guide for Scientific Travelers* dispensed entirely with instructions for earthquake observation.[68] If, by the turn of the century, scientists were no longer recruiting European travelers abroad as seismological researchers, this was in part because they believed they had found a replacement: the instrumental registration of distant tremors.

Earthquake-Watching on Samoa

Samoan legends collected by Europeans in the nineteenth century told of Mafuïé, the god of volcanoes and earthquakes. Mafuïé had broken an arm in a fight with a young warrior, leaving him only one arm with which to shake the earth. A British missionary recounted this story at the second meeting of the Australasian Association for the Advancement of Science in 1890 and drew a geohistorical lesson: "The testimony of the old natives is that the shocks of earthquakes were much more severe in olden times."[69] Other Samoan myths invoked the god Tangaloa, who cast his fishing line into the sea and pulled up an island. In fact, it is not implausible that strong earthquakes produced uplift of this sort.[70] Many such legends were collected during the last two decades of the nineteenth century, as British and German colonists began to settle on Samoa. The testimony of Samoans helped

Fig. 5.1. The seismic hut of the Samoa Observatory, under construction. Inside is a state-of-the-art inverted pendulum seismograph, designed by Emil Wiechert. *Ergebnisse der Arbeiten des Samoa-Observatoriums der Königlichen Gesellschaft der Wissenschaften zu Göttingen*, vol. 1 (Berlin: Weidmannsche Buchhandlung, 1908) (*Abhandlungen der könglichen Gesellschaft der Wissenschaften zu Göttingen, Mathematisch-Physikalische Klasse*, Neue Folge 7), plate 5, figure 1.

Europeans understand the past and present geology of the islands.[71] Then, in the new century, researchers began to arrive on Samoa in greater numbers. They no longer expressed interest in Mafuïe or Tangaloa. Instead, they spent most of their time in a coconut grove along the water, where native workers were building a set of huts, at great cost and with copious amounts of cement. The Europeans expected the Samoans to help operate the delicate instruments arriving by ship. Neither the Samoans nor the scientists associated this work with earlier European studies of the islands. In the minds of the Europeans, the Samoan geophysical observatory, founded by the Göttingen Academy of Sciences, was an institution of "pure learning." It was to be a window onto global physics, one that just happened to lie on a tropical island. The Samoans probably saw it as a profitable coconut farm (see figure 5.1).[72]

German geoscientists came to Samoa in search of "world quakes," *Weltbeben.* They reckoned that the little island would offer a prime view of the strong seismic waves arising in the Pacific region. By means of such world-shaking events, registered in a span of minutes by observatories around the world, they hoped to bring into focus a picture of the internal structure of the earth. For a landlocked young nation like Germany, a perch in the middle of the south Pacific looked like "a pre-arranged experiment" for determining "the manner in which the propagation of seismic waves is influenced by continents and oceans."[73] What the Germans hadn't reckoned with were all the quakes that were disappointingly unworldly: weak, nearby temblors, which might hold clues to local geology but would never be registered anywhere else. The seismic records of the Samoa observatory are littered with events marked "insignificant nearby quake [*unbedeutendes Nahbeben*]." This was one of the "shortcomings of Samoa from a seismic perspective—the abundant, mostly uninteresting nearby quakes."[74] The seismographic records required careful pruning to reveal the specimens of "world" quakes amid the overgrowth of "nearby" ones. Moreover, a few times each year, stronger nearby quakes fully disabled the seismograph, requiring up to two weeks to readjust it.

Not that the scientists ignored these events. They weren't allowed to ignore them: from the perspective of the colonial government, it was the scientists' duty to "reassure" the local population about volcanic and seismic activity. In the wake of an eruption on a neighboring island, the observatory director "now needed to place my quite modest knowledge in the field of vulcanology at the service of the official politics of reassurance."[75] Politics mitigated against either too much or too little attention to local nature. In 1906 the observatory's director claimed to have been "attacked from, to be sure, a biased party," because he had successfully predicted an eruption in the Samoan newspaper. He defended his conduct: "At first, as the residents of the adjacent villages, blind with fear, wanted to leave their dwellings and ceased work on their plantations, and even a few white settlers lost their heads, I tried to point out that the volcano was harmless for the moment; later, however, as concern and fear diminished with time, and a certain party proclaimed the imminent extinction of the volcano, I firmly drew attention to the latent danger, and—as the frightful eruption in July 1906 showed— quite justifiably so."[76] The same director established a service in 1905 to collect observations of local seismic events. At one point, the colonial governor was enlisted to cull reports from natives as well as Europeans, but the effort "remained entirely without results." A compilation of the reports was published every two to four weeks. "What was important to the public was the evidence that nearly all sensible earthquakes had their origin far outside the Samoan islands." On German Samoa, colonial scientists attended to nearby quakes only in order to mute the "blind fear" of the natives.[77]

How did the nineteenth-century quest for a global seismology lead to Samoa? What significance did this Pacific island hold for geophysics? Representatives of German science on Samoa must have asked themselves similar questions every morning when they awoke in a thatch hut, fifteen thousand kilometers from home. There was no ready answer. As one analyst of the Samoan seismic records warned, it was all too easy to mistake a local phenomenon for a global one, to misattribute a seismograph's jitter to a reported distant quake.[78] When a volcano erupted soon after the arrival of the observatory's first director, he simply assumed it was of global import: "For in 1902 the vulcanism of the earth had reached previously unknown dimensions and now let its fires smoulder on Samoa as well."[79] Geophysical observations on Samoa gained global significance by the same process that rendered the island's nearby earthquakes "insignificant." As we will see, this was a move to "purify" seismology by severing it from human bias—and human needs.

SIX

The Moment of Danger

In 1917 an American seismologist casually remarked that the psychological effects of earthquakes would form "an interesting thesis subject for some graduate student of psychology."[1] In fact, at a time when scientists were devising new means of measuring human emotions in the laboratory,[2] earthquakes had already drawn the science of fear into the field. The emerging disciplines of trauma psychiatry and criminology mined the testimony of earthquake victims for clues to managing the human aftermath of disaster. These episodes have been overshadowed in the history of the human sciences by military and industrial catastrophes. But natural disasters also played a role in shaping a modern understanding of the psyche and its defenses. The industrialized mind still recognized its own vulnerability to the natural world.

From Vertigo to Shock

In the late eighteenth century, witnesses to earthquakes were often diagnosed with vertigo, a condition of particular interest to eighteenth-century physicians and philosophers. As a disorder that did not belong exclusively to the mind or body, vertigo spoke to the Enlightenment fascination with the relationship between reason and sensibility.[3] Vertigo continued to fascinate the romantics because it blurred the distinction between the passive and active mind, between sensation and imagination.[4] When earthquake investigators reported cases of vertigo in this period, they were making a diagnosis at once physical and psychological. They did not fret over the relative contribution of reality and imagination, as later researchers would. They simply accepted that vertigo (and associated symptoms like nausea) were typical reactions to tremors, particularly where shaking was too weak to cause visible damage. In the wake of the Calabrian earthquakes of 1783, for instance, an Italian physician cited "squeamishness" and "vertigo" as among the symptoms of a seismic "influence on the animal economy," the result of an imbalance of electrical fluids and an excited imagination.[5] Self-reports of dizziness and nausea thus became trusted indicators of ground motion. Human fear likewise figured in scales of seismic intensity as a criterion of earthquake strength. (This is true of nineteenth-century scales as well as those in use today, including the Rossi-Forel [1883], Mercalli [1902], Sieberg [1912], Mercalli-Cancani-Sieberg [1932], and European Macroseismic Scale [1998].) The psyche, in this sense, was an essential component of the human seismograph.

Moreover, the frequency and intensity of earthquakes in a given region were understood as important factors in the physical and mental well-being of the inhabitants. It was not just that repeated catastrophes might induce barbarism, as Buckle suggested. People worried that even mild tremors, if common, might be insalubrious. In this respect, they were following the ancient belief that health is a property jointly of the body and of the landscape it dwells in.[6] For instance, after the 1855 earthquake in Visp, Switzerland (see chapter 4), the Zurich physician Conrad Meyer-Ahrens noted the variety of symptoms that ground movement could cause. In addition to the numbness, nausea, and vertigo reported by witnesses to the Visp tremors, he mentioned the "convulsive movements in the muscles, unusual movements of fetuses, shivering, heart palpitations, claustrophobia, [and] headaches" reported after the Calabrian earthquakes of the 1780s. "Naturally," he commented, "a portion of these phenomena, like the [miscarriages] and suppression of menstruation reported by Mignani as effects of this earthquake, were the result of terror and fear." A portion, but not all; some must have been due to the release of toxic subterranean gases, "which can be detected by a smell similar to that of wet clothes or rotting hay." For Meyer-Ahrens, the point was that earthquakes produced both direct and indirect effects on human health, physical and mental. Some of these effects could be long lasting, if they arose through "the transformation of the earth's surface, the destruction of vegetation serving animals as food, floods, changes in the weather, etc."[7] Earthquakes could thus harm people even in the absence of physical injury, by disrupting the complex dependence of human health on the environment.

It was only at the close of the nineteenth century that experts began to dissociate the mental effects of an earthquake from its environmental and physiological impacts. They began to describe the mental condition of earthquake survivors in terms of a new concept: traumatic "shock." The notion of shock has become fundamental to a modern understanding of the psyche, its defenses and its vulnerabilities. Historians have constructed a history of shock centered on industrial and military contexts, tracing the diagnosis back to the early days of the railway. Railway travel was typically the middle class's first encounter with industrial technology. Accidents occurred relatively frequently; when they did, some victims complained of ills that were similar to hysteria and could not be associated with any visible injury. These patients were diagnosed with "railway spine." At first, this was assumed to be a form of fatigue due to the mechanical jolts and tremors that the railway carriage inflicted on passengers. By the 1880s, though, this condition was increasingly understood as psychic in origin, a form of "traumatic shock."[8] In the First World War, shock became the principal framework for understanding the psychic effects of mechanized violence. Historians have thus understood the concept of shock as a product of the military and technological contexts of industrialization.

Originally, however, shock also described a reaction to *nature's* violence. Several exemplary early cases were connected to natural disasters. Near-miss lightning strikes, for instance, were to blame for several cases of male hysteria treated by the famed Paris psychiatrist Jean-Martin Charcot. Similarly, an earthquake that produced moderate tremors in Nice in 1887 was the cause of "nervous-pathological disorders." Charcot treated a witness to this event who "was preoccupied by the memory of the earthquake; at home, she was haunted by this obsessive idea that nothing is solid and that perhaps the ceiling is going to fall on her head."[9] Charcot was quick to associate these cases with other instances of "traumatic hysteria." Crucial in this respect was his observation that the hysterical condition could be triggered by a wide variety of natural and man-made events. "Whether it is a matter of a railway collision, of a nervous shock of any kind with or without trauma, earthquake, carriage accident, or on the contrary of an intellectual or genital overstimulation; of alcoholism, of saturnism; no matter, the nervous effect always remains essentially the same."[10] Charcot was led in this way to his fateful conclusion: nervous illness could arise in the absence of a physical or even neural injury, from a purely psychic wound.[11]

Among cases of traumatic shock, earthquakes seemed to produce particularly intense emotions. An article in a journal cofounded by Charcot observed: "There are no accidents that are accompanied by stronger emotions, by more intense fright than those that are due to these great telluric disturbances. Even if people have received no wound, not even a knock, they seem no less absolutely unbalanced. One would say that the earthquake has produced in their brains a disorder analogous to the disorder it produced in their homes."[12] As one of Charcot's associates noted, hysteria sometimes afflicted even those who escaped physically untouched. Even a weak tremor could turn normal people into hysterics and "liars." These conclusions were echoed by a Russian psychiatrist's report on the mental effects of an earthquake in a Polish-speaking town, namely, a "weakening of the cortical centers of inhibition and a significant elevation of nervous excitability. . . . The people gave the impression of automata, which fell into tonic-clonic convulsions as a consequence of the smallest noise."[13] Similarly, a Dr. Schwarz reported on the effects of the 1880 earthquake on the island of Chios: "With regret I must report that the majority of the young women fell sick after the beginning of the earthquakes, specifically some of epilepsy, some of spasmodic attacks. If a keen observer of human nature were to behold these suffering faces, tinged more blue than pink, he would certainly be astounded that fear and horror can cause such a transformation."[14]

Psychiatrists considered the possibility that psychic "shock" could be a direct result of the seismic "shock," rather than of the fear it induced. Eduard Phleps, an assistant at the psychiatric and neurological clinic in the Austrian city of Graz, admitted the difficulty of this question in a 1903 article entitled "Psychosis after Earthquakes." It was eight years after the Austrian Earthquake Commission had begun collecting and publishing the reports of lay observers (chapter 7), and Phleps noted that "in all lay observations [Laienbeobachtungen] illnesses are reported that, according to the form of the phenomenon, may very well be connected in part to the powerful natural disaster." The self-reports of survivors were essential, Phleps argued, since neither the scientific literature nor newspaper reports could be trusted: "There are always elaborate descriptions of general panic, the manic racing-around of those in despair, mortal fear, and the like."[15] Phleps drew his evidence primarily from the cases documented by the investigation of the 1895 earthquake in Ljubljana, less than two hundred kilometers from Graz. Phleps concluded that earthquakes did seem to correspond to a "well-defined clinical picture," distinguished by "headaches, dizziness, and nausea."[16] He was inclined to believe that the ills of earthquake victims had a unique etiology, one explained by the significance of gravity for the human psyche. He cited Humboldt: "'A bad earthquake destroys at once our oldest associations'; we could add our most solid associations, since the constant influence of gravity presents not only a constant stimulus for the molecular constitution of all organic and inorganic bodies, but may also be a constant and important factor for the formation of the totality of our purely psychic capacities, perhaps the most fundamental and important, the significance of which in psychology and physiology, it seems to me, has received too little attention."[17] The psychic effect of earthquakes could thus be reduced to that of "the sudden, completely unmediated onset of the stimulus, which produces a strong fright and as such represents the immediate cause of the illness."[18] In the end, Phleps never clearly drew the line between the physical and psychic impacts of earthquakes. But this did not detract from the power of his conclusion: "As far as the development of psychoses after earthquakes is concerned, it is probably sufficient to point to the uniquely powerful natural disaster, *which unsettles every party more or less profoundly.*"[19] In an earthquake's wake, no one, male or female, was immune to hysteria.

Not everyone, however, was convinced by the diagnosis of traumatic shock. The anti-Nazi criminologist Hans von Hentig, for one, believed that the ills of earthquake victims had a far more complex etiology. Hentig, who had studied with the renowned psychiatrist Emil Kraepelin before the war, suspected that the pathways by which earthquakes affected human health were both environmental and psychic. He thus fused the nineteenth-century tradition of medical geography with the new psychiatric findings. He felt that the environmental dimensions of human health had been obscured by the successes of modern bacteriology. The human effects of earthquakes allowed one "to glimpse an intermediate level between the bodily and the mental."[20] They therefore could not be deduced from even the most sophisticated mechanical devices: "The cause must be a different form of energy than the 'shock.' The experiences of seismo-pathology permit no other interpretation. While the most sensitive horizontal pendulums have always lured us farther into a one-sided analysis of the mechanical components of the phenomenon, the reagent of the living organism, above all of the nervous system, opens surprising perspectives for the earthquake researcher as well as for the psychiatrist."[21] In an age when seismology was increasingly confined to geophysical observatories, Hentig recalled how much could still be learned from the human seismograph.

Earthquakes thus deserve a place in the history of trauma psychiatry. They were, in fact, an ideal site for investigations of shock. When technologies went awry, it was always possible that people would claim injury in order to win insurance compensation. Earthquakes, on the other hand, were legally "acts of God." Victims had no financial incentive to simulate a malady, and the symptoms of shock were therefore all the more irrefutable. In fact, earthquakes often provided a point of reference for early industrial accidents. In her study of literary prefigurations of the concept of shock, Jill Matus notes, "Shock in the texts I have been quoting is likened to tremor and earthquake, a violent disruption, a clash, or physical upheaval."[22] One of the earliest instances Matus cites is a railway crash in Charles Dickens's

1846 novel *Dombey and Son,* in which the disaster site is compared to an earthquake zone. Comparing a technological catastrophe to a natural disaster hints at the Victorians' sense of the uncomfortable proximity between industrial progress and ruin. Henry Adams drew a similar comparison in 1904, "Every day Nature violently revolted, causing so-called accidents with enormous destruction of property and life, while plainly laughing at man, who helplessly groaned and shrieked and shuddered, but never for a single instant could stop. The railways alone approached the carnage of war; automobiles and fire-arms ravaged society, until an earthquake became almost a nervous relaxation."[23] It seems, then, that the emergence of the concept of traumatic shock depended on comparisons between the psychic effects of natural disasters and of military and industrial violence. The spread of industrial technologies did not eclipse Europeans' sense of vulnerability to elemental nature. On the contrary, Victorians' anxiety about the psychic effects of industrialization reflected in part their understanding of the psychic effects of natural forces.

Criminology and Cosmic Crisis

According to Freud's *Civilization and Its Discontents,* the Great War demonstrated that civilization's enemy was no longer external nature, but the unmasterable forces of man's inner nature. And yet the question of the geophysical determinants of civilization versus barbarism still loomed large in the interwar era.[24] Even after 1918, social scientists continued to study geophysical conditions as determinants of social breakdown.

Studies of the social effects of earthquakes had helped shape the emerging field of criminology in pre-1914 Europe. Lombroso's *Crime, Its Causes and Remedies* of 1899 had opened with two chapters on meteorological, climatic, and geological factors in the "etiology" of crime.[25] In the aftermath of the Messina earthquake of 1908, as the state struggled to impose order, Italian jurists had theorized a relationship between sovereignty and emergency that anticipated a pivotal concept of twentieth-century legal theory.[26] The Swiss psychiatrist Eduard Stierlin found the Messina disaster "of the greatest interest" as an opportunity "to compare how the frightful event affected survivors of such different races and social classes."[27] Two opposing tendencies were identified in the social response to earthquakes. As Cesare Lombroso wrote of the situation in Messina in his *Archivio di Antropologia Criminale:* "As for the effect on the moral sense, one may say that in this terrible catastrophe it was exaggerated in one direction or the other, as if human nature had wanted to show all that it could of the most sublime and the most savage." The catastrophe awakened what Lombroso termed "all the primitive and atavistic sentiments," religious and otherwise.[28] The *Archiv für Kriminologie* published a similar report on the San Francisco earthquake of 1906: "the *bête humaine* naturally held its orgies and there were plenty of looting mobs [*Schlachthyänen*]. On the other hand there were also traces of *homo nobilis*, and not just isolated cases, which should please the humanist. . . . It almost makes one think of an unconscious imitation, of a 'psychic infection' by example, which suddenly made heroes of average men . . . since indeed even in the worst something human slumbers."[29]

After 1918, criminologists were not satisfied with a simple analogy between the social effects of earthquakes and of wars; instead, they probed possible interconnections between war, revolution, and the social effects of "atmospheric and telluric disturbances." Thus, what observers described as widespread hysteria following the Tokyo earthquake of 1923 was attributed in part to the lingering effect of a world war that acted like a petri dish for mass panic.[30] Similarly, the 1928 earthquake in the Crimea was found to awaken "the whole already extinguished image of the war neurosis (at the moment of the catastrophe the image of the battle appeared; the sound of heavily loaded automobiles called for two series of associations—connected to the war and to the earthquake in the Crimea)."[31] In 1920, following Germany's November Revolution, Hans von Hentig published *On the Relatedness of Cosmic, Biological, and Social Crises.*[32] Hentig was interested in what happens when "an atmospheric or telluric infection [*Schädlichkeit*] reaches the light and porous substrate of moral feeling." Hentig concluded that modern states must recall what the ancients well knew: that even the slightest earth tremors could open the door to revolution. The authorities of ancient Athens would never have permitted crowds to gather on the day after a widely felt tremor—as occurred in Munich on the afternoon of 7 November 1918, when socialists staged a peaceful coup against the Bavarian monarchy. For information on the tremor of 6 November, Hentig was indebted to the Bavarian macroseismic service, founded in 1905. What clinched his argument, he believed, was that "one of the most excitable and most eloquent heroes" of the November Revolution—the anarchist and writer Erich Mühsam—was "by nature highly sensitive to earthquakes." He had in hand a letter from Mühsam, which he had received from the director of the Munich observatory. Though written over three years before the revolution, it was evidence of Mühsam's "earthquake-sensitivity"—a trait Hentig linked, predictably, to his Jewish origin:

Since about 2 o'clock I was unable to sleep and felt a strong nervous tension (as I almost always feel a very strong sensation of tension before storms). The nervousness, the cause of which I could not explain to myself, increased so much that I turned on the light. . . . I was then sitting upright in bed, when I suddenly had a sensation as if someone had grabbed the bedframe from beneath and were shaking it back and forth. . . . I understood immediately that it was an earthquake. . . . I note further that after the event my nerves became entirely calm again and I soon slept excellently, from which I conclude that the earthquake announced itself to my nerves ever more strongly for about 1½ hours before its start.[33]

Hentig believed that his discovery of the geo-psychic determinants of revolution would allow governments to extract revolution's "poison fang in due time."[34]

The Wisdom of the Body

Hentig's term "seismopathology" highlighted the difficulty of identifying a "normal" psychic reaction to earthquakes. Ideal conditions in which to study this question were created by the Crimean earthquake of 1927, at the height of the tourist season. Seismologists collected "normal" reactions from educated tourists, while doctors examined patients at the local sanatorium for "pathological" reactions. The analysis by Soviet physicians was keyed in subtle ways to the values of the state. Thus a Moscow psychiatrist paid particular attention to the drastic fall in the "work capacity" [Arbeitsfähigkeit] of the affected population; he noted too that the ability to withstand psychic trauma was supported by the "feeling of responsibility" proper to directors of institutes or state servants.[35] In all, this physician was able to define nine symptoms of "earthquake syndrome": "acceleration of the pulse, paleness, the sensation of ground oscillations, dizziness, nausea, general fatigue, restlessness (anticipation neurosis), insomnia, decreased productivity." A final symptom was a new perception of the immediate natural environment. As one survivor testified, "Nature has acquired greater significance, has let her power be felt; the horizon is significantly widened."[36] For many survivors, "The picturesque paths of the Crimea with the mountains looming overhead no longer offered any pleasure. . . . This blooming and vibrantly colored Crimea, bright of complexion and rich in the wondrous creations of nature, this dream of the ill and of those in need of rest, has in an instant become unpleasant, forbidding, malicious and menacing."[37] This "Crimophobia" struck even natives, many of whom moved permanently north.

These symptoms, then, constituted "normal" effects of the earthquake over the following days, weeks, or even months. But what of immediate reactions? What was the "normal" state of mind when the earth shook?

An assistant at the Moscow psychiatric clinic concluded that the earthquake cast doubt on the conventional wisdom that the human being is a "remarkably resilient creature." One fifty-one-year-old local physician, "strong and healthy," declared that his experience on the night of the Crimean earthquake "could not be described." In the moment of crisis, he "lost all courage, no longer even thought of my loved ones," and it took six to eight minutes before he "came to himself." The eighteen seconds of ground movement itself "meant a panicked terror, they were more horrible than death. The tongue did not obey, the body shuddered, only with difficulty could one use one's arms and legs. Not a single thought in the head. Paralysis."[38] Such an account could be explained neurologically: "The earthquake shock inhibited the influence of the cortex and thereby heightened the effect of the unconscious sphere of the instincts."[39] However, another commentator complicated this picture. "Normal" reactions to fear, he argued, were of two types, active and passive, though occasionally both reactions were combined in a single person. Active personalities, including those who had rescued or calmed others, "did not remain passive in the face of the disaster and displayed a reaction corresponding to their active, decisive character; they set their own wills to survival against the will of the elements." Passive characters, on the other hand, responded with fatalism and a sense of helplessness. As one survivor described it, at the moment of crisis "man has no will to battle with the elements, he is their slave." The passive reaction could even express itself as an "unconscious death instinct [unbewußte Todesdrang]." In general, "normal" reactions did not lead to "nonsensical, purposeless acts." To the contrary, "Terror in this way lent an unusual concentration of psychic energy, awakened acuity and clarity of thought.... The capacity to preserve the necessary concentration of attention for calm judgment is a property of a well-organized psyche, which does not permit sudden, reflexive, often purposeless reactions (jumping from the second or third floor)." Still other "normal" witnesses reported "a feeling of enchantment and wonder at the power of the elements that was close to ecstasy." During the weaker shocks, three even reported "erotic excitement."[40]

A most remarkable perspective was emerging from studies of "normal" reactions to earthquakes. Many survivors reported that, in the midst of catastrophe, their response was not panic but something like the opposite. Perhaps the most famous earthquake account of this genre was that of William James. As a young man, James had trained in the tradition of Humboldt and

Darwin as a naturalist on Louis Agassiz's Brazil expedition. He was undoubtedly familiar with their stories of their own first earthquakes. Indeed, he had
been disappointed that his voyage to Brazil did not bring much in the way
of adventure.[41] Moreover, James was a psychologist fascinated by the study
of the human mind under the extreme conditions that could be generated
in the laboratory; yet he had an abiding aversion to laboratory work. So
perhaps we should not be surprised that James's reaction, as his bedroom at
Stanford began to shake on 18 April 1906, was "delight." Beyond his sense
of good fortune, James's account departed from the traditional form of scientific earthquake observations. He did not describe the duration, intensity,
and direction of the quake, nor a scene of pandemonium in its aftermath.
"Little thought, and no reflection or volition were possible" as the ground
shook. Although he felt no fear, James had the sense that the quake was a
"demonic power," "an individualized being," a "living agent," which he
likened to "earlier mythologic versions of such catastrophes." James's narrative then turned to his observations in San Francisco later that day and in
the weeks that followed. Where others told tales of crowds gone "mad with
terror" and of "bacchanalian orgies,"[42] James stressed the remarkably sober,
cooperative behavior of the urban populace. James insisted that this ability
to organize in an emergency was not merely "American" and "Californian"
but also deeply "human." Cooperation was a firmly rooted feature of the
human response to catastrophe: "one's private miseries were merged in the
vast general sum of privation and in the all-absorbing practical problem of
general recuperation." James's account was characteristic of his efforts to
bridge materialism and idealism. It suggested at once the fundamental un-
predictability of the universe and the harmonies between world and mind.
The experience proved "how artificial and against the grain of our spontaneous perceiving are the later habits into which science educates us." At the
moment of crisis, James claimed, the keenest observations were the product
of instinct, not scientific training.[43]

James's account bears comparison to that of the German physician Erwin Baelz, who lived in Japan in the late nineteenth century. In 1901 Baelz
described his state of mind during an earthquake as "emotional paralysis":

> Suddenly, really absolutely suddenly, a complete transformation occurred
> within me. All the higher emotional activity was extinguished, all compas-
> sion for others, all sympathy for possible calamities, even concern for one's
> endangered relatives and for one's own life had vanished and left the mind
> completely clear; indeed it seemed to me as if I was thinking more easily and
> freely and faster than ever. It was as if a prior inhibition had suddenly been

removed; I felt like a Nietzschean master, not answerable to anyone, beyond good and evil. I stood there and observed all the terrible events around me with *the same cool attentiveness with which one follows a fascinating physics experiment.* . . . Then, equally suddenly, this abnormal state disappeared and made way for my prior self.[44]

Baelz reminds us of one filter through which Europeans experienced earthquakes in these years: a Nietzschean naturalism, according to which disasters would suspend everyday ethics and confer survival on the fittest. Baelz's "Nietzschean master" was clear-thinking and devoid of human sympathy. It was a new twist on the seismological fantasy of Immanuel Kant: the earthquake emptied of human content and reduced to a "fascinating physics experiment." Oddly, though, Baelz suggested that this attitude was not a scientific ideal but a spontaneous reflex.

Baelz's state of mind during the earthquake was diagnosed by one psychiatrist as "manic": "the different valuation of the associations under normal conditions is overridden by the excitement; the sanguine emotional state and the acceleration of thought, the elevated sense of self corresponded to the manic symptom complex."[45] Another psychiatrist, Eduard Stierlin, who had studied psychic reactions to earthquakes and mining accidents, considered Baelz's experience to be common: namely, "a total elimination of all sense of value" in the face of catastrophe. Stierlin speculated that the emotional response was "only 'repressed'"—and likely to return in nightmares or neurotic symptoms. From this perspective, Baelz was not so different from those earthquake survivors who succumbed to traumatic hysteria. Stierlin quoted Hippolyte Bernheim's opinion that everyone is, to a degree, hysterical. Indeed, hysterics had often been known to display a certain stoicism in the face of crises—their "emotivity was weakened rather than strengthened."[46] What emerged from this analysis was the unexpected difficulty of distinguishing between an attitude of scientific objectivity and the symptoms of traumatic hysteria.[47]

By the 1920s, the American physiologist Walter Cannon was studying fear as a reaction of the sympathetic nervous system: a physiological response to a hostile environment that produced, along with a faster metabolic rate, heightened mental acuity.[48] Among physiologists and psychiatrists studying humans under such extreme conditions, one story in particular came to be cited frequently. It was not an account of earthquake, for its author never experienced a major temblor—though he dedicated much of his life to studying them. It was, rather, the tale of an 1871 mountaineering fall that nearly cost Albert Heim his life. In a lecture to the Swiss Alpine Club in 1892 entitled "Death by Falling," Heim compared his memory of the accident to other accounts of near-death experiences. In all these cases, the senses were unusually acute; thinking was intense and accelerated; judgment was "objective." This response had nothing to do with "personal qualities," Heim argued. It was universal at such moments. Yet it could not be a "reflex," for it was fully conscious: "on the basis and in consequence of a complicated series of thoughts—clear in all its parts, although very rapid, but entirely consciously conducted. . . . A reflection such as appears on its own with absolute necessity when the human mind is maximally excited by terror, at the moment of the most extreme stimulation." Heim's state of mind while in free fall confirmed this seeming paradox: "Objective observation, thought, and subjective feeling took place simultaneously."[49] The fall occurred seven years before Heim penned his instructions to Swiss citizens for the observation of earthquakes.

At stake in representations of the mental effects of earthquakes was the basic question of the "reliability of witnessing."[50] These scientifically trained observers denied that their own capacity for clearheaded observation in the midst of catastrophe was the result of training and habit. They experienced it, rather, as a primal instinct—in Cannon's terms, as the "wisdom of the body." Erwin Baelz even went on to speculate that his "emotional paralysis" during the earthquake represented a form of "possession," and he called for "psychic research" to clarify the phenomenon.[51] James, Baelz, and Heim each framed the perception of disaster in terms of a preconscious synchronization of body, mind, and cosmos.[52]

Sensitive Instruments

Such cold-blooded representations of disaster have come to exemplify a modernist aesthetic ideal. Even as human observers were dropping out of seismology, literary writers began to claim that they were nothing more nor less than seismographs:

> Like a seismograph his sensitive nerves had already registered the underground shocks, while others were still entirely deaf to them.[53]

—Carl Georg Heise on Aby Warburg (1947)

> Both are very sensitive seismographs, who quake in their foundations when they receive waves and must transmit them.[54]

—Warburg on Burckhardt and Nietzsche (1927)

One holds the pen ready like a needle in a seismic observatory, and in reality it is not we who write; rather we are written.[55]

—Max Frisch (1950)

After the earthquake one lashes out at the seismograph. Yet one cannot blame the barometer for the typhoon, unless one wants to be classed a savage.[56]

—Ernst Jünger (1949)

The 'Storm of Steel" [by Ernst Jünger] demonstrates the course of events most powerfully, with the full force of the years on the front; without any pathos it depicts the dogged heroism of the soldier, recorded by a man who, like a seismograph, detects all the oscillations of the battle.[57]

—Erich Maria Remarque (1928)

The pen is only a seismographic pencil for the heart. It will register earthquakes, but can't predict them.[58]

—Franz Kafka (1920–23)

I . . . think of the publisher—how shall I put it—somewhat like a seismograph, who should aim to register earthquakes objectively. I want to record statements of the day that I hear, and, as far as they seem to me valuable to be heard, set them before the public for discussion. (To be a seismograph not a seismologist.)[59]

—Kurt Wolff to Karl Kraus (1913)

[The poet] is like the seismograph, which each quake, even thousands of miles away, sets vibrating. It is not that he thinks constantly of all the affairs of the world. Rather, they think of him. They rule him to such an extent that they are in him. Even his dull hours, his depressions, his confusions are impersonal states, they are like the twitches of the seismograph, and a gaze that reached deep enough could read something more mysterious in them than in his poems.[60]

—Hugo von Hofmannsthal (1906)

The poet appears as an indicator, as a seismograph, from which the moral state [*Gewissenszustand*] of his surroundings can be read off.[61]

—Hermann Hesse (1937)

Perhaps the emblematic literary seismograph was Ernst Jünger. A storm trooper in the First World War, Jünger famously portrayed a cyborg-like fusion of soldiers and their weapons. Such images have exemplified the contested aesthetic category of "reactionary modernism."[62] Yet like James, Baelz, and Heim, Jünger was also a naturalist: the son of a well-off pharmacist, he studied geology, zoology, and botany at university in the early 1920s. Indeed, Andreas Huyssen describes Jünger's writerly gaze as "entomological" in its sober attention to the smallest natural details in the midst of human catastrophe. Wounded by a bullet, Jünger's vision recalls the power of observation that Baelz had discovered through an earthquake:

> I throw away my rifle and jump with one leap down into the middle of the road. And though everything has gone well with me so far, not even the greatest of luck, which I have come to take for granted, will now keep me out of harm's way. While my knees are still bent from the impact of the jump, I receive a hard blow against my chest, which in a second makes me sober. In the midst of the tumult, between both raging parties, I stop and recollect myself. It was the left side, just on the heart, not much one can do about that. In a moment I shall fall full length as I have seen so many fall. Now it's all over. *Curious, this: while I stare at the ground I see the stones on the yellowish soil of the path, black chips of flint and white, polished pebbles. In this terrible confusion I see each one of them sharply, separately, and the pattern they make is imprinted in my mind as if this were now the most important thing of all.*[63]

At the moment of crisis, Jünger became what Huyssen calls an "armored eye." Or, perhaps, a mortal microscope, since the war, for Jünger, was equally about the mechanization of men and the enchantment of machines. It was a "storm of steel," a catastrophe both cosmic and man-made. And the writer fit to record it was, of course, a seismograph, writing "in agitated phrases, illegible, like the wavelines of a seismograph recording an earthquake, with the ends of the words whipped out into long strokes by the rapidity of the writing."[64]

The motif of the writer as seismograph reflected in part the modernist fascination with graphology and automatic writing, understood as techniques for the "objective study of human subjectivity."[65] But it was also a response to the question of the writer's responsibilities in the face of the human catastrophes of the first half of the twentieth century. The seismographic trope claimed for the writer an exceptional sensitivity to foreshocks and a cold-blooded acuity at the moment of crisis, even as it had the potential to divest him of moral responsibility for the ideas he expressed. At the same time, it severed the notion of the human seismograph from its origins in nineteenth-century natural scientific practice. Seismic observation had once been a widespread, acquired habit of empirical observation, often associated with women and with feminine nervous sensibility. The modernist writer as human seismograph laid claim instead to an innate and distinctly masculine artistic genius, a poetry of "iron nerves."

Conclusion

If one were only an Indian, instantly alert, and on a racing horse leaning against the wind, kept on quivering jerkily over the quivering ground, until one shed one's spurs, for there needed no spurs, threw away the reins, for there needed no reins, and could barely see the land unfurl.

—Franz Kafka[66]

With Jünger we arrive at the heart of modernism's fascination with earthquakes. The earthquake's invisible human effects—whether described as vertigo or as psychic trauma—mimicked the perceived effects of modernity itself. Nineteenth-century observers experienced the transformations due to modern science and technology as a "vertiginous violence" (Henry Adams); they induced "dizziness" (Mary Shelley) and "moral seasickness" (Max Nordau).[67] In the psychoanalytic language emerging at the fin de siècle, the earthquake pierced the modern psychic shield. Its effects were thus akin to the symptoms of neurasthenia and traumatic shock. Yet many of the earthquake narratives of this period described fascinatingly ambivalent reactions. Beside the tales of horror were others that insisted on a thrilling sense of liberation, a glorious dizziness. The earthquake gave modern men and women a tantalizing brush with a force they had trouble naming, something vast and primal. What their earthquake fevers shared was the fantasy of relinquishing a defining ambition of the modern age: the quest to master nature. The joy of the earthquake was the thrill of giving oneself up to a raw natural power. It was a fantasy, in Kafka's terms, of dropping the "reins" of progress. As Scott Spector points out, Kafka harbored an "explosive revolutionary element," the target of which was the materialism of his father's generation. That hostility expressed itself violently in his texts as a "contempt for the earth" itself, a "struggle to release the dormant spirit from within the comfortable but constrictive confines of the body."[68] In his fantasy of riding like "an Indian," the motion of the subject is transposed to the ground, such that the land itself "quivers" and "unfurls." Like an earthquake, Kafka's racehorse tears him free of the confines of gravity. Shedding his spurs, dropping the reins, he has shaken off modernity's claims on him. He gives himself up to what Michael Taussig calls "the wild abandon of sympathetic absorption into wildness itself."[69] The writer becomes, or longs to become, a body that speaks in the convulsive language of the earth itself: a seismograph.

At the same time, Kafka's fantasy of vertigo encompasses another fantasy: the discovery within the self of a hidden, inner equilibrium. In the moment that he releases control, the rider is "instantly alert." In this sense, Kafka writes in the tradition of Humboldt and Darwin, parallel to Baelz's "cool attention," James's "spontaneous perceiving," and Jünger's "entomological gaze." The earthquake, literal or figurative, was experienced simultaneously as an instant of wild abandon and of acute perception. Surely, this keenness of perception was a habit acquired through training, a product of these men's scientific backgrounds and of the nineteenth-century culture of popular science. Yet these trained observers insisted otherwise. They attributed their own acuity to a mysterious cosmic affinity.

By the 1930s, the human seismograph was no longer a geoscientific tool. The acuity of earthquake witnesses had been reinterpreted as a symptom of hysteria. The realism of felt reports came to be read as a mere literary exercise. This transformation was evident in the 1931 volume *Der Gefährliche Augenblick* (The Moment of Danger). This specimen of Weimar modernism was a collection of photographs and first-person accounts of various crises: a plane crash, the sinking of the *Titanic*, the assassination of Archduke Franz Ferdinand, battles of the Great War. The editor explained that the intention was to display instances in which "a highly primitive experience intersects with a very complicated one in a few seconds of decision."[70] First, as if setting a template for what followed, came natural disasters: photographs and narratives of earthquakes and volcanoes, from Vesuvius, Etna, and Mount Pelée, to San Francisco and Tokyo. Like the earthquake stories we have been examining, these narratives had eclectic origins: accounts of the eruption of Mount Pelée, on Martinique, for instance, came from a scientist, an imprisoned "Negro," a shoemaker, and a journalist. In the volume's introduction, Ernst Jünger described such cold-blooded visions of catastrophe as the "new style of language" of the modern age.[71] Readers have puzzled over these "moments of danger," which seem to be oddly detached and impersonal by the standards of bourgeois memoirs.[72] They seem incongruous only because they have been torn out of the context of nineteenth-century disaster science. As iconic of modernism as these accounts may now appear, they conformed to the scientific norms of an earlier modernity.

SEVEN

Fault Lines and Borderlands: Imperial Austria, 1880–1914

Our knowledge of a natural phenomenon, say of an earthquake, is as complete as possible when our thoughts so marshal before the eye of the mind all the relevant sense-given facts of the case that they may be regarded almost as a substitute for the latter, and the facts appear to us as old familiar figures, having no power to occasion surprise.

—Ernst Mach, *The Analysis of Sensations*, 1886[1]

How did the self-reports of ordinary people inform planetary physics? The Habsburg monarchy is an excellent field on which to track exchanges between local knowledge and international science. The circulation of knowledge could not be taken for granted in a state that included eleven major linguistic groups and encompassed the plains of Bukovina, the Adriatic coast, the Alps, and the Carpathians. Charged with understanding the monarchy's celebrated unity amid diversity, Habsburg scientists sought an integrated approach to the study of natural and human conditions.[2]

Austria-Hungary contained regions of moderate to high seismicity in the Alps and along the Adriatic. There is even evidence that seismicity in these regions was on the rise in the monarchy's final decades.[3] Particularly in the Balkans, earthquakes called into question the political framework that tied the monarchy's fringes to its two capitals: which level of the state's intricate web of governance would respond? Newspapers in Vienna, Prague, and Budapest printed wary reports of the camps pitched by earthquake victims to their south (see figure 7.1). The specter of homelessness raised fears of irredentist nationalism. Yet earthquakes also inspired humanitarianism. In 1895, when national disputes seemed to be crippling Austrian politics, a catastrophic earthquake struck the city of Ljubljana (German: Laibach),

Fig. 7.1. Camping in Jelačić Square after the Zagreb earthquake of 1880, creator unknown. Source: Jan Kozák Earthquake Image Collection.

capital of the crown land of Carniola. In Vienna, hundreds of writers and artists joined forces to produce a volume of engravings and verse to benefit the stricken city. In the accompanying poems, the contributors conjured up the monarchy as a "brotherhood" and Vienna as its "golden heart" (see figure 7.2).[4] The liberal press took a similar line:

> Where were Slovenes, where were Germans,
> Where were language strife and power thirst?
> Only humans, pale and fearful,
> Only humans, trembling, terrified,
> Only brothers, equal one and all—
> Whom now misfortune—unified.[5]

After the Ljubljana disaster of 1895, the renowned physicist-philosopher Ernst Mach presided (as secretary of the Vienna Academy of Sciences) over the formation of the Austrian Earthquake Commission. At its height this network encompassed over 1,700 observers reporting from all sixteen crown lands. The system's success reflected a culture of earthquake observing that took root in Austria in the last quarter of the nineteenth century. Habsburg scientists insisted that knowledge of a seismic event was possible only by combining the observations of numerous individuals, few of whom

Fig. 7.2. The iconography of imperial humanitarianism in the wake of a seismic disaster. Cover illustration by Alfred Roller for *Für Laibach: Zum besten der durch die Erdbeben-Katastrophe im Frühjahre 1895 schwer betroffenen Einwohner von Laibach und Umgebung,* ed. Genossenschaft der bildenden Künstler Wiens, 1895.

were likely to have any scientific training: "Probably in no other field is the researcher so completely dependent on the help of the non-geologist, and nowhere is the observation of each individual of such high value as with earthquakes. . . . Even when all conditions are favorable, one can do no more than ascertain the time as precisely as possible and observe the phenomena in the immediate vicinity. But thereby only one reliable report is produced, while for a correct judgment one needs a great many. Only through the cooperation of all can a satisfying result be delivered."[6]

A Perfect Earthquake

On the evening of 9 November 1880, readers of the *Neue Freie Presse* learned that, between 8:30 and 9:00 that morning, an earthquake had shaken much of the Balkan peninsula. Tremors were reported from as far away as western Hungary, Carinthia, and Lower Austria, and even from Vienna itself. The quake had killed at least two residents of Zagreb, wounded thirty others, and damaged more than three thousand houses.[7]

This was, in short, the perfect object of study for the seismology of its day. It was a "moderate" earthquake, neither so weak as to go unnoticed nor so strong as to leave observers in what was typically described as a state of "senseless panic or utter despair."[8] In addition, its impact was geographically extensive, meaning that observations could be collected throughout central Austria-Hungary.

The scientific response to the Zagreb quake embodied the dualist (potentially trialist) nature of power in the monarchy after 1867. Separate investigations were launched in the imperial capitals of Vienna and Budapest, as well as in the South Slav center of Zagreb. The imperial Geological Institute in Vienna solicited reports from eyewitnesses, while the imperial Academy of Sciences sent geologist Franz Wähner to Croatia to speak with witnesses and collect evidence of the damage to buildings. Independently, the director of the Geological Institute in Budapest was sent by the Hungarian Ministry of Agriculture, Industry, and Trade to inspect the area, but his inquiries were almost entirely restricted to assessments of damage to buildings.[9] Meanwhile, the South Slavic Academy of Sciences and Arts in Zagreb commissioned two researchers to report on the quake.[10] From mid-November to late December, the scholars from the Vienna and Zagreb academies covered approximately ninety square miles, interviewing witnesses and inspecting damage to buildings. The scale of the investigation dwarfed that of any previous earthquake in Austria-Hungary.

The Vienna geologist Melchior Neumayr conveyed the magnitude of Wähner's effort:

> To get an idea of the quantity of observations that are necessary for a correct assessment of an earthquake, we may consider the materials on the basis of which a recently published work by Wähner on the Agram earthquake has been composed. The author himself spent five weeks on location, occupied exclusively with this matter; several other geologists from Vienna, Pest, and Agram were similarly occupied, and it was possible to use their observations as well. Through the intervention of a few railway authorities, the reports of well over a hundred railway stations were furnished, the maritime authorities transmitted [reports] from all the port captains and lighthouse keepers of the entire stretch from Cattaro to the Italian border; in addition a great many private communications and newspaper reports arrived, such that observations were brought forth from approximately 750 different locales. Certainly there were still many holes, but overall on this basis an accurate insight into the nature of this earthquake was possible through the united effort of more than 1000 different observers, whose results were united in one hand.[11]

No doubt Wähner had achieved an impressive feat of public outreach, but the question remained: having collected over a thousand reports, what was a scientist to do with them? "The analysis of the collected reports, the assessment of their reliability, the separation of the worthwhile from the unusable, finally the synthesis and application of the information obtained, demanded a significant investment of time and effort."[12] It would ultimately take Franz Wähner three years to accomplish.

"To Discern the Phenomenon in Its Physical Elements"

Wähner insisted on reproducing observers' reports as completely and accurately as possible:

> Despite my attempt at brevity . . . I may indeed have reproduced the [witnesses'] reports in too much detail. But I believed I should not go too far with the abridgements, if I wanted to maintain complete objectivity and also make it possible for others to draw an independent conclusion from the reports. This cautiousness was particularly necessary, because in the course of my work I arrived at a judgment that diverges from the views that have prevailed until now.[13]

To be sure, Wähner's ideal of objectivity did not exclude being an active and critical editor of the testimony he collected. One witness reported: "A fearful earthquake. A few houses have partially collapsed." Wähner cautioned: "This is exaggerated." He omitted reports of churches being locked because "people seem to be quick to lock churches in the interest of safety." When another witness claimed that a church steeple had been damaged, Wähner cautioned, "No mention of this in the following report, thus to be considered incorrect." In another case he inserted, "The remaining comments not reprinted here prove to be somewhat exaggerated in view of the previous account."[14] In all these cases Wähner engaged with witness testimony in an intimate and critical manner more typical of the human sciences than the earth sciences.

As a native of northern Bohemia, Wähner depended on the Croatian colleagues with whom he traveled for translations. Yet he took pains to analyze the word choices of his witnesses. For instance, when discussing reports of earthquake sounds, he inserted the original Croatian next to German translations. He thus alerted his readers that the German *Getöse* corresponded to Croatian *tutnjava*.[15] In one case, he quoted an observer's seemingly awkward description, "The dreadful *Getöse* [*tutnjava*] began to shake stronger and stronger with the earth and the houses." He commented, "Despite the unusual manner of expression, very indicative of the character of the ground motion."[16]

Wähner's conclusions reflected this attention to the observers' own descriptive terms. In all the reports from the area of greatest destruction, he noted, "we look . . . in vain for the mention of an instantaneous strong movement such as would be termed, in everyday life as well as by a physicist, an 'impact.' The ground motion was, to the contrary, generally perceived as a long-lasting, continuous movement. . . . It seems to have been slower, gentler movements, though movements of great intensity, than would have resulted from a brief to-and-fro movement of the individual soil particles in a horizontal or diagonal direction."[17] Thus, in his final assessment of the nature of the Zagreb earthquake, Wähner concluded that it took the form of a transverse wave that shook the particles of the earth's crust nearly vertically, and which spread out not from a central point but rather from an extended area. In this way, by means of the language of "everyday life," Wähner lent support to Suess's tectonic theory and discredited Mallet's claim that earthquakes are longitudinal waves that propagate radially.

Wähner's respect for vernacular language was not uncommon among scientists in the Habsburg world. In contrast to Britain and Germany, where a distinct class of scientific "popularizers" arose in the late nineteenth century, recent research suggests that Austrian scientists took it upon themselves to communicate with the general public. Wähner, for instance, was praised for having "cultivated a popular writing style in the best sense."[18] Rather than seeing this peculiarity as a mark of delayed modernization in central Europe, we might view it as a conscious attempt to bridge scientific and vernacular discourses.[19]

The ultimate goal of Wähner's investigation was "to unite the individual observations, made and collected with the greatest objectivity, into a total picture and to seek to discover the law that lies therein."[20] As he wrote in the introduction, "In my view, given the current state of the discipline, it cannot be the task of a monograph on a large earthquake to investigate the final telluric or even cosmic causes of [the event]. It will be necessary above all to discern the phenomenon itself in its physical elements [Elementen], before it can be permitted to discuss causes hidden from observation, and I have convinced myself, that in this first area we still have a great deal to learn."[21] Wähner's sense of his task thus rested on an implicit phenomenalism: to identify the "physical elements" and unite them into a "total picture."

Wähner's study of the Zagreb quake became a model of the monographic method of earthquake investigation. But it also indicated the difficulty of translating this method from its original, landeskundlich context, to a culturally diverse territory. As he acknowledged of his Croatian collaborators, "I was greatly aided by their knowledge of the land and people."[22]

Nations and Regions

At first, Austrian seismology made little effort to overcome these barriers. Instead, it developed as a patchwork of independent provincial networks. In the early 1880s, geologists Rudolf Hoernes and Richard Canaval organized lay seismic observers in the Austrian crown lands of Styria and Carinthia, respectively. They also began research on historical earthquake catalogs for the two provinces. In these undertakings, both men were supported by local institutions dedicated to Landeskunde.[23] Hoernes anticipated objections to conducting seismology as a form of Landeskunde: "One might criticize me for undertaking first a discussion of a large area with political, that is more or less artificial, borders, instead of beginning by publishing a catalog of the earthquakes of the eastern Alps and then on the basis of the latter to consider the seismically active areas and fault lines of the entire region." This was a typical quandary for the earth sciences in the Habsburg world: whether to define the region of study according to physical or linguistic borders. In response to his hypothetical critic, Hoernes noted that he had, in fact, been collecting seismic observations from the entire eastern Alps; but the "critical inspection" of this material had "for many reasons" to be conducted piecemeal, "not according to period but according to place." Thus the evidence itself called for a *landeskundlich* approach: "The largest portion of the sources to be used for the critical annotation of the earthquake reports bear the character of contributions to the *Landeskunde* of the affected province."[24]

While Hoernes and Canaval were enlisting observers in the Alps, the South Slavic Academy of Sciences and Arts in Zagreb formed its own seismological committee. The Zagreb academy was a product of the movement for South Slavic unity at a time when the notion of a South Slavic identity was widely disputed. Under the direction of Michael Kispatic, the first PhD in the natural sciences at the University of Zagreb, the seismological committee collected observations of earthquakes in Croatia, Slavonia, Dalmatia, and Istria (in both Austrian and Hungarian halves of the monarchy) and in occupied Bosnia. Kispatic, the author of popular works on the geology and "natural-cultural" history of the South Slavic lands, also compiled a historical catalog of earthquakes in the region. To this end, he acquired medieval and modern documents from a circle of prominent intellectuals in Zagreb, who were in the midst of penning the first contributions to a Croatian national history.[25] His seismological committee gained significantly in prestige in 1893 when it was joined by Andrija Mohorovičić. Mohorovičić, already successful as a meteorologist, went on to discover the boundary between the earth's crust and mantle that now bears his name. According to Kispatic, Mohorovičić "found quite a number of *patriots* in Croatia and Slavonia who send in, on an almost daily basis," reports of seismological as well as meteorological phenomena.[26] The South Slavic seismological network, like those in Styria and Carinthia, pursued seismology as a patriotic form of regional natural history, as well as a means of geophysical analysis. It thus helped construct the South Slavic lands as an ostensibly natural region.

Elements of Observation

In Austria, as in Switzerland, nonexperts often proved to be excellent observers. Canaval adopted the questionnaires of the Swiss Earthquake Commission, and Hoernes adapted their observing instructions.[27] Hoernes suggested that earthquake observing was more a function of character than expertise: "People who have the *sangfroid* when an earthquake strikes to make useful observations on the nature of the motion will be able to ascertain this feature just as well as seismographs, even if they depend only on the perceptions of their senses. However only very few are able, with the approach of such an unusual and terrifying natural phenomenon, to direct their attention to the movements themselves."[28] Alongside the teachers, physicians, pharmacists, and civil servants who made up the Styrian observing network were occasional participants without titles. These observers in turn drew on observations from a far broader segment of the local population. In one case, Hoernes received a letter written by an untitled resident of Kappel, in the mountains of present-day Slovenia, which contained (in Hoernes's judgment) "very interesting data on the perception of the earthquake of November ninth in the Sulzbacher Alps."[29] The writer had heard of a tremor from two peasants, though they were unsure of its direction. He continued: "I sent my boy to Sulzbach in order to conduct a survey, since I had no success in writing. He made inquiries of several peasants, who had felt the quake in Sulzbach strongly—in two impacts. A time and direction of the impacts could not be determined, since the parish priest in Sulzbach, from whom I hoped for accurate data, had not felt the quake at all." Despite expectations, then, the priest proved a less valuable informant than the peasants, even if their reports were insufficiently "precise." In other words, seismological research was being conducted by a juvenile and peasants, at three removes from a scientific expert.

Earthquake researchers also collected oral testimony from witnesses as they made their way through an affected region.[30] The Carinthian geologist Richard Canaval gave a vivid picture of his own tactics in a report on his investigation of the Gmünd earthquake of November 1881. His interactions with eyewitnesses were far more intimate and reciprocal than written correspondence allowed: "I asked the observers with whom I was able to speak to lead me to the place where they had felt the earthquake, to place themselves in the position they had been in when it had occurred, and to show me the direction in which they had perceived the impact. With the help of a good compass the direction was then determined. It was possible in this way, with the aid of suitably applied questions (for example which wall appeared to be shaken earlier) and other data (direction in which objects swung, rolled away, or fell over, etc.) to correct many erroneous statements."[31] The British seismologist Charles Davison later pursued a similar strategy of reenactment: "Estimates of the duration of an unexpected and unusual phenomenon are, as is well known, almost always in excess of their true value. I endeavoured to check this tendency to exaggeration by suggesting in my circular-letter that the shock should be mentally repeated, the beginning and end being marked, and the interval timed by an assistant. My suggestion was carried out in some cases."[32] In cases like this, observers were taught to disaggregate their experience of an earthquake into basic factual statements of time, direction, and other physical "elements." Seismologists were then in a position to put these pieces together into an approximately complete picture of the physical phenomenon. It was only by combining the partial perspectives of individual observers that such a picture could emerge.

"Reassurance and Enlightenment"

Rather than drawing a dichotomy between "professional" and "popular" science, the practitioners of *Erdbebenkunde* contrasted two ways of bringing science to the people. The sensationalistic predictions of Rudolf Falb (chapter 3) represented the wrong way, stimulating panic rather than rational reflection. The alternative was to enlist members of the public as scientific observers, training them in the methods of modern science and teaching them to view earthquakes as a vital and natural part of the earth's life cycle.

Few were surprised when Falb rushed to the scene of the Zagreb earthquake and claimed to have predicted it. As he began to issue prophecies of catastrophes to come, professional scientists abandoned the wary skepticism with which they had initially responded to him. Now they were angry. None responded more aggressively than Rudolf Hoernes, a fellow Styrian. Hoernes denounced Falb's theory as "scientific humbug" and cast Falb as a charlatan and demagogue. He contrasted Falb's theory—speculative, dogmatic, and hubristic—with the modesty and empiricism of the Swiss Earthquake Commission. Falb had claimed that there "no longer remains in the whole, seemingly so inextricable earthquake question a single enigmatic point." The Swiss instead acknowledged that the causes of earthquakes were still uncertain and "multiple." Clarification could only come from "these precise studies of earthquake phenomena, founded on comprehensive observations."[33]

Meanwhile, back in the imperial capital, in the last weeks of November 1880, the private Scientific Club organized a pair of lectures by prominent geologists with the aim of raising money for the earthquake's victims. But the lectures had another explicit goal: "to bring reassurance and enlightenment from the standpoint of science in the face of the alarming theories of unauthorized people [*Unberufener*]." The first was delivered by the Vienna geologist and ethnologist Ferdinand von Hochstetter.[34] He lamented the fear caused by "false prophets" and denounced a certain theorist's reliance on "unproven hypotheses" as a distortion of the "scientific path to truth." At last, it had been possible "to make space for a calmer mood."[35]

The second speaker was Eduard Suess, who was then at a pivotal moment in his career. Five years earlier he had published *The Origin of the Alps*, in which earthquakes figured as evidence for a tectonic theory of orogenesis. Already, he had begun the work that would absorb him for the next thirty years—his great global synthesis, *The Face of the Earth*, with its integrated vision of earth, water, atmosphere, and life (the source of the term "biosphere"). Simultaneously, he was deeply engaged in civic projects like water provision and canalization. Like the Swiss seismic researchers Heim and Schardt, Suess was convinced that the geologist had a key role to play in the improvement of society. As a tribute to his renown as scientist and statesman, the audience for his lecture in 1880 included members of the imperial family, the minister of war, and other dignitaries.

Toward the end of his address, Suess urged his audience to take up the cause of scientific earthquake observation. He admitted that a scientific attitude was probably not compatible with mortal fear: "at the sites of primary impact themselves reliable observations are normally not obtainable, because this phenomenon occurs so suddenly and spreads such panic that few persons can be found, who in mortal danger have the *sangfroid* to record any observations of a reliable sort." But Suess did not conclude, as Mallet had, that eyewitnesses were unable to provide scientific testimony. To the contrary, he called for the "active participation of the entire educated population, without which a determination of the affected area [*Schütterkreis*] is not possible." Indeed, Suess was responsible for the collection of many of the observer reports used by researchers studying the Zagreb earthquake.[36] He also considered how "such earthquakes normally present themselves to the eye of the observer." At a distance from the epicenter, multiple seismic waves were superimposed, such that the strike might seem to come from any of several directions.

Enlisting the public as scientific observers was a means of transforming irrational fear, with its potential for social chaos, into scientific curiosity. Suess hinted that the damage from the Zagreb earthquake had been social as well as material. He described how " the individual, who is bound to hearth and family by a thousand threads, suddenly . . . sees these threads broken, like a plant torn up by its roots." Tellingly, however, Suess resisted the facile imagery of patriotism.[37] Where Hochstetter had used the language of *Heimat* and brotherhood, Suess couched the disaster in democratic and nonnational terms: "As in an instant all social borders, all differences of class fall

away and all are equal, equal in their degree of helplessness and of misery."
Suess stressed that social elites were as vulnerable as anyone else to both
the physical and psychological effects of earthquakes. "The dreadfulness
of the phenomenon lies in the suddenness of its occurrence and in that even
the educated person is likely to see in it something entirely unintelligible,
something entirely incomprehensible. Thus our response to the event must
be: sympathy for the victims and the conscientious advancement of our
studies." Unlike Kant in 1755, Suess portrayed the scientific and humanitarian perspectives on the earthquake as compatible, even interdependent. The
scientist needed to know how earthquakes "usually present themselves to
the observer," while the work of reconstruction required scientific knowledge of the distribution of seismic risk throughout the monarchy.

The Living Earth

If you want to discover the world, you need to pierce the eggshell.[38]

—Eduard Suess, 1888

In the tradition of Seneca, Suess set earthquakes in the context of "the all."
The loss of solid foundations could seem uncanny, even unnatural. By
bringing the earth as a whole into focus, as Suess did with his contraction
theory, geologists could instead show earthquakes to be a natural part of the
planet's life cycle. Suess set out to readjust humanity's sense of scale, both
spatial and temporal, by showing earthquakes to be the result of tectonic
shifts over the course of geohistory. Shortly after an 1898 earthquake in
Croatia, Rudolf Hoernes wrote in the *Neue Freie Presse* that Suess's planetary perspective "may comfort us when we recognize the essence of the
uncanny [*unheimlich*] seismic forces that are active on the Adriatic coast."[39]
The Ljubljana seismologist Albin Belar offered a similar image in the same
newspaper in the wake of the Messina earthquake of 1908: "The earth lives
and quakes—and a standstill means planetary death. We will no longer
shudder in fear at the word earthquake when we consider that it is merely
a balancing out of forces and tensions, similarly as in a thunderstorm."[40]
Indeed, Suess's view of earthquakes as symptoms of planetary evolution was
widely echoed into the early twentieth century. As a writer in the *Edinburgh
Review* put it in 1905, "Earthquakes are a sign of planetary vitality"; neither
"effete globes like the moon" nor "inchoate worlds, such as Jupiter or Sat-
urn" could be expected to suffer seismic waves, being too inelastic in the
first case and too "pasty" in the second. "Our globe is, by its elasticity, kept

habitable. . . . Just by these sensitive reactions the planet shows itself to be alive, and seismic thrillings are the breaths it draws."[41]

For Suess, the liberal statesman, the global perspective also stood for the interests of humanity as a whole, over and above those of individual nations or classes. In 1888, for instance, during a crest of anti-Semitism and nationalist conflict in the imperial capital, Suess called for broad mindedness: "Prejudices and egoism, above all the pettiness of the things with which we are accustomed to dealing with, have placed barriers around each of us which constrict our view. If they are removed, if we resolve to leave behind the narrow conceptions of space and time which bourgeois life offers us, and no longer to view the world from the base, self-centered perspective, which sees advantages here, disadvantages there for us or our species, but rather to admit the facts in their naked truth, then the cosmos reveals to us an image of unspeakable grandeur."[42] Suess associated the global perspective with modernization and ethical progress, but also with the bold empiricism of modern geology that grasped "facts in their naked truth."[43] He suggested that the mere recognition of the relative scale of the human and the cosmic could ground ethics in a secular, globalizing age: "There is nothing that raises the individual more completely and higher out of the narrow circle of egoism, as the thought of the infinite majesty of the universe and of its eternal laws."[44] Suess's global vision was not, however, a bid for transcendence. He never allowed the extraterrestrial perspective to become an end in itself. The human impact of an earthquake remained at the center of research—as a source of evidence and as a motivation for knowledge.

Seismology Comes to Ljubljana

The Zagreb earthquake of 1880 triggered the organization of earthquake-observing networks at the provincial level, but it took another fifteen years for scientists to begin to coordinate seismic observation across the Austrian half of the monarchy. Political resistance to scientific centralization was just one of the reasons for this delay. Another involved disciplinary dynamics: in 1880, the *landeskundlich* approach seemed perfectly suited to answering questions of crustal tectonics and orogenesis. It was not until the development of more sensitive seismographs over the following two decades that earthquake research would be reoriented around questions of the earth's internal structure, requiring larger-scale coordination of instrumental observations. Furthermore, the infrastructure that would make it possible to standardize observations on that scale was still under construction in 1880. Railroad time had not yet been coordinated, and many parts of the empire

Fig. 7.3. Ljubljana earthquake damage. In *Für Laibach.*

did not yet receive telegraphic time signals. As Hoernes found to his shock in 1880, even clocks at neighboring rail stations sometimes differed by several minutes.[45]

The event that raised the call for action in Vienna occurred at the end of Easter Sunday 1895 (see figures 7.3 and 7.4). In Ljubljana, the capital of the crown land of Carniola, most residents were snug in bed when an unusual "buzzing" disturbed their rest. This was fast followed by "powerful droning, rattling, thundering, and rumbling." Chimneys crashed to the ground, and church towers wavered. Over the course of the night, between thirty and forty further shocks ensued. The population fled the city and did not return until morning, when they found the city in ruins. Barely an exterior wall had survived without cracks, and interiors were littered with debris. Only two deaths were attributed directly to the earthquake, but 10 percent of the city's buildings were condemned to demolition.[46]

Because of the holiday, most of the monarchy only learned of Ljubljana's misfortune from Tuesday's newspapers. The *Neue Freie Presse* reported "panic" as far from the epicenter as Trieste, Fiume, and Abbazia. Eduard Suess's son, Franz Eduard, was commissioned by the education ministry to

investigate, and set off on Tuesday evening. Meanwhile, Eduard von Mojsisovics, the vice-director of the Geological Institute in Vienna, circulated
questionnaires and enlisted newspapers to print requests for information.
In all, the Geological Institute would collect more than 1,300 felt reports
from more than nine hundred locations, plus over two hundred negative
reports. Franz Eduard Suess's final analysis would mine all this testimony
alongside over five hundred published sources. Remarkably, the younger
Suess printed all these sources as an appendix to his final monograph on the
earthquake, which appeared in 1897 and ran to 590 pages. Two hundred
pages alone were devoted to the observer reports received by the Geological
Institute, which were necessarily "mostly reduced to keywords; only particularly elaborate and typical descriptions are reproduced word-for-word."[47]

Fig. 7.4. The Ljubljana earthquake of 1895, caricatured by a Viennese artist as a grotesque giant stomping across the Balkan peninsula. Note the prominence of the Sava
River, the tributary of the Danube that flows through the South Slav lands—symbolic of
the *natural* foundations of Habsburg unity. In *Für Laibach*.

For the city of Ljubljana, the earthquake was an occasion to modernize. The reconstruction effort laid bare a conflict between local and imperial direction, as Andrew Herscher has shown.[48] A similar tension was manifest in the scientific response to the earthquake. With the city still reeling from aftershocks, the Carniolan provincial government expressed interest in installing a seismograph at the imperial high school in the capital. There was, however, no one willing to supervise its operation until 1896, when the school hired Albin Belar, then an assistant at the Imperial Marine Academy in Fiume. Belar had studied with Eduard Suess in Vienna. He spoke German at home but also knew Slovenian, Croatian, and Italian.[49] In 1897 Belar drew up a petition to found Austria's first seismological observatory in Ljubljana. He later claimed to have coined the German word for such an institution, *Erdbebenwarte*. It would seem to be a miraculous stroke of fortune by which Belar's petition found its way to the direction of the Carniola Savings Bank, which promised the necessary financial support. But luck had less to do with it than connections. The bank's director was Joseph Supan (or Suppan), a member of the imperial court of justice and one of the foremost German nationalists in Carniola.[50] Indeed, the bank itself would become the target of Slovenian nationalist fury over the following years, with Supan even facing accusations of embezzlement. Supan's brother Alexander, meanwhile, was one of Austria's leading physical geographers and the editor of the influential journal *Petermanns Mittheilungen*, where he had discussed seismological questions.[51] Apparently, an intervention from Alexander Supan had something to do with the continuing support of the Carniola Savings Bank for Belar's observatory.[52] Conspicuously, the observatory also received the personal support of several Habsburg archdukes.[53] From the start, then, the Ljubljana observatory was implicated in the debates over imperial authority in the South Slavic lands.[54]

Belar's interest in seismology lay in instrumental design and the interpretation of seismographic records—in his terms, "the technical aspects of the science."[55] Yet much of his work after 1897 involved comparing the instrumental records of his observatory with human reports.[56] "It has become an urgent necessity to bridge the large gap that unfortunately still exists today between the earthquake observer with the help of instruments and the observer who relies exclusively on human perceptions, his own or those of others." Belar introduced a linguistic metaphor to describe this effort: it was necessary to correlate the characteristic curves of the seismograph with terms in common use for the sensations of ground movement: earth shock, ground vibration, ground shaking, earth movement, and so on. ("Erdstoß, Erzittern des Bodens, Bodenerschütterung oder Erdbewegung").[57] "Bridging the gap" between instrumental and human observations would depend on smooth communication between seismologists and the public. To this end, in 1901 Belar founded the monthly journal *Die Erdbebenwarte,* a unique forum that addressed both expert and popular audiences. It covered the latest research in seismological geophysics and geology as well as issues of seismic safety. Belar also lectured frequently to popular audiences in Ljubljana and elsewhere in the monarchy, and he served as a correspondent for newspapers in Vienna, Berlin, London, and New York.[58] He thus helped forge a seismological language that linked scientists, citizens, and instruments.

Imperial Science

Just ten days after the Ljubljana earthquake, the Academy of Sciences in Vienna ruled to establish a special commission for the "more intensive study of seismic phenomena in the Austrian lands." The initial form of the Earthquake Commission remained significantly decentralized: each crown land had one reporter responsible for recruiting observers; mailing, collecting, and compiling questionnaires; and investigating significant seismic events within his province. There was thus no direct contact between the commissioners in Vienna and the volunteer observers.[59] In practice, this scheme was further decentralized by the provision of one reporter each for the German- and Czech-speaking regions of Bohemia and the German- and Italian-speaking regions of Tyrol. According to Habsburg logic, these were geographical rather than social divisions, meaning that each reporter was still required to be bilingual. The commission's second task, that of collecting historical information on past quakes, was to be organized according to "appropriate regional sections, so for example the Alps, the Sudetenland, and the Karst region."[60] This too proved tricky to implement. The problem was, at core, the familiar conflict between territorial and linguistic divisions of the monarchy. Earthquake observing was uniquely susceptible to this tension, dealing as it did directly with both the land and with people's perceptions of the land.

For instance, the Czech-Polish physicist Václac/Waclaw/Wenzel Láska found that the preparation of a treatise on "The Earthquakes of Poland" was problematic. "I have called my project 'The Earthquakes of Poland,' but more accurate would be 'The Earthquakes in the Polish Historical Sources,' because in this I did not by any means think of political, but rather of historical borders, and those not so much of the land as of the sources." Poland, of course, was not a political entity at this time, having been swallowed by neighboring states over the previous century and a half. Although the Austrian kaiser was far more tolerant of publications in Polish than the Russian tsar, and while Láska benefited from the collection of the Lemberg (L'viv) Ossolineum, his task proved frustrating: "The acquisition of the necessary literature created great difficulties for me, because so many fundamental texts could not be found in the Austrian libraries. It seems superfluous to add another request for the kind loan of documents necessary to me, which in the present paper are marked as inaccessible to me."[61]

The commission's directors in Vienna devoted careful thought to the challenges of public outreach. They deliberated, for instance, about how to recruit observers at no pay. As compensation, they offered each observer a copy of the volume in which his or her observations were printed. In 1899 they resolved to send each regular observer in addition an annual note of thanks. They also struggled with the problem of turnover among the reporters for individual crown lands. At one point, for instance, there were complaints that the reporter for Carinthia took no interest in the job and failed to report on earthquakes that were clearly felt in neighboring Styria. Another reporter resigned, confessing that he was not really interested in seismology. On the other hand, six dedicated men served as reporters continuously from 1895 to 1919. Five years into the commission's work, the director Mojsisovics summed up the lessons learned:

> The collection of observational data from the observers scattered across the province is far more difficult than it might appear to those farther removed from the situation. It requires the unlimited attention and the constantly renewed initiative of the reporters, to ensure that the individual observers send in their observations. More than any other branch of natural science, we require the participation and cooperation of a broad class of the population. We must therefore make an effort to cultivate the awakening interest of the public; thus we make clear, by publishing the essential contents of the observations received and by listing the names of our collaborators, how valuable and important the prompt cooperation of the public is to the fulfillment of our task. As soon as the institution of earthquake-reporting has become more habitual [*mehr eingelebt haben wird*], all participants will be eager to serve the cause of our efforts, conscious of supporting a purely scientific enterprise and making contributions to its progressive development.[62]

The commission also faced a linguistic challenge. Originally, the questionnaires and instructions were printed only in German. In 1898, they were translated into Czech, Slovenian, Croatian, and Italian (though not Polish or Ruthenian, because of the low seismicity of Galicia and Bukovina). In 1901 Belar requested that the portion of the annual chronicle concerning Dalmatia also be printed in Croatian. According to his own reports to the commission, 382 of his observers wrote in Croatian, only twenty-six in German, and fifteen in Italian.[63] The commission responded that it would be impossible to translate this section of the chronicle: to do so would draw analogous demands from the other nationalities, and the cost of satisfying them all would be prohibitive. The commission recommended that Belar have the chronicle printed instead in a local newspaper.[64] A month later Belar complained to the commission that his efforts to recruit observers by means of articles in Croatian and Italian newspapers had so far been fruitless.[65]

The fate of the imperial Earthquake Commission would play out, like imperial politics overall, as a struggle to strike a balance between centralization and decentralization.[66] Seismology demanded centralization in order to synthesize data on midsize earthquakes, to respond concertedly to seismic disasters, and to facilitate communication among the new generation of seismic observatories. Yet earthquakes also required study region by region, by scientists fluent in the local language, versed in local history, and familiar to the local community of observers. After 1904, Vienna's Central Institute for Meteorology and Geophysics provided the necessary degree of central organization, while managing to preserve the lively intercourse between provincial scientists and the interested public. Its success can be measured, for instance, by the over two thousand reports collected in the course of two large swarms in Bohemia in 1900 and 1903.[67] Indeed, in the Habsburg world, "local knowledge" was often produced as "imperial science." Research conducted in the service of nationalism—even the "national schools of thought" celebrated by post-Habsburg historians—was often sponsored by the Habsburg state and articulated, in German, by Habsburg-loyal scholars as contributions to the ideology of unity in diversity.[68]

Conclusion

For Albin Belar, the dissolution of Austria-Hungary in 1918 meant the loss of his *Erdbebenwarte*, the perch from which, for twenty years, he had reported on each spasm of the planet. The new Yugoslav government, suspecting him of German nationalism, seized the contents of his observatory, dispossessed him of his apartment, and sent most of his instruments to Belgrade. While his children emigrated to the United States, Belar moved what remained of his instruments to the Triglav valley in the Julian Alps, where he owned a villa designed by his friend Max Fabiani, the architect of Ljubljana's post-1895 modernization. The Triglav valley was one of the "natural monuments" for which Belar had first sought state protection in 1903, one of the earliest proposals for a nature preserve in European history.[69] A conservation area was established in the valley in 1924, and a national park created in 1961, with no acknowledgment of Belar's contribution. On the door to Belar's villa hung a sign in Slovenian: "Silence, silence, for here resides a solitary man who uses his instruments to listen to the sound of earthquakes being born in the heart of the Earth."[70] This lonely enterprise bore no resemblance to Austrian seismology before 1918.

In *Contributions to the Analysis of Sensations* (1886), Ernst Mach famously argued that the goal of physics is to build up a description of the world out of the most basic components of human experience. To illustrate the goal of complete knowledge of a physical phenomenon, Mach provided two examples. One was straightforward, culled from the standard repertoire of laboratory physics: the analysis of white light by means of a prism. The other was more elaborate:

> Our knowledge of a natural phenomenon, say of an earthquake, is as complete as possible when our thoughts so marshal before the eye of the mind all the relevant sense-given facts of the case that they may be regarded almost as a substitute for the latter, and the facts appear to us as old familiar figures, having no power to occasion surprise. When, in imagination, we hear the subterranean thunders, feel the oscillation of the earth, figure to ourselves the sensation produced by the rising and sinking of the ground, the rocking of the walls, the falling of the plaster, the movement of the furniture and the pictures, the stopping of the clocks, the rattling and smashing of windows, the wrenching of the door-posts, the jamming of the doors; when we see in mind the oncoming undulation passing over a forest as lightly as a gust of wind over a field of grain, breaking the branches of the trees; when we see the town enveloped in a cloud of dust, hear the bells begin to ring in the towers; further, when the subterranean processes, which are at present unknown to us, shall stand out in full sensuous reality before our eyes, so that we shall see the earthquake advancing as we see a wagon approaching in the distance till finally we feel the earth shaking beneath our feet,—then more insight than this we cannot have, and more we do not require.[71]

The scientific observation of an earthquake involved all the forms of sensation explored earlier in Mach's treatise: movement, sight, time, even tone. In this way, seismology could serve as a test of the capacity of Mach's psycho-physical program to produce practical knowledge in real environments, beyond the laboratory, as his evolutionary epistemology demanded. In other respects, however, Mach's choice of example is confounding. Having spent most of his life in Bohemia, he was unlikely to have experienced a major earthquake firsthand. Moreover, how could he expect anyone, in the midst of catastrophe, to record all these details—and manage to survive?

Mach's choice of example seems to have reflected the ambitions of earthquake research in imperial Austria. As early as 1869, Mach likely heard Rudolf Falb discuss earthquake prediction at the Lotos natural-historical society in Prague.[72] As Mach would have known—whether from the Lotos society, Austrian scientific journals, or even the popular press—the monographic method of earthquake study was being held up as a model of empiricism and an antidote to popular unreason. Moreover, this method embodied Mach's principle of analyzing experience into its most basic elements. Only by combining the limited perspectives of individual observers could a total picture of the phenomenon emerge. Tellingly, the subject of Mach's passage on the earthquake is not the generic *Man* ("one") but the first-person plural *Wir*. Such a panoramic view of a seismic event could not be the property of a single psyche. Seismology was therefore a perfect instantiation of Mach's principle that physics would do well to abandon the fiction of the ego. For Mach, "The primary fact is not the I, the ego, but the elements (sensations)."[73]

Mach's denunciation of "self-centered views" in *Analysis of Sensations* was as much a political critique as an epistemological objection. He likened the metaphysical notion of the bounded self to the social phenomena of "class bias" and "national pride," and called on the "broad-minded inquirer" to renounce them. Behind this appeal lay bitter personal experience. Just as Suess resigned a university rectorship in 1889, so had Mach in 1883, both apparently worn down by the politics of nationalism. Both longed to escape the "egoism" of modern politics for a collaborative, supranational science.[74] To Mach's Austrian colleagues, his seismic example of complete knowledge in the *Analysis* would have evoked a thriving culture of earthquake observation—a culture that managed to fuse the provincial practices of *Landeskunde* with the transnational organization of a continental empire.

Both Mach's epistemology and the practices of macroseismic survey reflected a peculiarity of Habsburg science: the function of translation. Implicit in the work of the Earthquake Commission was a philosophical resistance to reduction. No single perspective on the event was privileged. Instead, the full array of impressions—visual, aural, tactile, instrumental—

were taken as irreducible elements of knowledge of the whole. The epistemic model was one of translation. The goal was emphatically not to reduce the observational reports to a single "language of nature," as the curves of self-registering instruments were often described in the nineteenth century.[75] To a North American seismologist of this period it seemed self-evident that "reliable seismograms furnish us the data for all our computations about an earthquake . . . in short, the seismograms give the story of the earthquake, just as the spectrogram gives the story of the distant stars."[76] For Mach, Suess, and their colleagues, however, no seismogram could ever tell the story of an earthquake—most basically, because the earth was not a distant star. The goal was not to explain the earthquake in terms of a single physical cause nor to reduce its description to a single variable. Instead, all possible perspectives on an earthquake, like the multiple languages of the empire, were in principle granted equal status. "Complete knowledge" of an earthquake corresponded not to a mathematical law nor an instrumental trace, but rather to a multilingual archive.[77]

Yet Mach's epistemological fantasy hid the immense work of constructing such an archive. The history of the imperial Earthquake Commission demonstrates, to the contrary, that translation was labor—and it was labor that the Earthquake Commission was at times unwilling to fund. Viennese seismologists were not even aware of the extent of the effort involved, since linguistic equivalences were constructed primarily through discussions between the provincial reporters and local observers. For instance, the German, Croatian, and Czech words *Getöse, tutnjava,* and *rachot* became the standard descriptions for the rumbling noises accompanying earthquakes, rather than other possibilities in each language.[78] Even less visible was the subsequent work of translating the official results of the commission from German into languages in which they were more likely to be read by potential observers. Arguably, the invisibility of such labor is endemic in the history of science.[79] By eliding the labor and politics of translation, Mach contributed to the discipline's enduring tendency to view translation merely as an epistemological metaphor.[80]

EIGHT

What Is the Earth?

The establishment of the International Seismological Association (ISA) in Strasbourg in 1901 was, in the words of its founder, Georg Gerland, a case of "nations uniting for collaborative, idealistic work."[1] Gerland even suggested that the ISA's members could "call themselves apostles of world peace."[2] As a Humboldtian quest to integrate local geology, global physics, and the anthropology of disaster, seismology would seem ideally suited to internationalism.[3] Yet the ISA represents a contentious episode in the *un*making of disaster as a scientific object. Gerland's version of scientific internationalism pried the earthquake apart into (1) a geophysical cause, accessible only to instrumental analysis, and (2) a human impact, the object exclusively of an "anthropomorphic" gaze. The story of the ISA manifests incongruities between the global visions of scientific modernizers and the local realities of communities at risk. Such tensions are part of what Jeremy Vetter has recently described as a persistent conflict between "the globalizing and universalizing ambitions of modern knowledge-making and the practical needs of a world of tremendously complex and variable environments whose knowing matters so much for human survival and sustainability."[4]

The ISA was founded at a pivotal moment in the history of what historians have recently labeled "the observatory sciences." Taking both the heavens and the earth as their subjects, these fields shared a Humboldtian focus on precision measurement, numerical data processing, and the representation of scientific information on a global or cosmic scale. At the turn of the twentieth century, the observatory sciences were uncertainly poised between a nineteenth-century dream of popular enlightenment and the imperatives of scientific professionalization and military-industrial development.[5] What form seismology would take as an international, observatory science was open to debate. What would a global seismology be a science *of*? Or, as Gerland put it, "What is the earth after all?"[6]

Matter

One way to answer Gerland's question was bottom-up. From this perspective, the globalization of seismology would entail the multiplication of regional studies on the model of Suess's monographic method. This would be a collaborative process, recalling Humboldt's vision of a "general physics of the earth." As one of Gerland's Strasbourg collaborators explained in 1908, studies of individual earthquakes are "building stones for the construction of a large edifice, the theory of the seismicity of the entire earth. He who wants to be active in this field must follow seismic activity through time and space, to a lesser degree as a historian, principally however as a geographer."[7] Such studies were in demand both to map seismic hazard and to corroborate Suess's tectonic theory, which was still contested at the turn of the twentieth century. Not all geologists accepted that the majority of earthquakes were tectonic in origin, resulting from movement along fractures in the earth's crust and often related to mountain building.[8] Gerland, for one, took pains to point out that the tectonic explanation of earthquakes remained very much a hypothesis—"always only theoretically assumed, nowhere really proven." (He could not imagine how this theory could explain earthquakes at sea, where there were no mountains. This was a curious opinion, since submarine extensions of mountain chains had been suspected to exist since ancient times, and nineteenth-century coastal soundings had already revealed submerged ridges and hollows.) Evidence for the tectonic theory would come from correlating the location and direction of individual earthquakes with features of local geology. Gerland termed this approach derisively "local observation."[9] It led, nonetheless, to one version of a global seismology. This was the path marked out by Montessus de Ballore, the French pioneer of a worldwide "seismic geography." For Montessus, the problem of globalizing seismology came down to the question: "How then to succeed in the observation of macroseisms [that is, earthquakes], small and large, across the entire surface of the globe?"[10]

Montessus initially hoped his seismic geography would generate a universal law relating seismicity to topography. The closest he came was the hypothesis that the highest seismicity was associated with geosynclines, the regions where uplift and folding produced the world's great mountain chains. But he also came to appreciate the limits of any such global generalization, based merely on statistical distributions and not on field studies. Such laws were "only valid for one and the same region, where they determine the direction of the largest ground tremors; they become incorrect however if they are carried over from one region to another. This is because earthquakes, apart from the geological aspects, also derive in part as *causa efficiens* from the mutual adjustment of matter on the surface. These laws thus have a far more narrow range than those that result from geological-seismic studies, which are undertaken on the whole surface of the planet and extended across it."[11] Without local field studies, seismology would be blind to the ways in which local conditions determined the development and impact of earthquakes.

Analogously, Montessus's investigations of earthquake-resistant construction methods showed him that there could be no universal rules of seismic safety. "No one knows how to establish general rules. Sometimes hard soils are to be preferred, sometimes soft soils. Here the heights are less vulnerable than the lowlands, elsewhere it's the opposite." Indeed, Montessus advocated the complete cessation of construction in certain areas, such as Ischia after its disastrous quake of 1883. "The point," he explained, "is that the entanglement of the conditions of location and composition excludes any general solution to the problem."[12] Like recent scholars, Montessus recognized that the study of earthquakes as disasters—as conjunctions of geological, geophysical, and human factors—often required a regional rather than global lens.[13]

Gerland's Observatory

Gerland envisioned an altogether different path to a global seismology. In his opinion, the methods of local observation were already being rendered obsolete by the progress of globalization in its widest sense: "Today, however, the earth is freely accessible, humanity rises and unites ever more to a common enterprise; and thus earthquake research has also reached a new level. What formerly a single country was for its inhabitants, such is the earth as a whole for us today."[14] But what indeed was the earth?

Gerland's answer was that the earth was nothing but a "gigantic totality of cosmic material." Such was the definition he had offered a decade earlier in a bid to redefine "geography" (*Geographie* or *Erdkunde*, terms he used synonymously) as the study of this totality—minus its human inhabitants. This was an astonishing move from a man best known to this point for his contributions to anthropology. Gerland's immediate goal was to establish

Erdkunde as an autonomous scientific discipline, unified by its own methods and laws. "For only through such a conception will *Erdkunde* stop being a vague field of miscellaneous knowledge, whose details, certainly valuable in themselves, in this association all too easily—indeed almost of necessity—lack a strictly scientific foundation; only through this conception, as a dynamics of the earth as a whole, will *Erdkunde* become a strictly methodologically unified science."[15] Gerland's rhetoric, with its stress on scientific foundations, universality, dynamics, and methodological unity, was intended to mark him as the long-awaited modernizer of geography.

In 1887, in the founding issue of his seminal journal *Beiträge zur Geophysik* (Contributions to Geophysics), Gerland firmly excluded humanity from the domain of geography. Geography was strictly the science of "the continuous mutual action between the interior and the surface of the earth."[16] Gerland rejected both crass environmental determinism and anthropogeography's more sophisticated models of nature-culture interactions. He argued, emphatically and controversially, that there was no way to bridge the study of man as a product of external physical forces, on one hand, and as an autonomous, conscious agent, on the other. Explanation could never run from human geography to physical geography; human activity could modify the environment in limited ways but could never alter the laws of nature. Since "inferences from the nature of the men to the nature of the land are impossible," a science of the human (if such could exist) had nothing to teach the science of the earth.[17]

In Gerland's globalizing enterprise, the phenomenon of seismicity took on "a remarkable position." Seismicity seemed to him quite different from planetary properties such as heat, elasticity, or magnetism, which could be studied as independent subfields. Seismicity instead depended on the "mixture, the aggregate conditions, the chemical-physical process in the earth's interior"—of which, Gerland noted, "we know almost nothing." Volcanism, to be sure, was also a composite phenomenon, but it was also "always a local phenomenon." Seismicity was a truly global property, "spread over and through the entire earth, at least in the movements called forth by it."[18] Seismology's goal was nothing other than "the discovery of the fundamental character and constitution of the earth" (*die Erkenntniss der Naturbeschaffenheit der Erde*).[19] And the study of the planetary interior was, in Gerland's estimation, the geographer's raison d'être, the path to "a causal—not a merely descriptive—earth science."[20]

The key to modern *Erdkunde*, in Gerland's view, was the seismograph. It was the telescope of the earth scientist, zooming in on the planet's hidden depths.[21] For Gerland, making seismology "modern" meant making it into an "observatory science." Yet observatories themselves were in flux circa 1900. Would observatories continue to anchor networks of lay observers or rely only on experts? Would they maintain the lively exchange with field-based research that Humboldt had inspired? Would they endure as public spaces as well as scientific workplaces? As historians have recently noted, the Observatoire de Paris and the Berliner Sternwarte invited the nineteenth-century public to train telescopes on the heavens, but their open-door policies were short-lived. Scientific observatories were increasingly divided from public ones and tied to military and industrial ambitions.[22]

Gerland's aspirations were embodied in Strasbourg's Imperial Central Station for Seismology. Unlike the stately facades of earlier observatories that announced their public function, the Imperial Central Station presented a squat, plain face of homely brick. Behind it was a state-of-the-art, triple-walled design of iron-reinforced concrete. The three types of seismometer—soon supplemented by five others—were steadied against nearby vibrations by pillars 2.3 meters deep and detached from the building itself.[23] In a self-conscious contrast with the older seismic observatories of Italy or Japan, Gerland intended that in Strasbourg "local research would recede before the great questions of the seismicity of the entire earth."[24]

Gerland's version of seismology was only incidentally concerned with the cataclysms that had been the science's subject matter since ancient times. The threat that earthquakes posed to human communities was at most marginal to his global seismology. Addressing a popular audience in 1898, Gerland insisted that seismology's goal must be pure rather than practical knowledge: "Seismic research thus has no practical benefit. All prophecies, even if now and then they seem to come true, are entirely groundless. All that one can say for now is that some regions and eras are more given to earthquakes, some soil types are more dangerous; and that certainly has no practical use."[25] Here Gerland went well beyond disputing the evidence for the tectonic hypothesis. He dismissed outright all the work his colleagues were doing to avert future disasters—the mapping of seismic intensity, the location of fault lines, the study of architectural damage. To the German public, Gerland flatly denied that seismology was of any practical value at all. Its goal was purely intellectual: to fathom the physical world. It was, he argued, an endeavor wholly in the tradition of Kant, who was drawn to earth physics because of "the majesty and novelty of the task, the immensity of the vision."[26] Ensconced in his windowless, climate-controlled fortress, hundreds of kilometers from the nearest active fault, Gerland promoted a vision of seismology in which the earth figured more as a distant planet than as the home of intelligent life.

Across the Surface of the Globe

The most notable absence at the ISA's 1901 meeting, next to Milne's, was that of Montessus de Ballore. It was Montessus who later pinpointed what was so controversial about the fledgling ISA: "If the seismographs register every slightly intense earthquake, wherever it is produced, there follows the unexpected consequence that, at least theoretically, a single seismological observatory would suffice to observe them all from a single point on the earth's surface. That is perhaps what the countries that abstained from the Strasbourg conference said to themselves."[27] The fear was not just that Germany's rivals would be cut out of seismological research, but that seismology itself would be transformed: that its diverse modes of investigation—geological, statistical, historical, and geophysical—would be reduced solely to the instrumental registration of remote events. Montessus appreciated the potential of seismographic research, but he feared an overreliance on it. To cast doubt on Gerland's project, he drew freely on geology's romantic, heroic image of empirical field research.[28] There was a risk of overdrawing this caricature. As Stephen Jay Gould once noted, geologists have tended to paint the history of their field in black and white. Hutton figures as the honest empiricist, Werner as the proto-lab scientist. Yet Gerland's critics took a more nuanced stance, acknowledging the complementarity of new instrumental techniques and field-based experience.

Montessus also pointed out the artificiality of Gerland's contrast between the "new" seismology and the "old." Gerland implied that these fields could be distinguished in part by their objects of study: "microseisms" versus "macroseisms." Microseisms were, by definition, imperceptible to humans and of uncertain origin, but they could hold clues to the nature of the planetary mass through which they traveled. Montessus objected: "The distinction between macroseisms and microseisms is artificial in the sense that it depends on the individual acuity of the senses of the different observers." Moreover, microseisms might be due to any of a potentially infinite array of causes. As Montessus attempted to enumerate them, these included "winds, tides, solar heat, barometric pressure, ocean currents, snow accumulation, avalanches, rockslides, movements of railway cars, explosions of mines, volcanic phenomena, etc., etc."[29] The category of microseism also included waves from strong earthquakes, which would register with small amplitudes and long periods after traveling great distances. "Hence," quipped John Milne, "the largest earthquakes have been called small."[30] Milne himself preferred the term "unfelt earthquake" to microseism. The distinction between micro- and macroseisms—one that has shaped the discipline's development ever since—was thus a clumsy attempt to naturalize the growing fissure between field-based and observatory-based methods.

In 1903 Montessus appeared at the second meeting of the ISA with a warning: "The study of seismological waves originating at a distance has the attendant drawback, that the earthquake phenomena lose their characteristic features on the long path through the earth. It is to be feared that their nature is veiled to us; the geological causes, the discovery of which is the ultimate goal of our work, are in danger of escaping us."[31] He made the point more baldly the following year, noting that "by an unfortunate consequence of this brilliant research, above all physical and mechanical, we have increasingly neglected the geological point of view, the only one by which it is possible to attack head-on the origins of the seismic phenomenon, and we have delayed studying in itself a movement that, as interesting as it may be, is not any less a merely secondary effect of earthquakes, such that these studies remain incapable of reaching them [earthquakes] in their genesis. Nonetheless a few brief considerations of the geology of the shaken regions, including the descriptions of the most important seisms, would have sufficed to remind the physicist-seismologists that they were directing their efforts at a truly ancillary problem, despite the interest—entirely indisputable, for that matter—of the results that they obtained." To attack the earthquake problem at its core, Montessus continued, seismologists had to get to the scene of the action. Those who sat tight in their observatories were like meteorologists who hoped to understand the trajectories of cyclones based only the "mechanical laws of the movement of fluids," "without attending to the atmospheric phenomena preceding and accompanying their formation at the very heart of the country where they are born."[32] Seismology could not be confined to the observatory, for its evidence was written in part on the face of the earth.

Montessus's point concerned not only the methods of seismology but also the identity of the seismologist. He should be "geographer and geologist" in addition to "physicist and engineer [*mécanicien*]." Most importantly, the seismologist must remember his primary responsibility: "Science must not only shut itself up in the ivory tower of its observations and speculations. Whenever it can, and it is certainly the case here, it must think of the well-being of humanity and of the relief of its sufferings." To Montessus, the purpose of the ISA was "not to predict earthquakes, but simply to mitigate the disastrous effects by a more profound knowledge of the seismic phenomenon, just as hydraulic services and foresters diminish those of floods."[33] Like Albert Heim, Montessus prioritized the geologist's duty to mitigate hazards. Implicitly, he questioned the moral consequences of an earth science defined, à la Gerland, to the exclusion of humanity.

While Montessus was piecing together a global seismic geography, the Austrian geologist Eduard Suess was pursuing an even more ambitious synthesis: *The Face of the Earth*, his sprawling three-volume survey of global tectonics, published over the course of three decades. No individual at the turn of the twentieth century had a clearer vision of geology as a global science, and earthquakes played a special role in it. They were movements of the earth's crust that made visible the activity of mountain building—the vital process of a planet dissipating its ancient heat and gradually collapsing in on itself.[34] Yet it was not as a progenitor of a global tectonics that Suess confronted Gerland. It was, fortuitously, in his capacity as president of the Austrian Academy of Sciences (an office he held from 1898 to 1911). In that role, Suess was consulted by imperial ministers when the Austrians were formally invited to join the ISA in 1903. Suess was then in the midst of arranging the transfer of the Austrian Earthquake Service from the private hands of the Academy of Sciences to the state-run Central Institute for Meteorology and Geomagnetism. As we have seen (chapter 7), Austrian scientists recognized the need to balance the centralization of seismic research and response against the decentralization demanded by Austria's linguistic and environmental diversity.

Writing in a confidential report to the education ministry in Vienna, Suess argued that "northern" Europeans (read: the Germans) were interested exclusively in the propagation of waves from distant earthquakes, which was "only a special question within the great task." Moreover, the instruments for registering such waves were still imperfect, as the Viennese physicist Franz Exner had concluded after inspecting several European seismological observatories. Exner also noted that the study of nearby earthquakes required different instruments than the study of distant ones (the Ehlert as opposed to the Vincentini seismograph). As Suess pointed out, the lands of the southern Alps and Mediterranean had "a far greater interest [than northerners] in studies of this [local] kind and also a more intensive knowledge of their nature." Suess also stressed the practical advantages of relying on domestic rather than foreign institutions: "the individual investigations, such as the earthquakes of Laibach [Ljubljana] and Agram [Zagreb], are carried out much better and with an incomparably greater measure of local knowledge and local interest by the individual states than by a central office." As a "practical example" Suess pointed to a seismic observatory in Bohemia that was run by the Academy of Sciences "at no small cost." This observatory depended on the cooperation of the imperial finance ministry, the railway ministry, the mining administration, and the Vienna Physical Institute. "I doubt that such an enterprise could be better carried out from Strassburg, and that it would be possible at all without the many domestic connections and sponsorships."[35]

Suess's critique of the ISA echoed in part the political discourse of decentralization in the waning years of the Habsburg monarchy. It also reflected the years he had spent scaling icy peaks and scouring provincial archives in search of clues to the seismic history of the eastern Alps. As he put it elsewhere, macroseismology "demands very precise tectonic local knowledge, does not permit of centralization, and seeks the causes of tectonic earthquakes on the dislocations themselves." For microseismology, earthquakes were merely "the incidental trigger of secondary phenomena, which are soon diffused among other questions of geophysics."[36] Characteristically, Suess refused to sever the micro from the macro. He recommended against Austria's joining the ISA.

Montessus and Suess were defending a Humboldtian approach to scientific internationalism against Gerland's purifying impulse. In addition to international cooperation, they argued for the value of local knowledge. In addition to building observatories, they urged seismologists to strap on their hiking boots and return to the field. And in addition to probing the inorganic stuff of Gerland's "earth science," they called on seismology to serve human welfare.

Man

Gerland's elimination of humanity from the purview of geography has long been misunderstood as an expression of positivism. He was accused by the geographer Hermann Wagner of being the kind of person who "despises all anthropogeographic problems, because they are not capable of precise solution."[37] In truth, Gerland's motivations were more complex. Paradoxical as it may seem, he was attempting to reinvent humanism for a scientific age.

Gerland was, after all, a human scientist before he was a physical scientist. Well before he took up the cause of seismology, he had won respect for his anthropological scholarship on the relationship between *Kulturvölker* and *Naturvölker* (civilized and primitive nations). In *Über das Aussterben der Naturvölker* (On the Extinction of the Primitive Races) he argued that colonized populations were not in fact doomed to extinction, and he condemned Europeans for their brutal treatment of natives. His insights were well noted by theorists like Darwin[38] and Boas.[39]

Yet Gerland's conclusions were ambivalent. On one hand, he recognized that Europeans' barbaric treatment of natives cast doubt on the very notion of "civilized peoples." Not only were colonialists at risk of becoming barbarians themselves. Even in the metropoles Europeans were threatened by symptoms of what would soon be termed degeneration: "We have grown pampered in our bodily existence, accustomed to a host of comforts that we can not do without; we are mentally far more sensitive, and a downfall of that which is holy to us crushes us as well." It was, he judged, time to "stop speaking of a particular incapacity for life of the *Naturvölker,* since we would succumb far faster than they to the disaster they have suffered." *The Extinction of the Primitive Races* was thus, in part, an allegory of modernization and its discontents. Gerland's ambivalence was underscored by repeated comparisons between colonial subjects and the Germanic tribes in the face of Roman culture. He could not help but agree with J. G. Seume's aphorism that "we savages are better men."[40]

On the other hand, Gerland did not doubt that the "struggle for existence" would end with the triumph of the colonial order. *Naturvölker* were vulnerable because of their exquisite adaptation (in a Darwinian sense) to a particular natural environment. By definition, these were populations who adapted to nature instead of controlling it; hence their catastrophic confrontations with modernity. At the end of his analysis, the fundamental distinction between *Kulturvölker* and *Naturvölker* remained intact: the former were able to transcend the conditions imposed by nature, while the latter were bound by them.

The New Humanism

What then did colonial anthropology have to do with geophysics? Gerland's reinvention of geography, *Erdkunde,* as a strictly physicalist, instrument-based, and highly mathematical discipline might seem typical of what Andrew Zimmermann has described as the antihumanist turn in German anthropology in the late nineteenth century. Like Zimmermann's antihumanists, Gerland replaced "hermeneutic notions of understanding and interpretive empathy with models of objective observations borrowed from the natural sciences." According to Zimmermann, this move "devalued the human both as an inquirer and as a subject of inquiry."[41] For Gerland, however, the abandonment of hermeneutic inquiry—the turn to a dehumanized geography—was meant to ensure that the human would *not* be thus devalued. Humanity must not figure in a true science of the earth, Gerland argued, precisely because (European) humans were not passively subject to natural conditions.

Gerland broke with anthropologists like Tylor, Ratzel, and Lasch, who were struck by the parallels between mythical and scientific explanations of earthquakes. Gerland argued instead that the history of geographical thought illustrated clear disjunctures in human intellectual evolution. He insisted on the distance between primitive anthropomorphism and modern science, with its capacity to transform "the entire range of telluric influences into psychic force." Gerland made this point in 1901, just as the ISA was taking shape, in a series of lectures at the University of Strasbourg on Kant's physical geography and anthropology.[42] It was there that he identified Kant as the first modern seismologist. Why, at the dawn of the twentieth century, would a geophysicist have taken up Kantian philosophy?

Gerland sought to prove that Kant's "pragmatic" geography had never been merely pragmatic. In a convoluted and often implausible argument, Gerland argued that Kant's geography and anthropology were merely "propaedeutic" to his critical philosophy. Apparently, the inventor of modern seismology never intended it to serve practical ends. Hence Gerland's insistence that seismology was not a practical but a "pure" science, and that earthquakes were not "disasters" but "microseisms." Gerland went so far as to attribute an "anthropomorphic" perspective, from which earthquakes did appear as disasters, to a stage of barbarism. To speak of geophysical cataclysms in human terms was to regress to a pre-enlightened age: the age before Kant.

Gerland stood in a tradition of central Europeans, including Gustav Theodor Fechner and Ernst Haeckel, who yearned to reconcile materialism and evolutionary theory with the romantic need to find meaning and purpose in nature. "The atomic-mechanistic point of view is not only *not* hostile to a religious and aesthetic conception of life and of cosmic evolution," Gerland insisted. "To the contrary, it leads to it, and only through it becomes complete, vital, indeed becomes capable of life—and reciprocally, this second conception of life is nothing without the first."[43] Gerland could not accept the pragmatic views of scientific knowledge that were gaining ground at this time, and which proved so fruitful to seismology. Indeed, what united various contributions to seismology as a science of disaster—from Eduard Suess to Harry Wood—was a certain suspicion of claims to absolute truth. For these seismologists, the primary criterion for valid knowledge was its utility for human welfare.

For Gerland, the resolution of the tension between materialism and humanism lay in the expectation that evolution would gradually loosen natural constraints on human freedom. "But the higher man climbs, the more he makes himself free of the coercive influence of the earth . . . thus this ever growing freedom and independence is an invigorating and uplifting outlook for the future that we are building for ourselves . . . and life does not end with life on earth, for 'All that is transitory is but a metaphor.'"[44] Darwinian anthropology had spelled the end of the old humanism by casting man's intellectual efforts as no more than a struggle for existence. Gerland was therefore on a quest for a new humanism, one in which even the most earthly knowledge would serve a purpose higher than mere survival. But his new humanism also introduced a radical elitism. By stigmatizing a human perspective on nature as primitive, Gerland's humanism abandoned the hope of a convergence between expert and lay perceptions of nature. In the words of George Sarton, founder of the history of science, only common folk paid any attention to "mere earthquakes."[45]

History

In Gerland's *Beiträge zur Geophysik* in 1900, Montessus de Ballore pressed for "the most complete seismic description of the universe as possible."[46] Montessus had already laid the foundations, having built a correspondence network that furnished him with felt reports and instrumental data on tens of thousands of seismic events. His library of seismic data eventually numbered approximately 170,000 events. At this scale, systematic organization was of the essence. Seismology's age of newspaper clippings was over.

Montessus's solution embraced a technoscientific trend of the fin de siècle: the *système de fiches* or card catalog, which, as he noted, "does such great service in bibliography." In preparing his global seismic geography, Montessus divided the earth's surface into fifty major geographic regions and 451 seismic subregions. For each of the latter, he arranged index cards chronologically with "all the necessary indications of time, place, sources, etc., the detailed history of every observed fact, which thus acquires a kind of individuality."[47] The discrete "fact" represented on each card often masked the difficulty of deciding whether separate reports pointed to the same seismic event. By 1907, these "facts" took up almost twenty-six meters of shelf space. (Their subsequent history is fascinating in its own right. Montessus left the cards with the Geographic Society in Paris upon his departure for Chile in 1907. During the German occupation of Paris in World War Two, they were safely hidden in the vaults of the National Library. There they were forgotten until the 1970s, when the fiftieth anniversary of Montessus's death inspired the director of the Strasbourg geophysical institute to inquire into their whereabouts. It took seven years for the Paris librarians to find the catalog, still in the basement where it had been placed in 1942. An inventory was finally published in 1984.[48])

As historians of information science note, it is difficult to appreciate in the Internet age the enthusiasm with which the concept of the card catalog was greeted circa 1900. Advocates of the card catalog saw it as the means of liberating information from its confines in books and organizing it for scholarly and public use. "Cards of a uniform size, on which standardized data were transcribed, housed physically in card drawers and related furniture, and organized conceptually by classification schemes of various kinds, in effect epitomized a new 'modernist' technology."[49] In seismology, the card catalog promised a new path to the nineteenth-century ideal of complete knowledge, guided by the internationalist values of efficiency and commensurability. It was the quintessential modernist solution to the globalization of knowledge.

The State of the Planet

Montessus and Forel urged the ISA not to neglect its responsibility for "universal" seismic description. Back in 1899, Gerland had assured his audience at the International Geographical Congress that the ISA would pursue "the publication of a chronicle, which would extend over a certain period [and include] a bibliography."[50] A catalog had also been discussed at the Strasbourg conference of 1901. Yet "inexplicably, it is never mentioned anywhere in a precise manner," as Forel pointed out at the 1907 meeting of the ISA.[51] In 1905, Gerland's Central Bureau of the ISA in Strasbourg had published the first volume of a putatively global catalog, two years after the events it chronicled. Forel evidently hoped to give the ISA a degree of control over future volumes. He was soon appointed to a commission to oversee their production. The ISA's catalog would be "more complete" than those of Fuchs, Perrey, and Mallet, and based on "more extensive information"; they would "be perfected from year to year, [and] become the fundamental repertory of the entire seismological science of the future."[52] The catalog, in short, was to be a distinctly modern object.

Forel sought a wide audience for the catalog, as for his limnological studies. This included geographers who wished to compare regions of stronger or weaker seismicity, or to trace the rise and fall of seismicity in a region over time; geologists interested in "the points and the lines of greatest fragility on the globe"; physicists and astronomers studying the nature of seismic waves or the constitution of the earth; seismologists tracing the periodicity of earthquakes; architects and engineers seeking "the rules to make human constructions safe from seismic devastation"; and "many others still." "Let us not forget," Forel cautioned his colleagues, "that the Catalog must be able to be consulted by people who are not seismologists." The bibliographical references would need to be complete enough to serve nonspecialists, and a summary table would "certainly be greatly appreciated by the general public."[53]

The inaugural 1903 catalog proved a disappointment, however. Some critics complained of the two-year delay in publication. Others, like the Italian seismologists Agamennone and Palazzo, charged that the catalog had been produced too quickly, without enough time for full studies of individual earthquakes. The key question was how to balance discursive detail with analytical convenience. Should the catalog be a chronicle or an overview? August Sieberg, responsible for analyzing foreign macroseismological reports at the Strasbourg observatory, saw both sides of the issue: "In order to allow others to verify the conclusions arrived at, it is indispensable to publish the observational material in extenso, but not, as used to be common, in long-winded descriptions, but rather in the concise and clearly organized [*leicht übersichtlich*] form of tabular compilations."[54] Following Forel, Montessus advised that the catalog should not be a "dry enumeration of observations," but rather a "readable description."[55] Not everyone at the ISA saw the trade-offs so clearly. Palazzo recommended that future catalogs restrict themselves to "the most interesting reports." "In this way one could immediately record the earthquake activity of the entire planet for a given moment and the distribution of earthquakes in the various regions." The result would be a catalog "at a glance."[56]

However, the variability of earthquake reports undermined the ISA's goal of producing a global seismic overview. Among the noninstrumental reports arriving at Strasbourg, five different intensity scales were in use. In some countries, no scale at all was employed, and "the observers instead attempt to express themselves in quite general phrases, for which a standard of comparison is lacking. In the reproduction of intensity determinations of this last kind, we have adhered strictly to the descriptions of the sources, even when these proved to be improbable or even incorrect. In order to make it possible above all for the user of the catalog to form an independent judgment of the degree of intensity of an earthquake, we have added any comments on the effects of an earthquake according to the sources in column 11 [of the table]."[57] The observers would still have their say.

In fact, Montessus anticipated that the progress of seismology would make a global catalog impossible. For example, the ISA's catalogs for 1903 and 1904 reported only a few dozen Chilean seisms. Meanwhile, improvements in the Chilean service meant that the catalogs for 1906 to 1908 would need to list 1,888 separate events for this country alone. "I am firmly convinced precisely by the experience I have acquired in this genre of research," Montessus attested. A global catalog would soon be a hopeless task: "Once the efforts of the International Seismological Association have produced their effect, that is to say that macroseismic and microseismic observations have developed on the surface of the globe with the necessary fullness of detail and generality, the work of the annual global catalog will become completely unrealizable; the central office will be literally overwhelmed and the ISA will no longer be able to meet the expenses. . . . So it will be absolutely necessary to renounce this publication. Moreover, and to use what seems to me an entirely natural analogy, can one imagine the International Meteorological Association publishing an annual catalog of rainfall on the surface of the globe?"[58] Montessus concluded that seismicity, like rainfall, was a phenomenon for which annual variations were meaningful only at the regional scale. By 1909 Montessus seemed resigned to the failure of his universal seismic description.

Standardizing Disaster

Internationalization requires agreement on standards of measurement, and establishing these was one of the broad achievements of nineteenth-century science. Standardizing the measurement of disaster, however, was nothing like standardizing mass, length, or electrical resistance. Unlike the wind-speed scale used to assess hurricanes, seismic intensity scales involve many parameters of many kinds. When introducing the Rossi-Forel scale in 1883, the Swiss Commission expressed the hope that it had "thereby created an international intensity scale."[59] In a literal sense, this was true: the scale was published in German, French, and Italian and immediately put to use in Switzerland and Italy. But were these really the same scale? In places, the Italian one hewed closer to De Rossi's original criteria than to Forel's French amalgam of the two.[60] There were significant differences among all three in the criteria of human response. Degree 1 required that the shock be "determined by a seismologist" in Italian, but "determined by a practiced [exercé, geubte] observer" in French and German. Degree 2 specified that the shock be felt by "few people at rest" in French, but there was no comparable criterion in Italian. Degree 3 required that the shock be reported in newspapers and perceived "by people who are not concerned with seismology" in Italian but not in French. Degree 6 specified in French that "a few fearful people leave their houses"; in Italian that people exited "due to fear or prudence"; the Italian alone further specified that in describing the shock people would mention "that fortunately there were no damages." Not surprisingly, the French version was the one most often cited as "*the* Rossi-Forel scale," thanks to the language's cosmopolitan status. Charles Davison judged it superior to the Italian scale, but that was probably beside the point. In 1903 the ISA resolved to develop an international scale of seismic intensity, but candidates emerged slowly.[61]

Few members had acquired as much expertise with the international measurement of intensity as August Sieberg, secretary of Gerland's Strasbourg observatory. By virtue of "processing a vast amount of macroseismic observations, collected from people of the most varied education levels throughout the world," combined with on-site inspections of earthquake damage in Italy, Sieberg felt ready to propose a new scale for international use in 1912. One drawback of earlier scales, in his view, was the brevity of their descriptions of observable effects. "I cannot fathom the considerations behind this terseness; in any case it has proved in practice to be the greatest obstacle to an effective use of intensity scales."[62] He also called for more detail in the questions posed to observers. The questionnaire should suggest typical, recognizable earthquake effects; otherwise novice observers answered the bare question "Effects?" with "None."

Sieberg cautioned against including human injuries or emotional effects as criteria, but he could not dispense with them. Indeed, his scale is remarkable for the care with which it describes psychological and sociological phenomena. For example, degree 1 stipulates a shock not felt by people, "however once the registration becomes known, individual people report that they purportedly felt the quake." Degree 2 notes, in keeping with Davison's research on "the effects of observers' conditions": "Only a few people finding themselves completely at rest, specifically those with sensitive nerves, noticeably feel the quake in the higher stories of houses, while on the ground barely anyone is conscious of it. The quiet of the night is also more favorable to perceptibility, if the observer is awake." Degree 3 specifies: "many only recognize afterwards, through the mutual exchange of ideas, that they were dealing with an earthquake." And degrees 4 through 6 distinguish among different levels of panic:

4: "This movement almost never induces terror, only if the residents have already been made nervous and fearful by other earthquakes."

5: "In isolated cases the inhabitants flee their dwellings."

6: "The earthquake is felt by everyone with terror, so that very many flee their dwellings; many feel like they will fall over."[63]

In these ways, Sieberg's scale exploited much of what nineteenth-century seismology had learned about human responses to earthquakes.

By 1939 there were thirty-nine versions of the Rossi-Forel scale in use. Some revisions expanded descriptive language, while others pruned it. In Wood and Neumann's 1931 revision to the Sieberg scale, for instance, the line "Plants, the branches and weaker boughs and trees sway visibly, as they do with a moderate wind" became "Trees, bushes, shaken slightly." A recent critic judges that the earlier version was "more vivid and clearer."[64] Indeed, devising intensity scales posed a rare challenge to scientists' communicative skills. Composing a scale was no less than a poet's quest for the evocative image, the telling detail.

Moreover, the descriptive criteria used to classify shocks were highly dependent on local cultures. It mattered how easily the masonry cracked, whether the houses had chimneys, how much of the population worked indoors, how given they were to panic. In Santiago, Montessus de Ballore found counterintuitively that "the observation 'felt by persons walking' corresponds to a greater intensity than the record 'felt by everyone.' This seems to be due to the more or less coherent nature of the subsoil at Santiago."[65] In the United States, Charles Rockwood noted that the report of "heavy" damage by an earthquake in the eastern United States was equivalent to "light" in reports from Peru.[66] When Edward Holden adapted the Rossi-Forel scale for use in California, he appended each degree with a word or phrase typically used by local witnesses, such as "slight" or "violent." Yet his assignments of intensity were unreliable because he was, inexplicably, "partial to intensity V."[67]

How then could earthquakes be compared internationally? Montessus again drew a comparison to meteorology, specifically to the Humboldtian methods of global physical geography: "When meteorologists draw their isobars or isothermic curves, they begin by reducing the temperatures or the atmospheric pressures to sea level by means of appropriate formulae. In other words, they calculate these elements according to what they should have measured in places equally distant from the earth's center. But as regards the earthquake observations of intensity, we cannot conceive of any method or formula which would enable us to reduce all estimates to uniform exterior circumstances, that is to say, to uniformity of subsoil, rigidity of buildings, temperament of the observers, etc. The mere enumeration of these requirements shows clearly the illusiveness of such an attempt."

Montessus added with a touch of sarcasm that an acceptable international intensity scale would have only two degrees: felt or not felt.[68]

In 1900 Charles Davison succeeded in devising a rough conversion scheme for eleven different scales, including corresponding values of maximum ground acceleration.[69] In 1933, he expanded this to include thirty-nine scales. By then, however, he had grown pessimistic about the possibility of settling on a single international scale. As he reflected in 1921, "After an experience of thirty years in constructing isoseismal lines, I venture therefore to suggest that an arbitrary scale of seismic intensity may be of very considerable service in the study of earthquakes, provided that each degree of the scale is restricted to one test only, and that personal impressions, such as those depending on the degree of alarm excited, are excluded. I feel doubtful, however, whether it would be wise to attempt agreement in the adoption of one uniform scale in all civilized countries."[70]

Indeed, a scale with criteria like "chandeliers swing" or "chimneys thrown down" will not help a seismologist evaluating damage to thatched huts on Samoa. In Japan, intensity is measured by damage to features of traditional architecture like stone lanterns and paper-covered sliding doors. The European Macroseismic Intensity Scale of 1998 assumes that for any degree of shaking, the damage will follow a normal distribution: the number of structures with greater-than-average damage will be roughly the same as the number that fall at the low end. Such a scale won't work in places where typical structures (say, California's wooden bungalows) either survive undamaged or are wrecked beyond repair. By the same token, in countries where quakes are typically stronger, scales must be extended at the high end; in regions of low seismicity, scales must be more finely graded at the low end. As Davison concluded, "Conditions vary from one country to another, and, so long as those conditions are known and so long as the scale is expressed in definite terms, it seems to me probable that the gain will be greater if each seismologist is left free to adopt that scale which is best adapted to the conditions under which he happens to work."[71]

Documenting the Earth

With its information overload, seismology circa 1900 looked poised to become a poster child for the benefits of the "documentation" movement. In 1895 Paul Otlet founded the International Institute of Bibliography in Brussels for the purpose of "unifying the bibliographic notices relating to writings of every nature, dealing with all matters, published in every era and in every country."[72] Otlet's institute was supported by international conferences and an annual *Bulletin*. By 1930, it had amassed a catalog of 16 million entries. At the 1907 meeting of the ISA, Georges Lecointe—the director of the Brussels observatory and an associate of Otlet—called for an international seismological bibliography. The idea received unanimous support, but debate quickly turned to the pros and cons of competing bibliographical systems. Some urged linking the association's work to the compilations undertaken by the British Royal Society, while Lecointe spoke in favor of Otlet's institute. The question was submitted to an expert commission—to which Otlet himself was appointed. It was decided that the bibliography would be cast as a standardized card catalog, to be integrated into Otlet's "universal bibliography."[73]

Yet macroseismology did not fit neatly into Otlet's scheme. Otlet aspired to release information from its linguistic trappings, to solve the problem of Babel. Not only were he and his followers warm supporters of the Esperanto and Edo movements. He also advocated "vocabulary control," such as restrictions on the words that could be used as subject headings. Otlet likened his system to the work of a knowledge factory, running documents through a giant grinder in order to extract "the purest matter useful for civilization."[74] The factory metaphor underlined the difference between Otlet's encyclopedism and that of the Enlightenment *philosophes*. Otlet sought to decontextualize information: to abstract it from its context of production, decompose it into standard units, and set it into circulation as a form of knowledge capital.[75] His attempts to obviate human language were, however, at cross-purposes with seismology's monographic method. Macroseismic information was inextricably language bound. It lodged in archival documents, newspaper reports, and the stories of earthquake survivors. It mattered whether an impact was described as a tremor or a jolt, whether a house was described as cracked or split. And the researcher had to develop an ear for such local turns of phrase.

Nor were macroseismology's goals well served by Otlet's ultimate aim: the "Universal Book." It was a telling irony that Otlet envisioned his final product as a bound volume: "It will constitute a systematic, complete, current registration of all the facts relating to a particular branch of knowledge. It will be formed by linking together materials and elements scattered in all relevant publications. . . . By gathering these leaves together, and classifying and organizing them according to the headings of a reliable and detailed classification, we will create 'the Universal Book of Knowledge.'"[76] How could one hope to enclose the young field of seismology—with its intuition of the infinite and evolving interconnections among terrestrial phenomena—within such a static, closed system?

The conflict between the ideals of the documentation movement and the methods of macroseismology was clear in their very different uses of the term "monographic." The monographic method in seismology, as we know, entailed a comprehensive narrative of a given earthquake, incorporating geological, sociological, and historical research. The documentation movement, on the other hand, adopted from Wilhelm Ostwald its own "monographic principle," denoting the "division, dissection and redistribution of items of information."[77] In the first instance, seismology's monographic method embedded physical facts in historical narratives and descriptions of the physical and human environment. Documentation's monographic principle, by contrast, sought to cut information free from narrative, from context, even from language itself.

This effect was vividly illustrated by the scheme for a seismological bibliography proposed by the ISA commission. It consisted of no fewer than 165 categories of phenomena. The primary division was based on the method of registration, whether human senses or instruments, that is, "macroseism" versus "microseism." Thus "earthquakes per se" were classed apart from the "microseisms" that corresponded to "distant" quakes.[78] Catastrophic effects and geophysical parameters were strenuously pried apart.

Such a bibliography represented nothing less than the systematic dismantling of disasters as scientific objects. As one commentator remarked at the time, the bibliographic problem in seismology was a "very complex question in the present case, a veritable spider's web, in view of the multiplicity of topics which seismology touches on and the perspectives from which they may be considered."[79] Applying the monographic principle of documentation to nineteenth-century seismology was like snipping the threads of the "spider's web" that represented its holistic approach to disaster.

The ISA had been created by an internationalist movement preoccupied with the present—most urgently with averting the threat of war. In the sciences, internationalists focused on cataloging the *present* state of knowledge. In seismology, they aimed to standardize data and organize it efficiently, in order to survey the present seismicity and physical constitution of the globe. To that end, the ISA came up with various schemes to purge felt reports of their discursive character. They seem not to have realized, however, that there is a trade-off between making data commensurable across space and across time.

The sciences responsible for helping humanity plan wisely for the future have the tasks of reconstructing the earth's past on the basis of existing data, producing "complete" data on the here and now, and archiving that data for future generations. As Geoff Bowker argues, the problem with computerized data in the environmental sciences is often that it includes too little of the context in which it was produced.[80] Without information about the physical, scientific, and social milieus in which it was generated, data loses much of its value to future researchers. Lack of context is evident in many ways. Sometimes it's a matter of language, as when a climate model exists only in an outdated code, or when weather has been described in the lost vocabulary of seafaring (how did a "moderate gale" differ from a wind that "blows hard"?). In this vein, seismologists have developed "lexicons" of the words typically used in other eras to describe earthquakes and tsunamis.[81] Sometimes, the problem is an incompletely recorded measurement process—likely something so basic to the scientist's training and everyday work that she would never think to write it down. Sometimes it's a result of theory change: seemingly unrelated phenomena might one day be judged essential to the process she is recording. Indeed, the relevant information goes beyond facts and protocols to include values, or what historian Christian Rohr calls "mentalities." In order to date floods of the Danube in the sixteenth century, Rohr argues, one needs to know that most floods were not perceived as disasters at all. They were normal, seasonal events, which were manageable within a system of warning and prevention. For that reason, they do not appear where a climatologist would look for them today, namely, in historical annals. Rohr underscores that environmental data is deeply embedded in historical cultures of perceiving, interpreting, and managing nature.[82]

Geoscientists look to the past and to the future. In order to interpret records of the earth's history, they need knowledge of the human past. In order to serve future generations, they need to index their data with elements of the present—the world they take for granted. These tasks require a sense for historical otherness. Like nineteenth-century seismology, they require a feeling for history, both human and natural.

Conclusion

The ISA collapsed during the Great War, and with it the hope of constructing an integrated international approach to seismic risk. Until the late 1960s, the human dimensions of earthquakes were studied, if at all, within the national frameworks of hazards geography and Cold War disaster sociology. International seismology endured as an observatory science, a science of insensible vibrations. Under the International Union for Geodesy and Geophysics (founded in 1919, the year of Gerland's death), the Strasbourg observatory continued to function as the central node of an international network of observatories. In 1922, the International Union created the International Seismological Summary, a continuation of Milne's microseismic catalog. Since then, international catalogs have been almost exclusively microseismic. When Beno Gutenberg and Charles Richter compiled the first edition of their landmark *Seismicity of the Earth* (1942), the ISA's catalogs for the years 1903–6 and 1908 were their only sources of macroseismic (sensible) data.

Over the first decade of the twentieth century, Gerland increasingly faced criticism for his neglect of the study of the destructive impacts of earthquakes. Though Forel supported the ISA's founding, he was, from the start, "somewhat skeptical about such resolutions; I wonder if we are not already going a little too far with internationalism."[83] Such concern was not out of character for a naturalist who helped develop the small-scale field methods of modern ecology. Over the next few years, as Gerland's intentions became clearer, Forel's hesitation grew into serious concern. The ISA seemed to be pursuing instrumental observation to the exclusion of the geological investigation of earthquakes and studies of their periodicity. Forel argued that seismology could not rely on seismographic data alone, not now nor in the future. There would always be a need for eyewitness reports and geological surveys. "The details of direct observation will always remain useful for the study of the phenomenon in itself: direction of the plane of oscillation, study of the focus in the depths of the earth's crust, plotting of isoseismic curves, determination of the lines of fracture, for all the tasks of geological seismology, etc."[84] Forel's point was that seismology must not be confined exclusively to the observatory.

It was Montessus who most clearly tied this methodological critique to a moral injunction. Beyond the pursuit of scientific theory, he urged his colleagues to "give some thought to the well-being of their fellow creatures."[85] Indeed, Montessus pressed the ISA to redefine its purpose. Against Gerland's plans, Montessus reconceived global seismology as a humanitarian enterprise. The emergence of humanitarianism as an international movement in the nineteenth century was predicated on the belief that science could be used to rationalize the provision of relief and make human existence more predictable.[86] In this spirit, Montessus believed it was the ISA's duty to guide "unstable" countries that had not yet taken responsibility for monitoring their own seismicity. "They will need to press, for instance, countries like France or Germany to publish the tremors in Madagascar or in German New Guinea."[87] Montessus thus insisted on global responsibility for local seismic risk.

In 1901 the eminent Göttingen geophysicist Emil Wiechert—"a most complicated scientific imperialist"[88]—appeared at the founding meeting of the ISA to discuss "Seismological Observations in the German Colonies." "For a long, long time," he reminded his putatively international audience, "we Germans have sat in our narrow Fatherland and regarded what lay outside its borders as a foreign world. Today that has changed; we feel ourselves to be inhabitants of the earth and send out our ideas and plans with the ships that sail the world. . . . Not only politically and economically, but also scientifically we will need to look for points of leverage in our colonies."[89] Wiechert was right: imperialism motivated and structured internationalism in seismology. Most seismic observatories founded beyond continental Europe and North America in this period—aside from the ISA's stations in Iceland and Lebanon—were colonial outposts. The Germans, for instance, sponsored observatories in Jiaozhou, China; Apia, Samoa; and Dar es Salaam, Tanzania.[90] Of these, only the Samoa observatory survived until World War Two (under New Zealand's sponsorship), and only as the ill-equipped and poorly staffed shell of the former institution.[91] Subsequently, plans for a dense and truly global network of seismometers have failed in the face of objections that such surveillance would compromise national security.[92] As Helen Tilley has recently argued of the human and environmental sciences in British colonial Africa, nineteenth-century seismology had a surprising potential to generate friction with the very project of empire. The give-and-take between observatory and field studies tended to alert scientists to the value of vernacular knowledge and the complexity and specificity of human-environmental interactions. When confined to the observatory, though, seismology became all too easy to assimilate to the simplifications of an imperialist worldview.[93]

NINE

The Youngest Land:
California, 1853–1906

Among the visitors to the beautiful coast the Earthquake is one of the oldest and most influential. . . . Had he, the architect of the mountains, not been active, there would have been no Santa Barbara, no California, indeed.[1]

In 1857 the English historian Henry Thomas Buckle described California as a desert land, "scorched into sterility" and unfit for civilization.[2] Had he been aware of the region's earthquakes—few were at the time, for reasons we shall soon see—his judgment would surely have been still worse. For, as we know, Buckle believed that earthquake-prone lands could support only barbarians, mystics, and gamblers. Americans, however, were intent on proving him wrong. California gleamed in the eyes of late nineteenth-century Americans as "the heritage of the future."[3] The Harvard geographer Nathaniel Shaler called the influx of migrants into California "the most rapid movement of population ever known." "Everyone who feels an intelligent interest in the future of our race must be concerned for the prospects of this region." "Soil, climate, mineral resources, relation to other great centres of population"—all promised ideal conditions for "the type of civilization" that the "Anglo-Saxon race" was developing. Yet California was also rumored to be prone to drought, fires, flash floods—and earthquakes. The question loomed: "whether society can there find a stable footing on a firm-set earth."[4] Shaler acknowledged, as Buckle had, that frequent earthquakes could destroy the possibility of cultural progress:

There can be no question that where a people is exposed to recurrent and overwhelming danger, such as menace the inhabitants of Peru, Venezuela, or Calabria, a danger which as yet is not foretold by science or effectively guarded against by art, the conditions will tell upon its character. "To the firm ground of nature trusts the hand that builds for aye," is true in a real as well as in a metaphoric sense. This trust in a stable earth is a necessary element in much that is noblest and most aspiring in the life of men. Expose a people to constant devastations from an overwhelming force, whether it be in the form of an human enemy or a natural agent, and their state of mind becomes unfavorable for the maintenance of a high civilization. The best conditions of the state can only be secured when the laborer toils with the assurance that his work will endure long after his own brief life is over.

Shaler, however, was no prophet of doom. He was a devout man with a providential view of the American continent, a view he expounded in a schoolbook that defined "the fitness of a country for civilized man" in terms of natural conditions conducive to production and trade.[5] He firmly believed in a human racial hierarchy determined by environmental conditions. The young continent of North America was not destined to be a "cradle of civilization," as he saw demonstrated by the backwardness of the native Indians. For the right settlers, however, Shaler was sure that California would indeed be a golden land. He set out to convince the American public that earthquakes would not stand in their way.

In a series of popular articles, Shaler reassured his readers that earthquakes were but a magnified form of normal geological activity. "The notion that the ground is naturally steadfast is an error—an error which arises from the incapacity of our senses to appreciate any but the most palpable and, at the same time, most exceptional of its movements." Recognizing the inherent instability of the earth's crust required, in his analogy, a minor Copernican revolution for geology—a shift of perspective akin to the discovery of the heliocentric system. Ground movement was not only common, he argued, it was a necessary condition for organic life: "Were it not that the continents grow upward, from age to age, at a rate which compensates for their erosion, there would be no lands fit for a theatre of life." Shaler dispelled the occult qualities of earthquakes by means of mechanical metaphors: earthquakes were to be studied in terms of "the machinery which produces them"; faults were "earthquake factories"; lava was driven "perhaps with a greater impulse than that which propels the ball from a canon."[6] There followed the lesson that the proper response to earthquakes was not fear but scientific study and solid engineering. "And when experience has taught them the simple lessons which it is necessary to practise in order to obviate a large portion of the dangers occurring from these convulsions, there is no reason why this region, despite the frequent light shocks to which it is subject, may not enjoy as happy immunity from their worst effects as any portion of the continent now occupied by our people." European scientists were making the same point at the same time, but Shaler spun it into a racialized vision of environmental destiny. California was exceptional, in Shaler's view. Looking north, he judged it "probably a fortunate thing that the inhospitable and unproductive character of the Alaskan region will prevent any extensive settlements of civilized man in the midst of the terrible convulsions which are there so frequently occurring." Alaska's destiny was wilderness, while California's was "Anglo-Saxon" civilization. Shaler thus placed his faith in the ability of his "race" to engineer a more stable foundation for prosperity in California.[7]

California earthquakes were an abstract threat to East Coast naturalists like Shaler. As Sheila Hones has shown, Shaler's popular writings on earthquakes were part of a wider conversation in the *Atlantic Monthly* in the 1880s about the threat of natural disasters to the nation. The authors focused on the need to respond to disasters with centralized, expert-led interventions. Hones suggests that this discourse reflected the perceived fragility of America's democratic experiment. Earthquakes figured in social Darwinian terms as trials of national fitness, as moments that would define the nation's indefatigable spirit.[8]

At closer range, this myth was harder to sustain. Praise for California's "ideal conditions" from East Coast writers could not put locals' worries to rest. Earthquakes understandably made real-estate investors and insurance companies nervous. Settlers themselves worried that the cumulative physiological effects of mild tremors might be harmful to their health.[9] The history of California seismology in the nineteenth century is a story of how citizens were taught to distrust their own senses.[10] In the spirit of the state's "boosters," the California press persistently understated the threat from earthquakes.[11] One San Francisco paper claimed in 1868 that "earthquakes are trifles as compared with runaway horses, apothecaries' mistakes, accidents with firearms, and a hundred other little contingencies, which we all face without fear."[12] With equal confidence, the *Annals of San Francisco* announced in 1855 that earthquakes were "the greatest, if not the only possible obstacle of consequence to the growing prosperity of the city, though even such a lamentable event as the total destruction of half the place, like another Quito or Caracas, would speedily be remedied by the indomitable energy and persevering industry of the American character."[13]

In this vision of California's destiny, science and engineering held a privileged place. In the late nineteenth century, geology became "a tool to recast the West into a region of economic opportunity."[14] California also presented geology with a rare opportunity. The first directors of the state's Geological Survey, John Wesley Powell and Clarence King, found in California a dramatic landscape of cliffs and canyons that set each man imagining quite different causes. Powell, following Lyell, emphasized the roles of erosion and sedimentation, while King focused on uplift and subsidence, in keeping with the contraction hypothesis.[15] Both men, however, knew from long experience in the field that this land would not easily be tamed for "civilization."

King helped Americans recognize the role that catastrophic land movements had played in the formation of California's high mountains. His celebrated *Mountaineering in the Sierra Nevada* described a terrain that was profoundly unstable, a "dizzy" landscape. Staring into a chasm at Yosemite, he described the "titanic power, the awful stress, which has rent this solid table-land of granite in twain." Yet in his next breath he assured the reader of "the magical faculty displayed by vegetation in redeeming the aspect of wreck and masking a vast geological tragedy behind draperies of fresh and living green." He raised this relationship between catastrophe and renewal to a general principle: "Movements of great catastrophe, thus translated into the language of life, become moments of creation, when out of plastic organisms something newer and nobler is called into being." King's message was ambiguous. The vertiginous, precarious landscapes he described seemed to urge modesty on scientists bent on transforming nature to their specifications. And yet catastrophe appeared in these landscapes primarily as an actor already vanished from the scene, having set the stage for the vitality of California in King's own day.[16]

Powell's efforts were more explicitly political. He saw from his fieldwork in this arid state that agricultural development would be perilous without a rational irrigation system. He railed against the conceit that "rain follows the plow," that the very act of agricultural settlement would improve the climate. He even tried to convince the public that much of the country's western lands should be protected from settlement. But even Powell said next to nothing about California's earthquakes.

From time to time, lone voices disturbed this canyon of silence. In 1868, at the end of his work on the California Geological Survey, the East Coast geologist Josiah Whitney faced the problem squarely: "The prevailing tone in that region, at present, is that of assumed indifference to the dangers of earthquake calamities, the author of a voluminous work on California, recently published in San Francisco, even going so far as to speak of earthquakes as 'harmless disturbances.' But earthquakes are not to be bluffed off. They will come, and will do a great deal of damage. The question is, How far can science mitigate the attendant evils, and thus do something toward giving that feeling of security which is necessary for the full development of that part of the country?"[17] Again, after the Owen's Valley earthquake of 1872, Whitney argued for the need to adopt new construction practices, such as using wood rather than brick and abandoning heavy cornices and chimneys. He also stressed "the desirability of a scientific record and examination of the earthquakes occurring on this coast."[18] Whitney found a sympathetic ear at one San Francisco paper, which complained that "everybody has become indifferent to earthquakes." The editor rejoiced that "Professor Whitney and his party are among us at last. They will try to probe this earthquake business to the bottom, and I doubt not will arraign Mother Earth at the bar of Science to give an account of herself and explain the wherefore of her fatally playful ebullitions."[19] Whitney's warnings echoed like a pebble tossed into a chasm of willful ignorance.

"A Very Needless Source of Alarm"

Even the first attempt to catalog California's earthquakes did little to dispel the state's Edenic image. It was undertaken by an unlikely investigator: John Boardman Trask, a Yale-educated physician who set out for California in 1850 as part of the "California company" of John Woodhouse Audubon, the younger son of the famous ornithologist. The expedition came in search of gold, of course. But its members were naturalists, keen to explore the organic life and humbler minerals of the West. Lest one think Trask an opportunist, an obituary made a point of describing him, in his medical practice, as "quite free of the acquisitive instinct."[20] In California, Trask discovered that his amateur mineralogical skills were in great demand. In 1853 he received the first of a series of commissions from the state senate to report on the geology of California. He touted his research as "an exhibition of the capabilities of some of our soils for the production of the necessaries of life, unexcelled in the history of the world." California's scenery, on the other hand, was not yet a selling point. Despite the "repulsive aspect" of the coastal mountain range, with its "naked and barren appearance," Trask argued, these hills were "covered with a luxuriant growth . . . affording extensive pasturage for flocks and herds."[21] The boosters in the state assembly could not have been disappointed with his results.

In 1853, Trask took part in another tribute to California's natural riches: the founding of the California Academy of Natural Sciences. It was in the *Proceedings* of the new San Francisco–based academy that Trask began to publish on earthquakes. Through "careful inquiry of the older residents," he gathered information on seismic events in living memory. His findings sufficed, he believed, "to correct some of the misapprehensions and statements which have appeared from time to time relating to the severity of earthquake shocks in this country during the earlier periods of its history." Indeed, he learned of "but one shock that has proved in the slightest degree serious," and that was the Southern California earthquake of September 1812, which was reported to have killed between thirty and forty-five people.[22] Trask went on to publish annual lists of earthquakes observed throughout the state and a catalog of tremors in California since 1800. Although he rarely cited sources, he was reported to have interviewed "old inhabitants and foreign traders." What his "statistics" showed, Trask claimed, was that "even in the mountain districts, where during the day there is much less of turmoil and noise arising from business than in the populous city, of all these noticed, none have been of sufficient intensity to attract the attention of the inhabitants during the hours of daylight. These facts, though few in themselves, are of importance, to disabuse the public mind in relation to the danger to be apprehended from the occurrence of these phenomena. The reputation which we sustain both at home and abroad, of being in constant danger of being swallowed up by these occurences [sic], and the idea that our country is but a bed of latent volcanoes, ready to burst forth at any moment, spreading devastation over the land, is a very needless source of alarm."[23]

Trask never paused to consider that perhaps his data were incomplete.[24] By his reckoning, if California's earthquakes "possess[ed] that severity so often attributed to them, the attention of the people would much more often be directed to them. Yet we find that their first knowledge of such an occurrence is usually its announcement by the daily press." Trask thus preferred to interpret his empty mailbox as a sign of the insignificance of California's earthquakes—not of the shortcomings of his network.

After a decade of this work, Trask abruptly turned his back on the academy and returned to full-time medical practice. Without him, research on California's earthquakes languished for two decades. This was not for a lack of shakers, to be sure. In 1865 San Francisco newspapers reported the most violent quake since the city's founding. "Every door, of every house, as far as the eye could reach, was vomiting a stream of human beings," wrote the young journalist Mark Twain. "Thousands of people were made so sea-sick by the rolling and pitching of floors and streets that they were weak and bed-ridden for hours, and some few for even days afterward.—Hardly an individual escaped nausea entirely."[25] Just three years later, on 21 October 1868, San Francisco was hit by an even stronger upheaval, which tore apart buildings on the "made ground" of embankments and landfills. The 1868 earthquake killed thirty people, the first deaths by earthquake the young city had known.[26] The 1868 disaster would rival 1906 for sheer violence in the memories of older residents, though no fires ensued to complete the work of destruction. Even in 1868, though, the dominant tone of the city's press remained vigorous optimism: "There was nothing of despair, discouragement, or even doubt of the future to be seen on the countenances of our citizens, but everywhere a fixed determination to repair losses and do better work than before." The paper reported that only buildings on "made ground" were damaged, "and nine-tenths of it was to old structures." The report concluded that "this earthquake demonstrates the proposition that, with proper care in the construction of our buildings, San Francisco is as safe a place to live as any on the Continent."[27]

Not everyone who lived through the San Francisco earthquake of 1868 was reassured by such reports. One of the city's more prominent businessmen called for an investigation. The committee he appointed has since become famous for its failure to reach conclusions of any kind. It has even been accused of sabotage. More plausible is that the committee refused to be shaken in its opinion that the earthquake risk in San Francisco was minimal. Most of the members had arrived in California well after the start of the gold rush. None but Trask had previously studied earthquakes, and only the mining engineer Thomas Rowlandson could claim so elevated a credential as fellowship of the Geological Society of London. Rowlandson was dropped from the committee as a result of a dispute with the chairman, but he went on to publish a *Treatise on Earthquake Dangers, Causes and Palliatives*. This treatise is one of the few sources to shed light on seismological thought in California at the time.

Since Rowlandson had been denied access to the observations collected by the committee, he was free to rely completely on his own memories of the quake. He recounted his experience in excruciating detail. "Owing to indisposition," he was in bed at his home north of the Pajaro River, "a position better calculated to remark the course of the two chief shocks than the bulk of people at midday." Moreover, his wife was "reclining on the bed in an opposite direction . . . her head pointing near due east, mine nearly due west," thus perfectly situated as a source of "corroboration." While their bodies served to measure the direction of the waves, their conversation—"which was not hurried" and which Rowlandson paraphrased with particularly unhurried pedantry—served to measure the duration. This scene perfectly served Rowlandson's purpose of "domesticating" the earthquake threat. Where one might expect to find death and destruction, he offered only this image of lazy domesticity.[28] For Rowlandson intended to go one better than the boosters. Instead of denying that California was seismically active, he would prove the beneficence of earthquakes. First, he argued that the seismic threat in San Francisco was mitigated, "if not wholly obviate[d]," by the city's geography, with a shallow bay to protect against landslips and a coast set back far enough to protect against tsunamis. Next, he furnished a scientific account of earthquakes, in the hope that the phenomenon would "cease to trouble timid minds. Timidity has certainly been carried to excess by those who have left or expressed a purpose of leaving this State on account of earthquakes."[29] But Rowlandson did not stop at demystification. Here, in the heart of the book, his argument became quite literally nebulous. Starting from the Laplacean principle "that universal matter . . . was originally in a gaseous condition," Rowlandson argued that earthquakes somehow brought precious metals trapped in the earth's "molten center" to the surface. "I have long held and still adhere to the opinion, that human access to these valuable metaliferous [sic] accessories to the luxuries, comforts, and necessaries of human life has resulted from earthquake influences. The earthquake, in fact, being one of the cosmical agents employed by the great DESIGNER OF ALL, for contribution to HIS final aims." Evidence for this startling claim came from the stratigraphical location of precious metals, which pointed to their having reached the earth's surface no earlier than the appearance of man. "They were not required by the wants of the then animal creation, and they were not needed until MAN made his appearance on the surface of the earth, endowed with faculties to extract, reduce and utilize them."[30] Rowlandson had found an ingenious way to reconcile California's earthquakes with the myth that America was destined by providence to bear wealth for white settlers: earthquakes were California's secret source of *gold*.

Remote Sensing

While Trask and Rowlandson were sweeping California's seismic rubble under a carpet of gold, interest in seismology was growing in more quiescent parts of the country. Local volunteer observing networks for both weather and earthquakes flourished in many states in the 1870s. These efforts demonstrated what scientists and citizens could accomplish when not encumbered by the denial of seismic hazard. The emergence of earthquakes as scientific objects elsewhere in the United States underscores the work necessary to produce seismic ignorance in California. It also points to the tensions between local and federal initiatives that would plague American seismology into the twentieth century. In imperial Austria, the arguments for and against the centralization of the earth sciences resulted in an effective compromise, a system balanced between the intimacy of local natural history and the efficiency of a central bureaucracy. In the United States, the result of analogous tensions was a shiftless path from one form of organization to another. More than elsewhere, the initiative to record earthquakes came from individual scientists, and the quirks of personality set the limits of communication between scientists and citizens.

Before the Civil War, a successful network of weather observers had threaded across the United States from Massachusetts to Missouri. It had been built in the 1850s by Joseph Henry at the Smithsonian Institution and had depended to a large degree on amateur "cooperative observers." As Jim Fleming has documented, the composition of this corps shifted in the 1850s and 1860s to include fewer professors and more farmworkers.[31] Weather observing was on its way to becoming a grassroots enterprise. However, the Civil War unraveled much of this network. Telegraphic connections were broken or overloaded with military messages, and observers were drafted into the army. The reconstruction of the observing system began in 1870, when Congress passed a bill establishing telegraphic weather forecasting under the auspices of the Army Signal Service. The lawmakers were convinced that military discipline was essential to a proper routine of meteorological observation. So too, however, was civilian expertise, and the new weather service was fortunate to recruit Cleveland Abbe, an astronomer who had established a regional telegraphic observing network out of Cincinnati in 1869.

Abbe later claimed that he had hoped from the start to include the reporting of earthquakes in the weather service's agenda. That plan was apparently foiled by political resistance to the further expansion of the Signal Service.[32] As the acting chief of the Signal Service explained in 1887, "the data received at this office is so voluminous and recent regulations of the War Department relative to printing so restrictive, that the late Chief Signal Officer [Albert Meyer, d. 1880] ordered no further publication by this bureau of earthquake information, as it was thought that such publication might give ground for the charge that it was interfering in work not legally within its proper sphere." The acting chief did "not deem himself authorized to make any definite change at present."[33] In the absence of federal support, the compilation of data on American earthquakes was undertaken at the private initiative of Charles Rockwood of Princeton (chapter 3).

When a temblor took the Eastern Seaboard by surprise on 10 August 1884, Rockwood was able to collect observations from eleven states. Recent studies locate the epicenter in Brooklyn, New York, and judge it the strongest earthquake on record for the New York region.[34] Along with information from the Weather Bureau's observers and newspapers, Rockwood amassed about 150 reports from private citizens.[35] By the end of the month, he was able to furnish a preliminary report in the *American Journal of Science*, describing an affected area with boundaries "along the coast states from Washington, District of Columbia, and Baltimore, Maryland, to Portland, Maine, and Burlington, Vermont; and on the west . . . a nearly straight line from Burlington to Harrisburg, Pennsylvania."[36] However, the initial reporting on the earthquake of 1884 put American science to shame.

The most extensive coverage appeared in the *New York Herald*, including a two-page spread on 12 August featuring interviews with most of the leading East Coast scientists.[37] They contradicted each other at every turn. John Wesley Powell, for instance, claimed that "earthquakes are probably independent of meteoric and to a large extent of astronomic conditions." Yet according to Simon Newcomb of the Naval Observatory, it was "impossible to say" "whether or not earthquakes are influenced by cosmic causes." Powell also claimed that the interior of the earth was undoubtedly molten, while William Harkness at the Naval Observatory said the latest evidence pointed to a completely solid earth. None of the scientists evinced any knowledge of the tectonic theory. One reporter offered the deadpan understatement: "Geologists do not agree as to the cause of earthquakes." The headline "Harvard's Wisdom" was an ironic introduction to a statement from the astronomer Edward Pickering, who admitted that he had no information to offer. According to Pickering, there was no system for observing earthquakes in the Northeast because they were so rare. James Dwight Dana, Silliman's successor at Yale, apparently responded to a question about the earthquake by saying that he "saw no occasion for it."[38] The *Herald* added to the impression of confusion with headlines and pull quotes such as "Nothing Known—Everything Known"; "No Little Indignation among Professors at the Unexplained Phenomena," "The Science Only Beginning"; "As ignorant of the cause as of any subject about which we know nothing."[39] The Columbia geologist John Strong Newberry joked to a reporter, "I want to show you my credentials." Out came an old sheaf of paper addressed to the "Professor of Earthquakes, Columbia College, New York." Newberry explained that it came from a witness to the Bay Area quake of 1865 "who had a theory concerning earthquakes that he wanted to ventilate."[40] The joke, of course, was the absurdity of the academic title, "Professor of Earthquakes." Readers would have come away with the impression that seismology hardly deserved to be called a science.

What must Rockwood have felt as he filed this publicity away in his scrapbook? No doubt it disturbed his colleagues too. A letter from Cleveland Abbe arrived on the nineteenth, containing two letters and two clippings. Abbe hoped to receive information from Rockwood, but he had heard that Rockwood was himself looking for reports. Something had to be done, Abbe decided. "I have always hoped that our office [the Army Signal Office] would pay more attention to Seismology but the present feeling is that it is too far outside of our legal business. So probably it will be left to the Geological Survey. However if as a private person I can help you to organize a general Seismological Association I shall be glad to do so. I will set up a lot of pillars and balls in one corner of my cellar to begin with." Abbe mentioned two colleagues who had lived in Japan and were "interested observers" and another who had set up a rough seismometer at Harvard. "I really think we can organize systematic records and observers in every Astronomical observatory east of the Mississippi. Possibly the Appalachian Club would undertake and study Appalachian Seismology."[41]

In mid-November, Powell, director of the Geological Survey, invited Rockwood, Davis, Abbe, Paul, and Dutton to a meeting in Washington, all expenses paid. "After looking over the ground pretty fully," Powell thought it would "be possible to inaugurate systematic observations, and it seems that we ought first to give attention to the character of the observations most desirable, the instruments to be used, etc."[42] A press release announced plans for a national seismic survey, to include a "large corps of observers" reporting to local directors.[43] A report on the meeting in the *New York Herald* hoped that "the observers will not have a chance to make a single observation for half a century at least."[44]

Only one participant in the November 1884 meeting seems to have taken any notice of the example of the Swiss Earthquake Commission. In *Science* the following spring, Harvard's William Morris Davis judged the Swiss far ahead of Americans in the study of earthquakes. Davis described the commission's publications as "entertaining" and "attractive" and noted with disappointment that they were not available in any American libraries. He discussed the Rossi-Forel scale, the significance of the commission's work for the tectonic theory of earthquakes, and its potential to clarify the geological interpretation of the Alps.[45] Perhaps it was Davis who brought Swiss seismology to Rockwood's attention. By January, Rockwood was in correspondence with Albert Heim, who offered details on the Swiss system and assured Rockwood that human observers had proved far more valuable than instruments.[46] But how could the United States adopt a research method designed for a nation smaller in area than West Virginia?

"Cooperatives"

What American seismology needed was a widely distributed corps of dedicated, attentive, and accurate observers: precisely what the US Weather Bureau could offer. The population of lay weather observers in the United States was more heterogeneous than in Switzerland and Austria. Its ranks included physicians, teachers, and clergymen, but also a significant proportion of storekeepers and farmers.[47] Many were women who took over the work of observing from their fathers or husbands, either temporarily during an absence or permanently following a death. Indeed, the task was more often assumed by daughters than by sons.[48] As in Europe, the cooperative observers were unpaid,[49] and their lack of compensation was taken as a guarantee of their trustworthiness. The bureau's weather records were accepted as evidence in law courts and were often used to plan major engineering projects. This meant that a great burden of trust rested on the shoulders of the volunteer observers. The bureau's meteorologist in Salt Lake City argued that the system was self-correcting. Only an observer who took pleasure in the work would stick with it, and "we cannot conceive of anyone finding a pleasure in a service that is carelessly done." "Unreliable" observers, he concluded, "find the work too objectionable to be continued long." Moreover, if the saying was true that "a bad workman quarrels with his tools," then the preservation of delicate measuring instruments was evidence of the quality of the observers.[50] Instruments, however, were not necessarily essential. It was not always the case that the bureau used instruments to inculcate bureaucratic discipline.[51] "Non-instrumental observations form the ground work" in meteorology, according to Gustavus Hinrichs, director of the Iowa Weather Service. "Hence we would also urge all those who take notice of special phenomena to report the same to us, as casual correspondents, describing the phenomena seen and stating time and place of the observation."[52] Seismic tremors were among the phenomena that observers were encouraged to register with their bare senses.

Based on a case study of the bureau's observers in Kansas at the turn of the twentieth century, Jeremy Vetter has recently come to the intriguing conclusion that the bureau succeeded only to the extent that it imposed a "rigorously bureaucratic form of top-down control over knowledge production."[53] By contrast, Jamie Pietruska argues in a forthcoming book that scientific and vernacular methods of prediction were mutually constitutive in the work of the Weather Bureau, producing a science that acknowledged the limits to its own certainty.[54] Consistent with Pietruska's conclusion about the mutual construction of scientific and vernacular knowledge, I have found evidence that the relations between observers and bureau officials were often flexible and personalized. Local officials of the bureau kept in close contact with the cooperatives. They gave careful thought to questions such as how to deal with a suspected error in a volunteer's observations. A South Carolina official stressed that the volunteers were as dedicated to scientific accuracy as the bureau itself: "The voluntary observers are, as a class, men [sic] who take up the work in a spirit of scientific investigation after truth, and in that spirit are as desirous as the Bureau possibly can be to have their observations and reports as nearly absolutely correct as painstaking care and thorough examination can make them." Therefore bureau officials should not "assume the role of censor," but "should frankly and courteously inform" the observer when an error has been discovered. Since this address was delivered to an audience of bureau officials, there is no reason to suspect that it was meant to flatter the cooperatives. Even if the intention was to reprimand more authoritarian types, one can only conclude that communication between scientists and cooperatives was often respectful and lively. As another bureau official described, the relationship between a section director and his volunteers was "of a dual nature, namely, official and personal, perhaps better expressed as friendly or social. The two are not inconsistent. The first, although always pleasant, is usually formal, the latter sometimes very cordial. The section director regards the voluntary observer as an indispensable aid in his work, to whom all encouragement should be given and every courtesy extended, and who is worthy and deserving of all that can be done for him, either officially or in any other manner in which the director can show his appreciation."[55]

The director of the Iowa Weather Service, Gustavus Hinrichs, had gone so far as to erect Iowa's central observatory in his own home.[56] He spoke fondly of his cooperatives. One, whom Hinrichs called "my friend," had taken up observing a full thirteen years earlier, when "his home was happy and wife and daughter aided him." This observer had since lost both family members, but his reports continued to arrive, now in a "masculine handwriting" that reminded Hinrichs of his loss. "Is it not worthy of special praise to keep up such public work under so depressing circumstances?"[57] Nipher, the director of the Missouri Weather Service, vouched for the accuracy of one of his observers, who set his clock to the telegraphic signal and was "a faithful observer." "I wish you would be kind enough to send them copy of any publication in which their names may appear," he added to Charles Rockwood. "It is well enough to encourage such men—they are scarce in this part of the moral vinyard [sic]."[58]

The bureau also valued cooperative observers for their ability to serve as intermediaries between scientists and local communities. As one bureau official wrote of an observer: "The citizens of his community rely implicitly on him for weather information, and through his careful educating of them in actual weather conditions for two decades they have largely abandoned their interest in weather folklore and general signs and now look to him for facts."[59] One cooperative observer in Memphis, a dealer "in all kinds of produce, flour, pork . . . lime, cement, etc.," informed Rockwood that "it would be an impossibility for [an earthquake] to be felt by any number of persons without my knowing something of it." He went on to explain how he had come to collect seismic observations and why he was to be trusted:

> I have taken more or less interest in the subject since 1855, which year I spent in New Madrid—the acknowledged great centre of internal disturbances in this region. I fully anticipated the shock of July 13th because the maximum average interval had passed between shocks—especially since the severe shock which occurred on the 6th of June 1862 at 10 a.m. the day Federal forces took possession of this city. I know of no way to forecast the future except from the experience of the past and if that is any criterion the time is only now at hand when we may begin to expect another shock. When we consider what occurred at New Madrid before this country was settled and what the consequences would be, were those scenes to be repeated, now, or in the future, when this country shall have become even more densely settled—and many costly habitation of man erected, and the consequent loss of human life, and destruction of property which would ensue, we can attain some idea of the importance of obtaining all the information, with a view of locating the great centre of these internal disturbances. You and I may not live to experience such a shock as would cause a loss of life or property, but the information now gained will survive us—and should such an event occur its importance will then be appreciated—for as I said before what has once occurred may as surely occur again. . . . [I] always carry with me as nearly absolutely correct time as it is possible to attain. And when the next shock occurs, I will send you as correct an account of it as it will be possible for human powers to obtain.[60]

This Memphis trader recognized that modernization was making Americans increasingly vulnerable to earthquakes. His efforts to observe tremors and collect reports from neighbors were a way to serve science and his community. "Correct" observation was the key to avoiding future disasters; it was a citizen's duty.

By the 1880s, then, the bureau's volunteer network was buzzing with activity. It was supported, however, almost exclusively by local officials. After the army's takeover of the weather service in 1870, the cooperatives were neglected at the federal level.[61] Moreover, the absence of consistent time zones before 1883 meant that comparing reported times of shocks across state lines was still a Herculean task. Indeed, little progress toward a national earthquake service was made in the two years after American scientists met in Washington in 1884. Systematic earthquake reporting in the Reconstruction Era would therefore be a strictly local enterprise, where it was attempted at all.[62]

Southern Discomfort

Consequently, not a single seismograph was in operation on the East Coast when disaster struck Charleston on 31 August 1886. The shock killed an estimated eighty people in the Charleston area. It was felt clear across the eastern half of the country, from the Gulf of Mexico to northern Michigan, and from Arkansas in the west to around five hundred miles beyond the Atlantic coast. The disturbed area measured 774,000 square miles, on par with that of the Lisbon earthquake of 1755.[63] The city of Charleston lost 25 percent of the total value of its buildings.[64]

Yet not even this disaster could convince Washington to organize a seismological network. The Geological Survey furnished Charleston residents with crude seismoscopes, but locals stopped paying them any attention within a few months. Although an estimated three hundred aftershocks occurred in the thirty-five years after 1886, a scientist searching for press reports of them in 1914 complained that seismic disturbances had lost their "value as news items as far as South Carolina newspapers were concerned."[65]

Several East Coast scientists descended on the stricken region to investigate. Some found the event well suited to Mallet's architectural methods of investigation, since it had thrown down walls, gables, and chimneys of many of the graceful wooden houses in Charleston and neighboring Summerville.[66] But the chief investigator, Clarence Dutton of the Geological Survey, considered Mallet's methods to belong to "the outermost pale of scientific philosophy."[67] Dutton was on his way to becoming the foremost proponent of instrumental seismology at the survey. However, of necessity and nearly despite himself, he founded his investigation of 1886 on a novel use of noninstrumental reports.

It was the first time that American scientists were pursuing the style of seismological field research pioneered by the Swiss and Austrians, and the investigation brought out some of the features that would set American seismology apart in years to come. With Rockwood's help, Dutton circulated a questionnaire and ultimately collected about four thousand observations from 1,600 localities. The journal *Science* judged of the Charleston catastrophe that "no earthquake of ancient or modern times has been observed with such care and fullness of detail. Besides the observations made by Professors in several Colleges, by hundreds of railway officials, and at signal stations, a large number of intelligent private citizens have given an account of their own experiences."[68] Dutton acknowledged the European precedent for this approach, but he perceived a difference in the United States: the questionnaires "were much fewer and more simple than those employed in Europe, because European investigators depend almost wholly upon the educated classes to answer them, while in this country the uneducated but intelligent and practical classes of the people must be the main reliance." Dutton's doubts about lay observation alert us to an important *perception* of cultural difference. Even as Dutton invoked the myth of American equality and meritocracy, he betrayed a hesitation about the consequences of democracy for intellectual culture. Dutton may not have appreciated that the "practical classes" in the United States were proving to be reliable observers in various fields of natural history, as well as in meteorology. More to the point, his pessimism about human observers was tied to his vision for seismology. For Dutton, seismology was a branch of physical science; its object was the precise measurement of physical forces and the elucidation of universal causes. He was far less interested in seismology as it was widely pursued by lay observers in Europe—as part of geohistory, local history, and the assessment of environmental risk.

Dutton was therefore skeptical of the felt reports from the Charleston quake. He treated them as a meager substitute for instrumental data. He was particularly dissatisfied when it came to measuring what was commonly called "intensity." The new Rossi-Forel scale was still relatively unknown in the United States. In 1884 Rockwood had been content with a rough qualitative scale running from "very light (noticed by a few persons but not generally felt)" to "destructive (causing general destruction of buildings, etc.)."[69] Dutton employed the Rossi-Forel scale, but worried that it was nothing short of "crude and barbarous." As such language suggests, Dutton recoiled from macroseismological methods without attempting to understand them on their own terms. What Dutton sought was an absolute measure of the "real energy" of the seismic wave at its origin below the surface. As he explained,

What is really desired is some reliable indication which shall serve as a measure of the amount of energy in any given portion of the wave of disturbance as it passes each locality. The means of reaching even a provisional judgment are very indirect, and qualified by a considerable amount of uncertainty. . . . *In view of the precise methods which modern science brings to bear upon other lines of physical research, all this seems crude and barbarous to the last degree.* But we have no other resource. Even if it were possible to obtain strictly comparative results from such facts, and decide with confidence the relative measure of intensity which should be assigned to each locality, we should have gained measures only of a series of local surface intensities, and not of the real energy of the deeply seated wave which is the proximate cause of the surface phenomena. Notwithstanding the indirect bearing of the facts upon the real quantities we seek to ascertain, and their apparently confused and distantly related character, they give better results than might have been supposed. When taken in large groups, they give some broad indications of a highly suggestive character.[70]

Dutton believed he had found a way to use isoseismals to calculate the depth of the shock's origin. He defined intensity as the energy of the seismic wave per unit area of the wave front; from this definition it followed that intensity varies inversely with the square of the distance from the origin. This method was as ingenious as it was unreliable. Dutton had confused intensity, the destructiveness of the earthquake shock as measured by the Rossi-Forel scale, with the energy contained in the seismic wave. As the British geologist Richard Dixon Oldham explained, Dutton's formula for the depth of the origin was virtually meaningless:

One of the most important problems of seismology is the determination of the depth at which earthquakes originate. One method after another has been proposed, only to be abandoned as its failure to give a true answer became apparent, and Major Dutton has himself invented a method which he believes to be sound, and which would be sound if its application were not vitiated by a logical fallacy. In the formula he uses the term "intensity" in the sense of energy per unit of wave-front, but in the application in the sense of a degree of the Rossi-Forel scale, which, like every other scale proposed in place of it, is miscalled a scale of intensity, being in fact a scale of acceleration, or, more simply, violence of shock. The two are very different, and differ in their rate of variation with distance from the origin, so that we are still left with no certain and trustworthy method of determining the depth at which earthquakes originate, but have, on the other hand, a lesson in *the danger of misusing words.*[71]

Dutton had stumbled into a hole that would ensnare many more American seismologists in the future. He had failed to appreciate that felt reports contain more than just geophysical information; they are not mere substitutes for instrumental observations. Oldham was right to fault Dutton for "misusing words." Exploiting felt reports required attention to the observers' own language. It depended on the consistent use of a scientific vernacular.

"The Trouble with Holden"

Among the small number of scientists who shaped the development of seismology in California, astronomers were remarkably well represented. They brought with them an expertise with measuring instruments, a distrust of naked-eye observation, and a disinclination for field research. These individuals included George Davidson, W. W. Campbell, Charles Burckhalter, and A. O. Leuschner and, last but not least, Edward Singleton Holden. Holden was a Saint Louis–born astronomer with no background at all in geology. In 1885, at the age of thirty-nine, he moved to California to become president of the University of California, and in 1888 he became the first director of the Lick Observatory, near San Jose. It was astronomy that piqued Holden's interest in earthquakes because of their tendency to disturb his telescopes. (In much the same way, astronomers were led to study meteorology in order to foretell favorable observing conditions.) At the Lick, Holden decided that it was essential "to keep a register of all earthquake shocks in order to be able to control the positions of the astronomical instruments."[72] To do so, he furnished the Lick and several other locations in the San Francisco area with Ewing duplex seismometers—an apparatus held stationary by two coupled pendulums, relative to which horizontal ground motion was automatically recorded on a smoked glass plate. He also began to compile a list of California earthquakes reaching back to the eighteenth century. He drew not only on newspapers and published catalogs (including those of Rockwood and Trask), but also on informers like the California state mineralogist H. G. Hanks, the San Francisco instrument maker Thomas Tennent, and the historian H. H. Bancroft, and on "verbal accounts from various gentlemen."[73] Like several of these predecessors, Holden seems to have embarked on his research already convinced that California's earthquakes were nothing to be feared. "When we take into account the whole damage to life and property produced by all the California earthquakes, it is clear that the earthquakes of California have been less destructive than the tornadoes or the floods of a single year in less favored regions."[74] Holden, however, conceived of a new way to bolster this conclusion: an "absolute" measure of earthquake intensity.

Holden's contributions to American seismology were shaped by a paradox: he craved public attention but had no talent for making friends. His long list of popular writings show him to be a prolific and "facile" writer.[75] Alongside scientific popularizations, he translated poetry and penned magazine articles under a pseudonym. He was better known as the founder of the country's "only truly successful" association for astronomy in this period, the Astronomical Society of the Pacific.[76] Holden wanted the society to be "popular in the best sense of the word."[77] He saw the ASP and its journal as a forum for communication between professionals and amateurs, to the benefit of both. "The few professional astronomers in our midst will here lose that sense of intellectual and professional isolation which is a drawback and a danger. . . . The opportunity to communicate the results of one's work readily and quickly is of the highest value; and 'the end of all observation is communication.'"[78] Like astronomy, seismology struck Holden as a field peculiarly suited to amateur participation. As he became interested in earthquakes as geophysical phenomena (and not just as a disturbing factor), he recognized a new opportunity for popular science. In an 1889 article for *Century* magazine on "Earthquakes and How to Measure Them," he appealed to "occasional observers, whose records may be of the greatest value." Rather than soliciting felt reports, however, Holden recommended the purchase of a duplex seismometer. "The instrument is extraordinarily simple and inexpensive, and requires next to no attention. It will give me pleasure to advise with any one [*sic*] who may feel willing to undertake observations of this kind."[79]

On the surface, then, Holden's efforts on behalf of seismology resembled those of the Swiss Earthquake Commission. He reached out to the public, even to "occasional observers," and sought to educate them about the nature of earthquakes. Like the Swiss, he saw this as a bottom-up project: "If results can be attained at all, we must begin by studying the statistics of small regions."[80] Unlike the Swiss, however, Holden defined his goal for seismology narrowly: a "mechanical" account of the propagation of shock waves. Holden tried to distinguish further between the "geologist's" interest in the *cause* of seismic shocks and the "mechanician's or physicist's" interest in how shock waves *propagate*. He struggled to relate this distinction to a presumed contrast between "popular" and "scientific" language:

There is little or no confusion of meaning produced by the use of the term earthquake in ordinary speaking and writing, but the moment an accurate scientific definition is attempted the term comes to include much more than is ordinarily meant. Any mechanical disturbance whatever, either on or within the surface of the earth, sets up a state of elastic vibration which is propagated to all adjacent parts of the crust by elastic waves which may or may not be evident to human senses. This motion constitutes an earthquake. Scientifically, therefore, an earthquake is the result of any elastic vibrations in the earth's crust, whether they are produced by volcanic eruptions, by the sliding of great strata . . . or even by the tread of a foot. In popular language, however, we are in the habit of restricting the use of the word earthquake to comparatively violent motions of short vibration-period which extend over a considerable area, and especially to such motions as are produced by some-what obscure causes. It is necessary to our popular use of the term earthquake that the cause, while natural, should be somewhat obscure.[81]

Holden's contrast between "scientific" and "popular" language remained elusive. Apparently, the term "earthquake" was "popular" because it referred to "obscure" causes. In what sense, though, were these causes obscure? Holden did not explain. Likely, he would not have been able to. He was searching the seismology of his day for a clear line between a "scientific" and a "popular" perspective; but that division simply did not exist.

From Holden's "mechanical" perspective, eyewitness reports of earthquakes were at best a supplement to the spotty coverage of seismographs—at worst, a source of exaggerations and distortions. Holden, a humanist and language lover, did not fail to appreciate earthquake narratives as a literary genre. But he drew a firm divide between earthquake "narratives in which the human element enters," proper to the domain of poetry, and earthquake "statistics," containing information of use to science.[82] Holden would rely on human observations only if they could be reduced to their instrumental equivalents. "In the investigation of the effects of any particular earthquake it is of prime importance to escape as soon as possible from the exaggerated accounts of special correspondents or of the inhabitants who wish to magnify the importance of their neighborhood or of themselves, and to obtain a numerical and quantitative basis for such comparison."[83] Like generations of astronomers before him, Holden hoped to replace unreliable observers with self-registering instruments. To do so, he intended to deduce mechanical equivalents for each Rossi-Forel degree of intensity. He began by defining intensity as the maximum acceleration of the seismic wave.[84] (This was a correction to his earlier statement that intensity could be measured by the length of the line traced by the duplex seismometer, i.e., the maximum displacement.)[85] If defined as the maximum acceleration of the ground wave, intensity could be calculated from the maximum period and displacement of the largest wave, quantities that could be directly recorded by existing instruments. But Holden never paused to justify his claim that the *maximum acceleration* really expressed the *intensity* of a shock.[86] Seismological practice suggested otherwise. Mallet, for instance, had assumed that intensity could be measured by the distance an object was thrown by a shock: that was proportional to the *maximum velocity* of the ground.[87]

Seismology's challenge, as the American engineer T. C. Mendenhall soon pointed out, was to clarify the term "intensity." Mendenhall observed that his colleagues tended to confuse two fundamentally different meanings of the term. One was the "destructiveness" or "destructivity" of the seismic wave: this measured the earthquake's "power to destroy" and was appropriately associated with the maximum acceleration of the ground, as in Holden's definition of intensity. The other meaning was Dutton's: "the magnitude of the [earthquake's] subterranean cause," or "the energy involved in an earthquake." If the seismic wave was assumed to be approximately harmonic, like a vibrating string, one could apply to it the mathematics of nineteenth-century acoustics. True, the amplitude and period of the seismic wave at the surface varied greatly from one point to another, but this was likely the result of variations of the elasticity of the ground, not of the subterranean wave. On this acoustic model, seismic intensity would be defined in terms of the time-averaged rate at which the wave transfers energy through a region of space. This, in Mendenhall's judgment, was "the most important sense" of intensity in seismology, and a meaning far removed from the "destructivity" of an earthquake. Mendenhall went on to apply his insight to a dizzying series of calculations. Using data from Tokyo seismographs on the Japanese earthquake of 1887, he determined—though admittedly only to within an order of magnitude—the "mechanical value of a cubic mile of earthquake." This, he showed, was equivalent to the free fall of "a cube of rock one thousand feet on each edge, the mass of which would be 75,000,000 tons, through a vertical distance of about 166 feet." He even took the time to translate that figure into horsepower.[88] As his delight in these numbers makes clear, Mendenhall's definition of seismic intensity as "the energy involved in an earthquake" was part of a nineteenth-century obsession with measuring, harnessing, and exploiting natural sources of energy. But the significance of his calculations for the history of seismology is more specific. What Mendenhall, Holden, Dutton, and their colleagues had begun to do, in the guise of forging an objective language for seismology, was to drive a wedge between two perspectives that had previously been fused: the geophysical question of the origins of an earthquake and the practical question of the earthquake's effects.

Holden would not be the one to carry this research program forward. In the late 1880s and early 1890s, just as he was getting his seismological project off the ground, he found himself presiding over a fast-growing community at the Lick. Scientists brought wives, children, and pets, and with them new challenges for Holden as director. In the close quarters on the mountain, Holden's strident manner became a tragic flaw. Even an admiring colleague observed in a biographical note that Holden "never seemed able to make an intimate and lasting friendship."[89] In one case, an astronomer's son developed malaria, which the father blamed on poisoned well water at Mount Hamilton; Holden refused to have the well cleaned, even after testing determined that it was contaminated. Among the colleagues Holden managed to turn against him was George Davidson, another astronomer with an interest in earthquakes who might otherwise have been an effective collaborator. Most significantly, the fights that erupted on the mountain could not be contained there. Holden was attacked in the California press in ways that could not have helped his efforts to cultivate relations with amateurs. He became known, for instance, as "The Emperor of Mt. Hamilton" for restricting public access to the observatory, which the papers duly noted was supported by taxpayers' money.[90] Another sympathetic colleague explained: "The trouble with Holden was that his personal traits offended and antagonized everyone that he had to deal with intimately. They had always done so."[91] Holden was forced to resign from the Lick in 1897 and never obtained another scientific position. He managed to support himself with his pen until his death in 1914, but his seismographic network collapsed along with his scientific career.

Into the Wild

By the time of the founding of the International Seismological Association in Strasbourg in 1901, it was clear that the United States was trailing far behind other countries in seismology. Harry Reid, a professor of geology at Johns Hopkins, attended the ISA's meetings as the American representative. In 1903, he began keeping a record of earthquakes in the United States, drawing primarily on newspaper reports and any instrumental data he could get his hands on. Apparently at Reid's request, the Weather Bureau urged its observers to send more detailed reports on felt motion directly to the director of the US Geological Survey. As of 1905 the only regularly operating seismographic stations in the continental United States were those maintained by Johns Hopkins in Maryland, the Lick Observatory in California, and the Weather Bureau in Washington.[92]

Meanwhile, a truly remarkable seismic event escaped scientific notice entirely. In 1905 Ralph Tarr and his young colleague Lawrence Martin arrived in Alaska to conduct a general geographical survey of Yakutat Bay. The bay lay beneath steep white mountains, on the border between Canada and Alaska, where glaciers stretched from some of the highest peaks down nearly twenty thousand feet to the sea.[93] "Before going to Alaska in 1905 we had seen one account of an earthquake in Yakutat Bay, in 1899, but many of the alleged facts were grotesque and failed even to encourage us to expect earthquake phenomena in the region. It was a thorough surprise to us, therefore, when, early in our work, we came upon clear evidence of recent uplift, in barnacles attached to ledges high above the reach of the present tide and among land shrubs." Suddenly, Tarr and Martin began to look at the beach from a new perspective. Those blue flowers they had earlier seen on rocks high above the water? Only now did they see that these were mussel shells. Now they could make sense of the extensive "benches" that provided them with convenient rocky surfaces for walking beyond the reach of the tides. They continued to collect evidence of recent uplift "until practically every foot of a shore line 150 miles in length had been examined." Several years of subsequent field studies showed how profoundly the earthquake had transformed the face of the land, even prompting the recession of nearby glaciers in the absence of climatic change.[94] It was an investigation driven by the enduring question of the American West, the question of the physical determinants of wilderness versus civilization.

However dramatic the physical evidence, it was their conversations with native canoemen that convinced Tarr and Martin that the 1899 earthquake was responsible for a dramatic transformation of this landscape. Their investigation is a story of scientific forensics and of a slow but sure appreciation of the value of eyewitness testimony. The scientific expeditions to the region in 1895 and early 1899 provided only "strong presumptive evidence" that the changes were of a more recent date. But the natives "state[d] definitely that the uplifts took place in connection with the earthquakes of the fall of 1899 and that there were no similar recent movements before or since." "Natives," wrote Martin in a preliminary report, "notoriously tell you whatever you want them to, especially if they can not speak your language well."[95] There were, however, "several reasons for considering [the natives' testimony] trustworthy. Our questions put to the natives were never in a form to suggest the answer desired. One of our canoemen in 1905,

J. P. Henry, a Sitka native long resident at Yakutat, was able to speak, read, and write English well and understood thoroughly the necessity of accurate information. After he knew of our interest in these phenomena he repeatedly indicated to us, before reaching certain places, that uplift had occurred there, and we never found such statements of his as we could verify to be untrue or exaggerated; moreover, he and other natives know the shores of the inlet intimately, for they canoe there every spring in search of seal and would certainly know when such striking changes occurred." Tarr and Martin therefore accepted the natives' statements without further reserve: "Even if other lines of evidence did not so convincingly point to the same conclusion, we feel that there should be no hesitancy in accepting this testimony of the Alaskan natives as to the date of uplift." The natives also "described the great shaking of the earth, the water waves, the fish killed or left stranded by these waves, the appearance of new islands, the uplift of sea caves and beaches (one of these the beach on which the natives camp each year in the sealing season), the formation of the whitened bryozoan film, and the avalanches." The natives were, in short, "keen observers of nature." The local "white men," on the other hand, residents of Yakutat Village, were likely not familiar enough with the bay to have taken note of such changes—"engaged in fishing or lumbering, [they] concerned themselves little with the wonderful fiord at their doors."[96]

The investigation also depended on the testimony of a group of eight prospectors who happened to be hunting gold and platinum in Yakutat Bay in 1899. Subject to the most violent effects of the earthquakes, these men "were too frightened to notice any change of level or saw too much of more spectacular things to report it." Two of the prospectors published accounts of their experiences in Alaska newspapers. Tarr and Martin interviewed a third, whom they found to be "a very intelligent man, then working as a carpenter in Yakutat." Only the most fortunate conjunction of events allowed the prospectors to escape with their lives. The men had been camped on the edge of a glacier in a fjord of Yakutat Bay when the shocks began. The first strong one came on 3 September, after which the men strung up a seismoscope out of hunting knives, clanging against each other as the tremors began. It was a "rude" instrument, Tarr and Martin commented, "but more delicate than their own perception." On 10 September the knives recorded fifty-two separate shocks between 9 a.m. and 2 p.m. The strongest one lasted nearly three minutes and threw the men to the ground. One of the men described the motion as simultaneously "circular" and "waving up and down like the swells of the sea, only with considerably more energy." They ran from their tents just in time to witness the bay roaring up from below on one side, while the mountains crashed down on the other. It may have been the most extreme display of sheer tectonic power ever seen by human eyes. "About 20 yards back of the beach and above us about 100 yards was a lake about 2 acres in area and 15 to 30 feet deep. This lake broke from its bed and dashed down upon our camp while we ran along the shore and escaped its fury. Everything went before it or was buried by the thousands of tons of rock that came down. This deluge was almost immediately followed by one from the sea. A wall of water 20 feet high came in upon the flood from the lake and carried all debris back over the undulating morainic hills." It seemed to the men that the glacier itself "ran out into the bay for half a mile," though Tarr and Martin judged more likely that a mass of icebergs had been thrown into the sea. What the prospectors knew for sure was that the ground was "cutting some of the queerest capers imaginable."[97] While three of the men made it safely to higher ground, the others were caught running back and forth between the furious waves. Eventually, the earth settled down. To get some rest, the men tied themselves by their clothes to the trees on the mountain side, to avoid "being carried away." Had they not been able to recover one of their boats and rescue a lost canoe, they would have been trapped for good. As they forded their way through the icebergs and toward the village of Yakutat, they saw marks left by the tidal wave "fully 60 feet up the bluffs."[98]

Tarr and Martin judged that the prospectors' story "contributes little of scientific interest."[99] Nonetheless, in their final report, they told this tale as the "thrilling story" that it was, devoting three pages to it and mostly letting the men speak for themselves. The authors did not turn their attention away from the prospectors until the men had made it safely to the village four days after the main shock. By contrast, the authors devoted a mere paragraph of their final report to seismographic records. As Grove Karl Gilbert pointed out in his preface to the final report, few "great shocks" since the advent of seismography had been "adequately" and "directly observed." In other words, teleseismic analysis as yet had little value in the absence of field studies. Gilbert attempted to differentiate between "two directions" in the study of earthquakes: "In its relation to man an earthquake is a cause. In its relation to the earth it is chiefly an incidental effect of an incidental effect."[100] Yet Tarr and Martin's analysis frustrated this distinction at every turn. They showed how an analysis of the earthquake as a geophysical "effect" was inextricable from an appreciation of the event "in its relation to man."

Roughly eight years after the earthquake tore through the region, Tarr and Martin sent a questionnaire to about six hundred addresses in Alaska, the Yukon Territory, British Columbia, and the western United States. The authors targeted a remarkably wide section of the population—including newspaper offices; private individuals suggested by various sources; US and Canadian officials, such as Weather Bureau observers and postmasters; "ministers and missionaries of all churches, and to managers of all canneries, salteries, etc., in Alaska"; "to the secretary of each Alaskan Brotherhood Lodge"; "and to others." An introductory note ended with the request "that you send some reply, even if it be that you know nothing of the matter."[101] Despite the time elapsed, the scientists reaped about 140 substantive responses. Fewer than forty of these reported that no earthquake had been felt, the rest "contained valuable information, either specifying places where we had not previously known certainly that the shocks were felt, or verifying information already at hand, or correcting mistakes printed in sensational contemporary newspaper reports, or referring to still other persons who had valuable information." These replies were "invaluable in determining the boundaries of the region where the shock was sensible to persons and in verifying, correcting, and rewriting many sections of the text, a few of the better replies being quoted in full. Those who filled out and returned the printed circular or showed it to others who did so have conferred a real favor upon all interested in the advancement of knowledge concerning earthquakes."[102]

The Yakutat Bay earthquake came to be cited as "one of the most remarkable for revealing the nature of the earth movements to which the quakings were due."[103] As Martin later observed, "In contrast with practically all the great earthquakes of historic times, the Yakutat Bay shocks of September 1899, stand conspicuous for the absence of loss of life and destruction of property. This was because they took place in an area largely wilderness at that time and because the frontier inhabitants lived in tents, in log cabins, or in low frame buildings."[104] The extent of the reported damage was the "shifting of an uninhabited log cabin," "the cracking of a few chimneys," and "slight damage to a wharf." These shocks were "in the unfortunately small class of world-shaking disturbances of which one may read without turning with a shudder at the loss of human life." Martin might have added that only in such "wilderness" conditions could eyewitness testimony have survived of such cataclysmic convulsions. Human constructions would surely have made the quake lethal.

In 1906, Tarr and Martin were putting the finishing touches on a preliminary publication of their research when a "world-shaking disturbance" struck the booming city of San Francisco. Writing in haste on 19 April, just before putting the manuscript in the mail, the authors added a postscript.

They invoked the American faith that the presence of civilization in California, rather than Alaska, was environmentally preordained. Yet they could muster less conviction than Shaler had in the 1880s. "It seems evident . . . that the [San Francisco] shock is the result of a normal process of mountain-building here as in Alaska." The great difference, they added fatalistically, was that in San Francisco "other geographical conditions have determined that a large city must develop."[105]

TEN

A True Measure of Violence:
California, 1906–1935

"1906 MARKED THE DAWN OF THE SCIENTIFIC REVOLUTION" proclaims a website of the US Geological Survey.[1] No event has received more credit for having jump-started the science of seismology in the United States than the San Francisco earthquake of 18 April 1906. The investigation is cited as the most thorough ever undertaken of a seismic disaster. The evidence collected is hailed as the source of the elastic-rebound theory, the cornerstone of the explanation given by seismologists today for most earthquakes. Yet 1906 would hardly have looked like a revolution to anyone who had followed the European history of seismology since the 1870s. The San Francisco investigation was, above all, an exercise in what Suess and his European colleagues called the monographic method, and the conclusions were essentially an elaboration of his tectonic theory. Nor did the earthquake shake the confidence of California's boosters.[2] No, the designation of 1906 as a scientific revolution expresses an aspiration rather than a reality. Since 1755, the earthquake has been the symbol of choice for an intellectual rupture that produces a "modern" discipline. But the foundations of a scientific discipline are rarely so easily shaken.[3] Well after 1906, indeed into the 1930s, "modern" seismology relied heavily on the eclectic methods of nineteenth-century *Erdbebenkunde*. That very continuity disturbed seismology's newfound American promoters. The history of seismology in California from 1906 through the publication of the Richter scale in 1935 is a story of attempts to distill a "pure" science from one that was deliberately hybrid.

It is also the story of Harry O. Wood, an unlikely hero for this book. Wood was an East Coast boy—born in Maine, a graduate of Harvard College. He failed to earn a PhD and struggled to find academic work in California. All his life his colleagues addressed him as "Mr. Wood," while he addressed them with academic titles. Wood seems to have brought something of a Puritanical spirit with him from New England to Southern California. He was a serious, conscientious man, and an independent thinker. "I am a good deal of an individualist myself," he once wrote, "and so I like to be good-naturedly tolerant of such traits in others."[4] He never married and seems to have had few confidants. He made no mention of family nor friends in his will. Instead, he left everything he owned to the Carnegie Institution of Washington, to support seismological research.[5] "Shy to the point of mortification around women,"[6] Wood was not cut out for public relations. "Wood did not like publicity," comments one historian, "at least in the beginning."[7] His interest in the public and their observations grew gradually, as his research showed him the limits of instrumental research. "Seismology owes a largely unacknowledged debt," attested Charles Richter in 1980, "to the persistent efforts of Harry O. Wood for bringing about the seismological problem in Southern California."[8] But this tribute came fifty years too late. It was in the early 1930s that Richter, Wood's junior collaborator, began to steer California seismology in a direction that Wood resisted. After 1934, Wood watched from the sidelines, the victim of a virus that ate away at his spinal cord. Though he lived until 1958, his subsequent work was, as he put it, "done little by little, with innumerable interruptions and intermissions."[9] He feared that "my own work and effort will have been wasted largely."[10]

Counting Seconds

The earthquake of 18 April 1906 tore a rift in the ground 290 miles long and caused an estimated $20 million worth of damage to downtown San Francisco. For roughly a minute, buildings shook like ships at sea. When the ground came to rest, most structures were damaged but still standing. The real horror was yet to come. Where the quake had cracked gas mains and chimneys, flames began to lick the ruins; within hours, firestorms were engulfing the city from multiple sides. The fires raged for three days, exacerbated in part by the fire department's misguided use of dynamite to create backfires. Crushed, trapped, singed, or asphyxiated, about three thousand people perished in the disaster. An unknown number were killed not by earthquake or fire but by some of the tens of thousands of military troops that descended on San Francisco to prevent the crowd from becoming "turbulent," with instructions to shoot "looters" and anyone else engaged in criminal activities.[11]

In these circumstances, it might have seemed absurd to expect San Franciscans to furnish seismological observations. What's more, the earthquake struck when most residents were still sleeping. The scientists in charge of the investigation complained, "So few people were awake at the time the shock began that but a small proportion of the replies come from people who were in full possession of their observational faculties at the beginning of the disturbance; and of those who were suddenly and rudely awakened, few were sufficiently alert for deliberate perception at the time and had to rely upon a somewhat confused memory for the character of the shock."[12] And yet, accounts of the earthquake are remarkable precisely for their character of "deliberate perception."

Several scientists were immediately alert. Grove Karl Gilbert, for one, counted himself lucky to have witnessed the temblor:

> It had been my fortune to experience only a single weak tremor, and I had, moreover, been tantalized by narrowly missing the great Inyo earthquake of 1872 and the Alaska earthquake of 1899. When, therefore, I was awakened in Berkeley on the eighteenth of April last by a tumult of motions and noises, it was with unalloyed pleasure that I became aware that a vigorous earthquake was in progress. The creaking of the building, which has a heavy frame of redwood, and the rattling of various articles of furniture so occupied my attention that I did not fully differentiate the noises peculiar to the earthquake itself. The motions I was able to analyze more successfully, perceiving that, while they had many directions, the dominant factor was a swaying.[13]

The astronomer A. O. Leuschner was no less precise in his report. His estimate of the duration of the shaking was, improbably, "based on counting seconds while carrying my small children out of the house." Leuschner counted seventy-five seconds, but feared he could have overestimated by ten, it being "safe to assume that I counted seconds too rapidly in the excitement of the moment."[14]

Today, it is easy to poke fun at the "pedantic precision" of these observers.[15] Such jokes made the rounds in 1906 as well. At Stanford, the statue of America's first great naturalist, Louis Agassiz, was thrown headfirst from atop a column of the zoology building. "Many stories were told about Agassiz's natural instinct that when the earthquake came he decided to stick his head underground to find out what was going on in the earth below and with his finger pointing saying, 'Hark! Listen!'" But to mock Agassiz's living counterparts is to miss the ways in which the earthquake called into question what it meant to observe an earthquake scientifically. (See figure 10.1.)

No one had worked harder in this vein than Alexander McAdie, chief of the San Francisco Weather Bureau. The *San Francisco Chronicle* reported

Fig. 10.1. The statue of Louis Agassiz on the Zoology Building at Stanford was knocked down by the earthquake of 1906; Humboldt, next to Agassiz, stayed put. http://commons .wikimedia.org/wiki/File:Agassiz_statue_FN-32903.jpg.

on 24 April that McAdie "stayed by his post in the Mills building until the structure caught fire and he was the last man to leave the building; he has not stopped work. His instruments are destroyed, but the records of the last sixty years are believed to be intact in the safe. He has been able, through the cable station at the [Ocean] beach and through wireless messages sent from Admiral Goodrich, to keep in constant communication with Washington, and only three observations have been lost in thirty-six hours." Even more remarkable were McAdie's personal observations of the earthquake. He began by noting that his "error" was "1 minute slow" according to the time signals received at the Weather Bureau,

with which my watch has been compared for a number of years. The rate of my watch was 5 seconds loss per day. . . . I would say perhaps that 6 or more seconds may have elapsed between the act of waking, realizing, and looking at the watch and making the entry. I remember distinctly getting the minute-hand's position, previous to the most violent portion of the shock. The end of the shock I did not get exactly, as I was watching the second hand and the end came several seconds before I fully took in the fact that the motion had ceased. The second-hand was somewhere between 40 and 50 when I realized this. I lost the position of the second-hand because of the difficulty in keeping my feet, somewhere around the 20-second mark. I suppose I ought to say that for twenty years I have timed every earthquake I have felt, and have a record of the Charleston earthquake, made while the motion was still going on. My custom is to sleep with my watch open, note-book open at the date, and pencil ready—also a hand electric torch. These are laid out in regular order—torch, watch, book, and pencil.

How then was McAdie to explain the fact that his reported time was about a minute later than most? He continued: "However, there is one uncertainty; I may have read my watch wrong. I have no reason to think I did; but I know from experiment such things are possible."[16]

Training, precision, experimentation, and skepticism: McAdie underlined the building blocks of his scientific attitude at this moment of crisis. He held himself to a heroic ideal of earthquake observation. As we have seen, it was an ideal with an illustrious heritage in the memoirs of Humboldt, Darwin, and Muir. McAdie, a founding member of the Sierra Club, once imagined what Muir would have said of a newly reported landslide in the Pamirs: "We would have had a description, both accurate and eloquent, for he would have written into it not only what the eye beheld, but much that other men must have failed to note, because they failed to feel." It was then that McAdie recalled Muir's famous account of the earthquake at Yosemite: "Mr. Muir often described the scene to the writer and fellow members of the Sierra Club. It is plain that after the first two or three seconds of doubt and trepidation, Muir realized what was happening and enthusiastically welcomed such an opportunity for close observation of the swaying trees, and the piling up of the talus by the torrent of rocks from the cliffs, forming a luminous bow as they fell. His intense interest and forgetfulness of self were not assumed, but the natural expression of a spirit all eager to observe and interpret, if he could, the shaking earth and allied phenomena."[17] Perhaps, on that April morning when McAdie's home on Clay Street began to heave, he conjured his friend Muir. At Yosemite, as McAdie pointed out, Muir "was probably the one man in the valley who kept his head."

McAdie's tribute to Muir implied that the act of earthquake observing was not simply a matter of scientific discipline. Muir's attitude was "the natural expression" of his "spirit," not a trained habit; it constituted a "forgetfulness of self" that was "not assumed." Perhaps these descriptions sound familiar. They recall similar testimony, from men like Erwin Baelz, Albert Heim, and William James, of the peculiar objectivity of the human mind in the midst of disaster. Like them, McAdie implied that this demeanor was more a primal instinct than a product of scientific education. Indeed, he hinted that Muir's qualities as an observer were precisely those that were missing from the science of his day—he was not just "accurate," but also "eloquent," expressive not only of "what the eye beheld," but also of what he managed "to feel." In this way, even the most "scientific" accounts of the earthquake raised questions about the nature and origins of acute perception in the face of an elemental catastrophe.

Scientists were not alone in staking a claim to an alert, expansive, and impersonal perspective on the 1906 earthquake. The event seemed custom made for youth with "literary aspirations" like the writer Kathleen Norris. Like five of her young literary friends, Norris experienced the earthquake as an "unmitigated delight." "How I wish that to every life there might come, if once only, such days of change and freedom, so deep and intoxicating a draught of realities, after all the artificialities of civilization and society."[18] On the afternoon of 19 April, she and her friends sent off their first stories of the disaster: "We realized that here was our golden opportunity, and we lost no time."[19] The writer Gertrude Atherton, who lost her home in the fire, likewise seized on the earthquake as a literary windfall. *Harper's Weekly* soon quoted her advice that there was "no better 'cure'" than an earthquake "for those that live where nature has practically forgotten them." The earthquake reappeared in Atherton's novel *Sisters-in-Law*, which opens with a young girl out alone after midnight for the first time. The anticipated scene of sexual awakening turns into quite a different form of liberation:

Alexina was a child of California and knew what was coming. She barely had time to brace herself when she saw the sleeping city jar as if struck by a sudden squall, and with the invisible storm came a loud menacing roar of imprisoned forces making a concerted rush for freedom. She threw her arms about one of the trees, but it was bending and groaning with an accent of fear, a tribute it would have scorned to offer the mighty winds of the Pacific. Alexina sprang clear of it and unable to keep her feet sat down on the bouncing earth. Then she remembered that it was a rigid convention among real Californians to treat an earthquake as a joke, and began to laugh. There was nothing hysterical in this perfunctory tribute to the lesser tradition and it immediately restored her courage. Moreover, the curiosity she felt for all phases of life, psychical and physical, and her naïve delight in everything that savored of experience, caused her to stare down upon the city now tossing and heaving like the sea in a hurricane, with an almost impersonal interest.[20]

In the course of this passage, sexual imagery—the trees "bending and groan-ing" in Alexina's embrace—gives way to something more radical: a natural world that calls on the young girl to abandon the conventions of femininity. Her laughter expresses "courage," not "hysteria," and her attitude of "curios-ity," "naïve delight," and "impersonal interest" marks her as an observer in the tradition of Muir and James. What unites their earthquake accounts and Atherton's is the "impersonal" quality of their attention, their "forgetfulness of self"—that state of pure objectivity that Albert Heim described as charac-teristic of near-death experiences.

The Report

"There was no hysteria, no signs of real terror or despair," recalled Arnold Genthe, who photographed San Franciscans gaping at the spectacle of the fires.[21] What to do with their testimony, however, was not immediately clear to the experts who set out to investigate the earthquake—commissioned by the state government and funded by the Carnegie Institution of Wash-ington. "Many of these replies are rather questionable scientific evidence," the commission judged, "inasmuch as many of them were in response to a leading and suggestive question, and very few of them have been subjected to the clarifying process of cross-examination."[22] The commission was de-termined to distance its research as much as possible from the sensational reporting of the press. Commission-member J. C. Branner explained: "The picturesque and sensational features of earthquakes are abundant and en-tertaining, but to the geologist these features have only a passing and acci-dental interest. For example, if a chimney top, broken off by an earthquake, should fall on a man in such a fashion as to go right over his head and leave him standing unhurt in the flue, it would be a striking, and to the man a very important, fact; but, from the geological point of view, its only impor-tance would lie in the fact that the shock was severe enough to throw down the chimney."[23] Like Kant in 1755, Branner initially attempted to divide the "geological" point of view from the human one.

Working with felt reports was a learning process for the commission members, several of whom were new to seismology altogether. At first, they found much of the testimony hard to believe. For instance, they doubted widespread reports of "visible undulations of the ground." By their esti-mates, the velocity of seismic waves in the crust—approximately two to three kilometers per second—was "so swift that they would scarcely be ob-served visually." Still, they found "considerable testimony, of a consistent and independent character, that much slower undulations were observed . . . a great deal of it is positive and unequivocal as to what seemed to be the fact. The evidence suggests that there is a type of wave in the ground, in the region of high intensity, which has not yet been sufficiently recognized, and the origin of which is obscure."[24] Based on the ratio of horizontal compression to vertical expansion at the earth's surface, the ordinary longitudinal seismic waves could indeed be expected to produce surface waves of approximately 1.33 inches. However, "it is not necessary to believe that the amplitudes of surface waves are nearly as large as they appear, for it must be remembered that an observer being shaken by the strong vibrations of a violent earthquake is in a difficult position to make good observations on the phenomena about him, and particularly to distinguish between the movements which are actually taking place and those which he apparently sees, but which are really due to his own oscillations."[25] The question of the reality of visible surface waves and their "obscure" origin was a persistent motivation for the analysis of felt reports over the following decades.

Reports of earthquake sounds also aroused skepticism at first. In an early publication, McAdie doubted that these sounds were a genuine seismic phenomenon: "in our judgment most of these sounds can be explained by the noise due to violent shaking of dwellings."[26] In the course of the investigation, however, the commission concluded otherwise. Of eighty-one people reporting having heard a sound accompanying the shocks, forty stated that the sound preceded the jolt, and "evidence as to the character of the sounds is consistent and uniform." These were low vibrations, "below the range of audibility of some people." This limit would explain the fact that the earthquake seemed not to have been heard in some places where its effects were seen. The commission went so far as to inquire of a professor of psychology at Johns Hopkins "in regard to the limit of sound." They were told that individuals varied as to the lower limit and were referred to Helmholtz's classic treatise on sensations of tone.[27] Eventually, scientists were willing to believe that San Francisco residents might indeed have seen the earth undulate and heard it roar.

For the harder hit regions, the distribution of intensity could be ascertained from surveys of structural damage alone. It was on this basis that Harry Reid developed his seminal theory of elastic rebound. He employed the "absolute scale of destructive earthquakes" that Fusakichi Omori had constructed from experiments with a shaking table. Assuming that most ground motion is horizontal, Omori determined the ground acceleration necessary to overturn or fracture brick columns of varying dimensions and quality. He correlated these laboratory results with field evidence from the Mino-Owari earthquake of 1891, producing a scale of seven degrees—from a "strong" earthquake that slightly cracks brick walls "of bad construction" (maximum acceleration no more than 300 mm/sec^2) to the upper limit of a "violent" earthquake that destroys all buildings, "except a very few wooden houses" (maximum acceleration above 4,000 mm/sec^2).[28] Reid used Omori's scale to translate damage reports into values of maximum ground acceleration. He then used the method Dutton had introduced in 1886 to calculate the approximate focal depth of the shock—a shallow fault of only twenty kilometers. This estimate of the depth figured directly into his calculation of the "work done by the elastic stresses."[29] Reid thus worked primarily with data at the high end of the intensity scale, where he could rely on architectural damage as evidence.

In order to trace fault lines beyond the hardest hit regions, however, it was essential to map the distribution of weaker effects. This meant working at the more ambiguous, lower end of intensity scales. In Sacramento, intensity was estimated between 6 and 7 on the Rossi-Forel scale (6: "general awakening of those asleep," "some startled persons leaving their dwellings"; 7: "general panic"), but at Santa Barbara it dipped down to 4 ("felt by persons in motion"). Effects were reported as far as 340 miles east of the fault. "Farther east the most notable feature of the reports is that wherever the effects of the earthquake were made evident, the physical signs, such as the swinging of suspended objects, etc., were described almost to the exclusion of direct physiological effects." George Louderback of Berkeley explained this contradiction to the Rossi-Forel scale by noting that the region was sparsely inhabited and its residents largely asleep when the shock occurred; "the few who were up were moving about at active work and were in general not of a sensitive type." A few people engaged in irrigation noticed slight waves in the water, an odd sight on a still morning. But those who mentioned it to others were met with "sallies of wit at the expense of the reporters."[30] There was thus no shortage of frustrations for researchers measuring weaker intensities east of the Sierra Nevada.

Indeed, it became evident in the course of the investigation that the commission was uncomfortable with the very concept of intensity. Lawson's introduction cautioned that intensity as measured by field observations was to be regarded as mere "apparent intensity," in contrast to the "real intensity" corresponding to the "energy" of the "earth-waves." "Inasmuch as we have to deal primarily with observable effects and record these as a basis for inference, it has been found convenient to use the term 'apparent intensity' in a technical sense throughout this report." "Intensity" as used in the report was "arrived at by applying *literally* the criteria of the Rossi-Forel scale," a remark echoed later in the discussion of the plotting of iso-seismals.[31] Clearly, the investigators did not appreciate that the Rossi-Forel scale and its relatives exist solely in order to be applied *literally*; their criteria derive from the reports of ordinary observers, as codifications of typical vernacular reports. One researcher was even assigned by Lawson to determine a coefficient for converting "apparent intensity," as reported on alluvial surfaces, to a "real intensity" corresponding to some ideal terra firma.[32] It was a quest that California's seismologists would pursue fitfully for the next three decades.

Despite such hesitations, the commission's final report consists in large part of the stories of survivors. It is two volumes and over six hundred pages of field observations, eyewitness testimony, photographic plates, and theoretical synthesis. As the British seismologist Charles Davison noted in 1925, "No report on any previous earthquake has been issued on so liberal a scale." The first volume, all 450 pages, is sold as a paperback today with a full-color cover view of sailboats on a rocky bay. The report synthesized observations from no fewer than three hundred individuals. As Davison observed, "Whenever possible, they have been allowed to speak for themselves in short notes and papers, so neatly worked into the text that, in reading it, there seems to be no breach of continuity." In this way, the report was able to follow the effects of the quake all along the San Andreas fault from the Mexican border to the Pacific. As Perry Byerly would note years later, "the earthquake itself had a certain simplicity—one unbelievably long fault which over much of its length was a single surface break."[33] The Lawson report has since acquired iconic status. It is perhaps the single best instantiation of Ernst Mach's ideal of complete knowledge of an earthquake.

Rebuilding a City and a Science

Lawson's commission had every right to expect that their massive research effort and meticulous survey of structural damage would lead once and for all to a reform of construction practices in the Bay Area. What they didn't foresee was the continuing strength of the boosters. In the logic of capitalism, destruction was conceived as an opportunity for modernization, not for pausing to take stock.[34] The policies of the insurance industry also promoted the continued denial of seismic risk. Insurers refused to pay claims for damage resulting from the earthquake, defined as an "act of God." Property owners therefore claimed that damage was the result of fire. Already on 21 April the *San Francisco Examiner*'s headline announced "The Water Front Destroyed / To Resume Business at Once." The conservationist Mary Austin noted a sign in one ruined building: DON'T TALK EARTHQUAKE / TALK BUSINESS."[35] Austin was an eloquent critic of this rush to rebuild. As she pointed out, "the greater part of this disaster—the irreclaimable loss of goods and houses, the violent deaths—was due chiefly to man-contrivances, to the sinking of made ground, to huddled buildings cheapened by greed, to insensate clinging to the outer shells of life . . . for most man-made things do inherently carry the elements of their own destruction. How much of all that happened of distress and inestimable loss could have been averted if men would live along the line of the Original Intention, with wide, clean breathing spaces and room for green growing things to push up between?"

Austin's critique reflected lessons she had learned from Native Americans, whom she credited with knowing how to "live off a land upon which more sophisticated races would starve, and how the land itself instructed them."[36] She was not entirely alone in seeking out the perspective of Native Americans after the devastation of 1906. Anthropologists interviewed the few surviving members of the Wintun tribe, located about eighty-five miles north of San Francisco. The tribe was of two minds on the meaning of the earthquake and its many aftershocks. To some it seemed to be the onset of the "great levelling" of the world, which would flatten the mountains and possibly destroy all life in the process. Others believed it betokened the "stretching" of the world by Old Coyote Man, in order to make room for the growing numbers of whites. Yet the Wintuns agreed that "ultimately there would be a great upheaval and levelling which would obliterate all things at present upon the earth." The anthropologists seemed to sympathize with the Wintuns' tragic vision.[37]

Natural scientists tended to be more optimistic. The geologist T. C. Chamberlin, better known for his theory of the greenhouse effect, suggested that the earthquake might liberate the public from unwarranted fears: "If, for instance, it shall later be shown, as I think not improbable, that the earth is now in a general way receding from a period of special deformation into one of relative quiescence, and that catastrophic action is on the decline, it will be a contribution of no small value to the comfort of mankind. The public is now very generally depressed by needless apprehension of great impending disasters, if not a universal and final catastrophe, apprehensions derived from the narrow and pessimistic views of the past. From my point of view, which is doubtless a partial one, a contribution of supreme value to the happiness and well-being of mankind is likely to grow out of rectified views . . . derived from the prosecution of the earth sciences."[38] Like ancient natural philosophers, Chamberlin viewed science in ethical terms as an antidote to fear.

The members of the Lawson commission shared this cautious optimism. They might even have been mistaken for boosters themselves, since they claimed that the quake had released the stress on the San Andreas fault for a long time to come.[39] Geschwind has characterized them as typical Progressives, valuing "objectivity, efficiency, and expert guidance."[40] However, Geschwind also makes clear that it took time for these scientists to settle on what he calls their "progressivist" strategy—meaning "deference to expertise" rather than "grassroots organization and protest," and the incorporation of seismologists into the "regulatory-state apparatus."[41] Otherwise known as technocracy, this solution was not a foregone conclusion.[42]

The Seismological Society of America

From 1906 to 1933, these scientists were experimenting with various approaches to the politics of seismic safety. At first, they turned hopefully to the public for support. In July of 1906, Lawson, Reid, and Leuschner, joined by Weather Bureau officials McAdie and Marvin, began plans for what would become the Seismological Society of America. The society was formally founded in late August, with the aim of "the acquisition and diffusion of knowledge concerning earthquakes and allied phenomena and [enlistment of] the support of the people and the government in the attainment of these ends." The annual dues were set at just $2, cheap compared to those charged by the Astronomical Society of the Pacific and the Sierra Club at the time ($5 and $3, respectively).[43] Scientists made up fewer than half the members, with the rest drawn primarily from engineering and architecture. Indeed, McAdie later credited the idea of the SSA to his next-door neighbor, W. R. Eckart, an engineer at the Union Iron Works whose meteorological instruments McAdie borrowed during the fire of 1906.[44] John Muir joined the society, as did other members of the Sierra Club. From the start, the plan was to organize an instrumental network to record both local and distant earthquakes, as well as a network of "200 to 300 cooperating observers who, every time they felt an earthquake, would report the quake's time of occurrence, duration, intensity, and other pertinent information to the central bureau. In this way . . . a complete catalogue of earthquakes on the Pacific Coast might be assembled."[45] The term "cooperative observers" suggests that the SSA modeled its network on that of the Weather Bureau.

Previous catalogs of California earthquakes had been designed to prove the insignificance of the seismic threat. The SSA's founders hoped instead to convince the public of the need for seismic safety measures.[46] They did not mean to cause alarm, nor to halt development. Instead, they aimed "to supplant any element of terror or helplessness which results from imperfect knowledge by an interest in natural phenomena and a sense of security resulting from familiarity with the facts and the taking of reasonable precautions."[47] Or, as J. C. Branner put it as president of the SSA in 1913, it was necessary to locate seismic faults "so that we can keep our houses, bridges, dams, pipe lines and other structures off them, or, we can do our engineering so that, when the next earth-slip comes, the damage will be negligible." In the society's *Bulletin*, the reports of the cooperating observers were "probably read more widely by the general membership of the society than the 'scientific' articles."[48]

The scientist in charge of collecting the volunteers' observations was Alexander McAdie, the man who slept with his watch, notebook, and pencil at the ready. McAdie's research interests lay in such practical areas as storm prediction and agricultural climatology. He was "an able interpreter of his science to the public," as a colleague commented.[49] In 1899 he published an article in *Century Magazine* on "Needless Alarm during Thunder-Storms." There McAdie distinguished between the "depression of spirits which is physical and real, brought about by some as yet unknown relation between the nervous system and conditions of air-pressure, humidity, and purity" and the "unnecessary" fear that was "largely the work of the imagination." McAdie hoped to teach the public to make rational distinctions between, in his terms, appropriate and "needless" fear.

McAdie and his fellow members of the SSA did not consider fear to be an irrational response to geophysical hazards. On the contrary, the society was invested in documenting an emotional response that was understood as "physical and real." At the same time, the SSA sought to serve the needs of industrial development. By no means did its members abandon the principle, long associated with Buckle, that fear of earthquakes could discourage investment and cripple a capitalist economy. As J. C. Branner would put it during his presidency of the society, "The more we know about them [earthquakes] the less harm they can do us, and the less reason we shall have to fear them."[50] The key, then, was to assess just how much fear was appropriate.

Branner's Society

By early 1911, when J. C. Branner took over as the SSA's president, the society appeared "moribund."[51] Branner immediately launched a campaign to revive it. In just a few months, he succeeded in doubling the society's membership. He added, "The Berkeley crowd seems to have wanted the society for personal use. I want it to awaken and keep alive an interest in seismology."[52] Though he failed to meet his goal of one thousand members, he had soon raised the number from 143 to four hundred, where it remained until the late 1920s.

Branner was, in his own estimation, a child of an oral culture. He had spent his early years on a Tennessee farm in the 1850s. He had little access to books as a boy, and instead delighted in the stories told by slaves. In 1921 he published a collection of these "How and Why Stories," transcribed in full dialect. In a preface he explained that it "seemed best to write them down as nearly as possible in the spirit and language in which they were told me without concerning myself with inconsistencies of which the narrators themselves were not aware."[53] As a product of the antebellum South and a critic of Reconstruction, Branner was also adept at casting suspicion on the federal government and positioning himself as a defender of the people. More than any other California seismologist of his era, he had a politician's instincts. At the time of the 1906 earthquake, his rhetorical skills were on display in a conflict with the director of the US Geological Survey, Charles Walcott. Walcott was "a skilled scientist-politico who was wont to breakfast with a congressman, a senator, or a President."[54] Walcott informed Branner that the USGS planned to take over the survey of Arkansas coal fields (which Branner had led in the 1890s), and forbade him from publishing anything on the topic. In a series of letters to *Science*, Branner framed Walcott's move as an "invasion" by the federal government of the province of a state survey: the issue was "not a question of geology, but a question of the administration of a public bureau."[55] Branner styled himself as a representative of "the people of Arkansas" against an expansionist federal office. He went so far as to accuse the USGS of being a trust. "Trusts and trust methods are in the air," he declared, but so are "protest, rebellion and resentment against these high-handed methods."[56] Branner's dispute with Walcott would return to haunt him after he took over the leadership of the SSA.

Branner was convinced from the start that the society would need to enlist a large number of cooperative observers. Perhaps he was swayed by his first experience with a seismoscope, a year after the Charleston earthquake of 1886. The journal in which he had intended to record tremors ended up with the title "Troubles with a Seismo, by J.C.B., 1887–8."[57] In 1909 he argued that locating faults could not be accomplished instrumentally, at least not for a "young" organization with limited funds. There was, however, "one excellent kind of a seismograph . . . that we can all use to great advantage if we will only set about it, and will take the trouble to put the records on paper and send them in. I refer to our own bodies."[58] It therefore seemed natural to Branner that seismology should be administered by the Weather Bureau, with its network of volunteer observers.

Soon after becoming president of the SSA, Branner pressed this view on the Weather Bureau's chief, Willis Moore. The bureau, Branner wrote, with its "large working organization covering the entire national domain, its permanent stations, and its intelligent observers trained to make observations and reports every day only needs a slight broadening of its field and the necessary equipment, funds, etc. to make it immediately the most effective organization in the world for gathering and using seismological data."[59] To Senator Frank Briggs, member of the Committee on the Geological Survey, Branner urged that "seismological investigations, in order to be efficient, must have a large number of observers, and that those observers must be accustomed to making, recording, and sending in certain observations." Emphasizing the value of lay volunteers, he noted that "these observers do not need to be geologists, but it is of the utmost importance that they be numerous, and that they be widely distributed over the area to be studied. It is only by such means that it is, or can ever be, possible to locate the seismologically active faults, zones or centres." Otherwise, he warned, "the machinery of the Weather Bureau would have to be practically duplicated at great expense and with much delay."[60] Of course, the Weather Bureau had collected seismological observations for years in an unsystematic way and without expending any significant portion of its budget to do so. Making this function official would hardly seem like a matter for controversy.

In 1909, however, Charles Walcott had succeeded Samuel Langley as director of the Smithsonian Institution. Walcott was soon lobbying Congress to establish a national seismological service—under the auspices of the Smithsonian. Branner feared that his opposition to Walcott's plan "is liable to be regarded as a purely personal affair which it is certainly not."[61] In 1910 the Treasury Department explicitly barred the Weather Bureau from applying its funds to earthquake research. The *New York Times* reported the verdict with a jeer: "The Controller says seismology relates to what is underground, while meteorology relates to conditions above ground. Therefore the study of seismology is not comprehended in the work of the Weather Bureau." An editorial described it as a decision made "in a moment of truly departmental inspiration." Surely the Treasury's controller would now be invited to join the world's elite scientific societies: "Possibly they will first

want to know, as a mere formality, how he discovered that there is no relation between weather and earthquakes . . . the Controller's competency to
decide the matter being universally recognized."[62] This was just the kind
of political control over scientific research that American scientists of the
Progressive Era could not abide.[63]

Finally, in 1914, Congress reversed the 1911 decision and approved the
Weather Bureau's bid to nationalize seismological research. Branner wrote
a relieved letter to C. F. Marvin at the bureau: "I should tell you frankly that
I have hitherto found the English language quite inadequate to express my
indignation at the attitude of Congress toward the Weather Bureau's work
on earthquakes. . . . You can count on the cordial support of intelligent
people out here, I am sure. . . . I think you are quite right to begin with
non-instrumental reports. I am confident that the results will be well worth
while. Instruments are too expensive to undertake in the early part of the
work."[64]

The cooperatives had won the day. Each of the more than four thousand
volunteer observers across the nation was furnished with cards for reporting ground movement, to be collected by their regional section directors.
Farmworkers and women were still well represented among the cooperatives, as in the late nineteenth century. The Weather Bureau hoped that
an instrument would soon be available that could "easily be cared for by
inexperienced individuals and that will give a trustworthy measure of the
intensity of local shocks." Such an instrument, however, was "more or less
completely unavailable at the present time."[65] Felt reports were thus the
bureau's priority. In California, 167 cooperative observers agreed to begin
reporting earthquakes to the bureau in 1915, in addition to the twelve regular weather stations.

Their reports went to Andrew Palmer, chief of the San Francisco office
of the Weather Bureau. Despite the disruptions of the war, the number of
cooperatives participating reached 350 by 1919.[66] Palmer vouched for the
integrity of these volunteers. "As it is recognized that psychological factors
play an important part in the recording of sensible earthquakes, the character of the observers deserves special consideration." Palmer praised the
cooperatives in the terms traditionally used at the bureau: "Nearly all of
these observers render both climatological and seismological reports without compensation, and this fact alone indicates their interest in and capacity for the work. Actuated largely by public spirit, these observers are almost
without exception leading citizens in the various communities which they
represent. Furthermore, the care required in the daily meteorological observations is a form of discipline which soon makes one exercise good judg-

ment in the recording of natural phenomena."[67] However, like so many California scientists before him, Palmer thought he knew all the results already. At the end of 1915, he was confident that no sensible earthquake in California had gone unreported. More shocks had been noted on the state's 5 percent of the total area of the United States than in the other 95 percent.[68] Yet his report would be "incomplete" if he did not add that California's earthquakes were less hazardous than hurricanes and tornadoes elsewhere in the country. Indeed, his "inevitable conclusion" after five years of macroseismic surveys was that California's tremors possessed "a constancy from year to year," indicating that "these slightly but constantly recurring tremors may well be regarded as a safety valve in efficient operation."[69] Palmer offered the public little incentive to report on the weaker movements that could have helped locate faults.

At the national level, oversight of the bureau's seismological program was handed to William J. Humphreys, an atmospheric physicist who had trained in the high-precision physics laboratory of Henry Rowland. Humphreys defined "modern seismology" in 1914 as a highly technical, instrumental science. It was "so very modern as to require considerable liberality in conceding it an age of even 30 to 40 years." "Modern seismology" had originated with the first instruments capable of detecting otherwise insensible tremors, as well as the geophysical expertise and "none too easy mathematics" necessary to interpret the instrumental records. Yet Humphrey worried that "modern seismology" had thereby acquired an image problem. "From this it might seem that seismology is an ideal subject for the private diversion of the abstract scientist, as indeed it is. Those who attempt difficult problems for the mere exhilaration they afford, or revel in the luxury of intricate equations, can find in seismology every excuse for endless self-indulgence." Clearly, such a field would be hard-pressed to attract either government funding or popular participation, especially with a war brewing in Europe. But there was another side to seismology, Humphreys remarked, one that could in fact engage "the engineer" or "the man of affairs": the location of active faults. This was a topic of immediate relevance to "a careful engineer" contemplating the erection of a bridge across a fault, or to "a properly informed and prudent banker" asked to invest in such a project.[70]

Under the Weather Bureau, then, earthquake observing was defined narrowly at both the state and national levels. It was a practical effort, divorced from the "modern" science of seismology, and serving the needs of engineers and businessmen. In California, it was expected to produce only further evidence of long-term environmental stability.

"Making It Local"

Even as he fought political battles in Washington, Branner pursued the other prong of his seismological program: enlisting local observers. His correspondence from 1911 and 1912 is filled with letters soliciting new members for the SSA, as many as 2,500 of them.[71] These efforts flagged only after 1913, when his new duties as president of Stanford University took precedence. Branner expected that his correspondents would not immediately recognize the value of felt reports. He explained that the investigation of an earthquake should collect "any facts that can be had of residents. Such notes seem worthless at first glance, but it often happens that we are enabled by them to determine the area and intensity of shocks."[72] Branner quickly perfected a pragmatic, antielitist tone and a grassroots rhetoric. One letter to a J. S. Rossiter in Pasadena announced, "As president of the Seismological Society of America I am trying to get the people of this state interested in earthquakes. To that end we want to get as many members as possible so that we can after a while organize some sort of systematic collection of data. . . . if you give me the names of a few others who will help I shall be greatly obliged to you. No particular skill or knowledge is required."[73] When Rossiter agreed to join and mailed his two dollars the following week, Branner replied, "It is pleasant to run across people who take a rational view of earthquakes. We cannot conceal from ourselves or from others that we have them, and the reasonable thing seems to be to study them and to find out how we can prevent their doing serious damages. That is what sensible people do about disagreeable things of all sorts. If we can get the cooperation of a large number of persons on this coast we shall soon have the earthquake business run into its hole."[74] A letter to a mining engineer expressed a modest view of seismological expertise: "None of us knows much about earthquakes, but if we all try to find out we hope to know something after a while."[75]

By the summer of 1911, then, a preliminary network was in place. On 1 July, a strong earthquake struck central California—strong enough to throw the seismographs at the Lick Observatory and at Santa Clara College out of order. As Harry Wood put it, it proved impossible to obtain an "unmutilated" seismogram of the event. It would have been an ideal test case of Branner's observing network, but for the fact that Branner was out of town, on a research trip to Brazil. Without his prompting, the network failed to spring into action. The collection of reports did not begin until Branner's return in late August, when he assigned the task to one of his graduate students. The conclusion was that the earthquake had reached a maximum intensity of Rossi-Forel 8–9 and was palpable over a region of about four hundred miles at its widest. There was no good reason for it not to have been reported by the SSA's observers.[76]

By early 1913, Branner was complaining publicly about the failure of his observing network. "The collection of information on the west coast of North America in regard to earthquakes is not as simple and as easy as it looks at first glance." A first problem, he explained, was that weak shocks were so frequent that "most people are accustomed to them. . . . It is a common experience to hear a remark like this in the middle of a conversation: 'By the way, did you feel that earthquake last night?' and after a yes or no, the conversation goes on without further interruption." Most earthquakes did not seem to most people to be "worthwhile" to report. But there was also a more insidious problem at work: the "deliberate suppression of news about earthquakes."[77] Branner was not the first seismologist since 1906 to have raised the charge of seismic denial, and he would hardly be the last.

Branner wanted California's residents and scientists alike to acknowledge what they did and did not know about earthquakes. He stressed that "no one needs apologize for any fact he sends in. To our requests for information about earthquakes we are frequently told apologetically that 'I don't know anything about earthquakes.' There is but one reply to be made to such remarks, and that is that 'we know precious little about them ourselves; we are just now trying to find out, and we want your help.'"[78] One respondent freely admitted his ignorance: "Would you kindly notify me, and I presume the information would be of interest to the other lay members of the Society, just how to recognize the disturbance occasioned by an earthquake. That is, how to differentiate it from similar disturbances. Not infrequently we have a little jar which we wonder about, as to whether it is an earthquake or the result of some blasting or explosion."[79] In another case, an engineer wrote to Branner asking for information on earthquakes in two states where his company planned to construct dams. Branner replied that he was sorry he had no information. He used this exchange in a 1913 article for the BSSA to "urge upon these very engineers the great importance and necessity of their own cooperation. . . . 'Help us and we shall gladly do all we can to help you. If you feel an earthquake, report the time, place, and intensity to the Seismological Society of America.'" It was crucial to "keep up our observations right straight along, year in and year out, whether the earthquakes are big or little."[80] Nonetheless, earthquakes continued to strike without being reported. The entire program was still experimental in 1914. It was still a matter of "try[ing] out the plan of locating the epicenters of our California earthquakes from personal observations."[81]

Branner's team resorted to canvassing the affected areas on foot. They found that people were often more willing to broach the subject of temblors in person. After a quake in the Santa Cruz Mountains in late 1914, Carl Beal noted that "only about twenty observations and notes were sent to the Seismological Society," likely because of the low population in the mountainous region of highest intensity. "The writer therefore spent four days in the field, and gathered from the people themselves most of the information contained in this brief account." One "interesting and valuable contribution to the subject" came from a young woman at Stanford who was phoning a friend in San Francisco: the friend said she would have to hang up, there had been an earthquake; it was only after a pause that the woman at Stanford exclaimed, "Here it is now!"[82] In January 1915, Beal spent nine days collecting information on a quake in Santa Barbara County. The locals were mainly ranchers, "who cooperated with the writer in the most courteous and efficient manner." He also interviewed fellow passengers during his train ride south, and contacted others via "the long-distance telephone," since some of the mountain roads were impassable in winter. Beal's published report cited witnesses' statements in detail, noting cases of dizziness, fainting, and nervousness; he noted effects on animals, "peculiar things" such as the starting of a stopped clock, and the fact that "near Santa Rita all the cream was spilled off the milk in a number of pans on a large table."[83] When a severe quake struck southwestern California the following June, Beal knew what he had to do: "It was impossible to get trustworthy information in any way except by going into the district affected and gathering it directly from those who experienced the disturbance"—even though Beal would have to rely on the military governor to give "sufficient guarantees of safety to permit the writer to travel through the northern part of Lower California."[84] Beal's adventures resulted in a preliminary fault map of Southern California in the spring of 1915.[85]

In February 1920, Los Angeles entered a period of heightened seismic activity. Over the following eight months, more than one hundred shocks, mostly of low intensity, were recorded in the city. In June a severe earthquake struck Inglewood, ten miles to the south; on 16 July a series of shocks caused minor injuries and light damage to buildings in Los Angeles's business district. "It has been many years since Los Angeles was subjected to earthquakes of as high an intensity," remarked Stephen Taber, a visiting geologist from the University of South Carolina who had trained with Branner. "People rushed from the buildings after each shock; many women fainted and some had hysteria. Business was practically at a standstill after the afternoon shocks, and many stores and offices closed for the day."[86]

With funding from the SSA, Taber spent a total of seven days investigating these two events. His experience was highly discouraging. He had "considerable difficulty" during this investigation "in securing such data, some people refusing to give us any information, and others giving us incorrect data, probably thinking it none of our business; and that the apparently trivial things we were asking about could be of no possible value."[87] Taber got around this problem in a creative way, using drugstores as his primary indicators of the distribution of intensity. As he explained, there was a drugstore roughly every half-mile throughout the affected area. "These stores are all on the ground floor, and they all have many bottles of different shapes and sizes similarly arranged on shelves." One couldn't ask for a quicker means of measuring relative intensity than counting overturned bottles. In this way he was able to locate the epicenter of the strongest July shock at the prominent fold in Elysian Park. The folds and faults in this area were known at the time as obstacles to prospecting in the surrounding oil fields. Taber saw their significance differently. This was a spot where mountain building had occurred "on a grand scale" for millions of years, and it did not look likely to stop in the near future.[88]

Meanwhile, Los Angeles's population was growing fast. In the SSA's *Bulletin* Taber stressed the need for geologists and engineers to work together to track earthquakes and their damage. Together, they could rationalize insurance premiums and devise proper means to protect the swelling city. "The society has not had the assistance of those people who would be most directly benefited by it," Taber charged. "Instead of being helped by the people of this state, we have had chiefly opposition." Thanks to this "ostrich policy," Taber argued, American seismology was "behind other countries which we are accustomed to regard as backward and uncivilized."[89] As so often before, the scientific study of earthquakes was being invoked as a measure of civilization, yet one that now threatened to class the Japanese ahead of Americans.

Taber's research in 1920 was aided by several Los Angeles engineers who helped collect felt reports and evidence of damage. Indeed, it was becoming clear that the SSA's tactics had shifted. They were no longer counting on a permanent observing network of ordinary citizens. Instead, the society was pinning its hopes on one small segment of California's population: engineers.

Engineers had a unique perspective on earthquakes in a fast-growing, water-starved region like Southern California. As Diana Di Stefano has recently noted, industrial workers in this period were not necessarily tools of the "capitalist exploitation of nature." In some cases, they were the most trusted repositories of knowledge about environmental risk. In her study of the "avalanche country" of western North America in this period, DiStefano shows that railway men were regularly called on as expert witnesses to determine whether or not damages caused by an avalanche could have been prevented. Her research identifies an unfamiliar moment in the transition to the "risk society"—before environmental risk became a matter of abstract, quantitative expertise.[90]

Southern California's engineers had a great deal of experience with seismic faults; the question was what they would do with it. In the aftermath of 1906, structural engineers and architects had sent the public mixed messages about the lessons of the San Francisco earthquake. Some lent credence to the state's boosters by attributing damage to the fires, or simply to shoddy workmanship.[91] In the face of these tendencies, the SSA helped redirect the seismic knowledge of California's engineers toward disaster mitigation. Those involved with mining and the construction of dams and reservoirs often had the keenest sense for locating faults and estimating seismic hazard.

Within weeks of the Los Angeles earthquake of 16 July 1920, the Southern California section of the American Institute of Mining Engineers met to discuss the earthquake threat. The meeting was organized by Ralph Arnold, a former Branner student whose research on the California oil fields had turned a nice profit. The featured speaker was William Mulholland, the chief engineer of the Los Angeles aqueduct. Back in 1906 Mulholland had downplayed seismic hazard. Earthquakes threatened nothing "beyond the possibility of repair in reasonable time and at moderate cost." He cheered the residents of Los Angeles who would take "their chances on earthquakes or other abnormal though seemingly inevitable happenings, rather than see the welfare of this fair country languish from lack of water."[92] In 1920, however, Mulholland struck a far more cautious note. He began by explaining that what he had to share were "observations not made in a scientific way at all, but the mere observations of a practical engineer, accustomed to stresses and strains in structures . . . the talk of a layman, pure and simple."[93] His knowledge came from long experience with building and maintaining the city's waterways. He had "known for many years of the existence of a fault or fold of the crust [by Inglewood]. . . . Along that hill or ridge the first earthquake I experienced in California occurred . . . in 1878. I was working at that time in a sewer pipe works near Santa Monica." The 1878 shock had been similar to that of 1920, Mulholland recalled, and in the intervening years he had felt "four similar shocks" on this same ridge. In 1918, Mulholland had built a water tank there. "When I constructed that tank, I knew the seismological character of the country, and I kept telling my young engineer assistants that it would be the most vulnerable part of the Los Angeles Water Works." Vulnerable, indeed: "Nothing is more inelastic and less contrived to accommodate itself to sudden shocks, than a great big tank full of water, for water is the most inelastic substance in nature. . . . There are major fractures, grand faults, that run through the country. . . . I have worked,—done lots of engineering work,—along that fault; made explorations at different points along it, and I know that it is a major fracture." These faults had cracked a reservoir of his own construction on Salano Hill, a reservoir "obliquely crossed by a fault line of minor character; but it broke the walls, and threw the walls about an inch and a half out of level. I repaired that, but came back in six months and found it had faulted the other way." In short, Mulholland had come a long way from his optimism of 1906. Instead of giving Californians a slap on the back for their pluck, he expressed humility. Rather than calculating acceptable damages, he was ready to rule out certain projects altogether. Mulholland compared his experience with earthquakes to that of his colleague Homer Hamlin, who had abandoned construction of an outfall sewer near the same ridge outside Santa Monica. Earthquakes, as Mulholland put it, were a "subject that gives a waterworks man a whole lot of worry, for those things are unavoidable. You cannot get around them. There are some things we have to face. . . . If an earthquake comes, there is nobody on earth that can build an aqueduct, or a building, that will be proof against it. You can build it so the damage will be slight; but it is going to rupture."[94]

Homer Hamlin had made a similar appeal to this same audience in January 1918, describing his efforts to "collect data and records pertaining to earthquakes in southern California." Hamlin was the autodidact director of the Engineering Department of the city of Los Angeles. Three months after the earthquake of 1906, he had written to Andrew Lawson to offer his support to the investigative commission: "I am of course much interested and would like to be in the field. Would such data as the location of some of the principal fault zones in Southern California be of interest or value to you. Two years ago I went down the Colorado River to the head of the Gulf of California and then down the west side for about 40 miles. There are many evidences of great faulting there. Will be pleased to write you about them if the data will be of any use."[95] Hamlin was just the figure the fledgling SSA needed. Following the disappointing performance of the volunteer observers, Branner was all too happy to fall back on Hamlin's knowledge and connections. When an earthquake struck the northeastern corner of Los Angeles County in October 1916, Hamlin was responsible for collecting most of the reports that informed Branner's published study. Branner even cited Hamlin's fieldwork on the Tejon Pass fault. Hamlin had also begun distributing questionnaires following each tremor near Los Angeles. "At the outset I thought we might have eight or ten shocks," he explained, "if it were a good year for earthquakes. In all, over fifty shocks have been recorded; but even these are not all, for I am sure that many slight shocks have passed unnoticed, or were not considered important enough to report. The effort is worthwhile and the work should be carried on even if it is rather strenuous at times, for we shall never solve our own earthquake problems until we study our own territory."[96]

Hamlin's informal network was tested three months later by an earthquake near the San Jacinto Mountains, which seriously injured several people. Hamlin and Sydney Townley, the secretary of the SSA, jointly investigated the affected region three days later. In all, Hamlin collected 166 reports; other researchers only managed to collect eight. Townley seemed genuinely surprised by the usefulness of these observations. "It is well known," he remarked, "that only a very few people are able to estimate a short time interval in a way which even roughly approaches scientific accuracy"; fortunately, two of the observers used their watches to estimate the duration of the shaking. "The moment the shock started," wrote one witness, "I pulled out my watch and caught the exact duration of the shock, which was one minute and thirty-one seconds."[97] Another witness took pleasure in describing the aftershocks—one was a "corker," another "twisted things up proper." This man had a good laugh at his wife, who "grabbed the baby and rushed outside" each time she heard a car approaching, mistaking it for an earthquake.[98] Hamlin's research began with such mundane details of the earthquake. But it culminated in meticulous knowledge of the damage and of the topography and geology of the San Jacinto fault.[99]

At the close of their 1920 meeting, the engineers voted to form a "Southwest Section of the Seismological Society of America" with Mulholland as chairman. Arnold likened the task at hand to the "organization of a fire department. Its members may sit around for a month or two months or even a year, and not respond to a fire call; but they go on practicing every day, so that they will know how to fight a fire when it does occur. When the necessity arises for their activity, they are prepared and ready to respond." Within four months, the SSA's first local section had won an additional forty-five members.[100] Arnold thanked the engineers who had volunteered for "organizing the work and making it local."[101] By 1920, then, "making it local" in Southern California meant appealing to a technical elite, not the general public.

"Dear Fellow Co-ops"

Ironically, just as the SSA was narrowing its outreach to target professional engineers, the Weather Bureau was expanding its own network of volunteer observers. Meteorology had proven its value for aviation in World War One,[102] but its practitioners found themselves hesitating between two divergent courses at war's end. Historians typically see the war as a watershed for technocracy, as military, industrial, and economic questions were redefined as problems for scientific experts. One sign of the growing political authority of scientists was the founding, in 1916, of the National Research Council, which directed postwar efforts to organize large-scale, collaborative research—including Harry Wood's efforts on behalf of seismology in California.[103] Yet even the NRC was aware, for its part, that future scientific funding was largely at the mercy of the public. Voters would have to be convinced that "pure" research was the key to military and industrial advance. In this sense, the war also pointed American science in a second direction, toward public outreach.[104]

The American Meteorological Society was founded in 1919 specifically as a means of outreach. As the AMS's first *Bulletin* explained, the "extension of meteorological knowledge and its applications require cooperation between amateur and professional meteorologists on the one hand, and teachers, business and professional meteorologists on the other hand." The AMS could soon count an eclectic membership of six hundred, about half of whom were either amateur or professional meteorologists, with the rest drawn from a variety of other occupations. The society planned to reach an even wider audience through "educational work," newspapers, trade journals, and direct mail. Laypeople would be encouraged to form committees that "might cooperate with the Weather Bureau . . . in gathering data and pursuing original lines of investigation."[105] For its part, the Weather Bureau recognized the AMS as a valuable tool for recruiting new cooperative observers and facilitating communication between them and professional meteorologists. In 1922 the *Bulletin* introduced a "Co-operative Observers' Department for Voluntary Weather Observers in the Americas."[106] It contained a letter from one Cola W. Shepard in Colony, Wyoming, addressed "Dear Fellow Co-ops":

> For many years we have been contributing our little mites toward the advancement of meteorological knowledge by accumulating daily records of the weather, you in your small corner, and I in mine. Like soldiers in the

German army, we have known little of the results of our work, but have un-
questioningly read our thermometers and measured the precipitation, occasionally recording a "thunder storm," or "Lunar Halo." We have received
the monthly and annual summaries of climatological data for our respective
sections, where our figures were printed beside those of other observers, and
our names appeared in type. We are not expected to know very much about
meteorology, and most of the more important observations are left to the
regular stations of the WB. But we are real meteorological enthusiasts. . . .
And it is possible also that the WB may through this magazine become more
confidential with us and less distant, for the Society includes in its membership most of the personnel of the WB, and, outside of being government employees, they are scholars and gentlemen and "regular fellows."[107]

This letter lets us glimpse the motivations of one participant in an ambitious experiment in "citizen science." The writer derived satisfaction from
seeing his name printed in the bureau's publications and from knowing,
like a German soldier in the trenches, that his work gained significance as
part of a larger whole. Most meaningful, however, was contact with the
bureau's scientists. Shepard confirms what European scientists had come to
suspect: that the public craved contact with "scholars and gentlemen and
'regular fellows,'" not "government employees." It was for this reason that
central European scientists had worried about the transfer of a volunteer
observing network from a scientific academy to a state bureaucracy. In the
United States, cooperatives hoped to interact with scientists who behaved
like "regular fellows," but who were, in their eyes, "scholars and gentlemen"—a status still at odds with the identity of a "government employee."
The cooperative observer system promised to forge just this type of bond
between scientist and citizen.

That promise faded quickly, however. In 1922 the AMS made the fateful
move of raising its annual dues from one to two dollars (on par with the
SSA). Most amateurs dropped their membership, and, ever since, the society
has been a typical professional scientific organization.

"From a Commercial Viewpoint"

The final blow to Branner's scheme for a cooperative seismic observing network came in 1924. It was then that responsibility for collecting instrumental and felt reports of earthquakes was transferred from the Weather Bureau
to the Coast and Geodetic Survey (CGS).[108] The transfer represented a shift
in the perceived significance of seismology at the federal level. Where the

bureau had emphasized the "practical" value of seismological observation broadly, the survey stressed its "commercial" value. The bureau had called for a "scientific" approach to earthquakes, whereas the survey now called for an "engineering" approach. Since World War One, the survey had been in the hands of Colonel E. Lester Jones, who pressed aggressively for the expansion of the agency's role.[109] In a programmatic government pamphlet issued in 1925, Jones wrote, "It would be an indictment of modern civilization and of human intelligence to say that it [the earthquake problem] cannot be solved. In a large measure the problems belong to engineering, and it is therefore not inappropriate that the Coast and Geodetic Survey, an engineering bureau which also makes investigations of physical phenomena, would take a part in earthquake investigation."[110] Likewise, Thomas Maher, the director of the CGS station in San Francisco, explained that the significance of an earthquake could be measured according to its disruption of commerce: "From a scientific standpoint, earthquakes may be great or small, depending on the extent and magnitude of earth movement; from a commercial viewpoint, they will be great or small depending on the damage done, and the problem is becoming more and more one for the engineer and for the man interested in industrial development."[111]

It was soon decided that the survey would rely for felt reports not on members of the public, but on "large public-service corporations of California." The explanation was that such corporations operated around the clock, had an expansive network of plants, and were supervised by "intelligent men." Speaking for the survey, Maher assured the SSA that "a very intelligent class of men are co-operating with us. From the public-service corporations, the reports are mostly by engineers, and such reports are generally accurate, without exaggeration, and without the suppression of important facts. Others are from business men who realize the danger of sensationalism and yet the necessity for having information of value."[112] As a "commercial" problem, macroseismology became a matter of ensuring the continuing profitability of California big business. The survey's new system would avoid "sensationalism," but at the expense of driving a wedge between the public's experience of earthquakes and scientists' interpretations of them.

Harry O. Wood and Regional Seismology

Although the 1906 earthquake was no "scientific revolution," it did mark a minor revolution in the life of Harry O. Wood. In 1906 Wood was a poorly paid instructor of mineralogy at Berkeley, grateful to find work with the Lawson commission. The disaster proved revelatory for him. He devoted

the rest of his life to understanding California's seismic threat. Most immediately, he conceived the ambition of surveying all of California's past
earthquakes, from the eighteenth century to the San Francisco megaseism.
This was to be no mere catalog in the style of Trask or Holden. It was to be a
"synthetic study of recorded shocks"—a work of analysis, not mere compilation. Wood would correlate each event with a known fault. For such a vast
job of data analysis, his goal sounded remarkably modest: "merely to bring
out clearly the suggestion that there is a causal association of earthquakes
with fault zones in this region—a relationship fraught with significance for
human affairs."[113] Wood's "synthetic study" would occupy him for nearly a
decade and make him the American scientist best versed in the subtleties of
exploiting felt reports.

Wood soon began to hone an argument for the value of noninstrumental
observations and for the integration of geophysical, geological, and social
analysis. Like his nineteenth-century predecessors, Wood viewed macroseismology as a hermeneutic challenge. He devoted twenty pages to a critical discussion of his sources, including the errors arising from incomplete
records and from the exercise of his own judgment in interpreting them.
Based on this experience, in 1911 Wood drew up a set of instructions to
earthquake observers for the first volume of the SSA's *Bulletin*. The problem
with seismographic observatories, he noted, was that their goal was "the increase of scientific knowledge," rather than what "the practical public deems
of greatest importance." More to the point, such observatories, "even if established in fair abundance, will not afford any detailed or precise knowledge of the size and shape of the area in which the shock is 'felt,' nor of the
way in which its intensity varies over this area, nor of the character of the
manifold attendant phenomena,—such as damage to structures or disturbances in the soil and rock." The public urgently needed to know "when and
where will strong shocks occur in the future, and what conditions, which are
subject to human control, tend to mitigate their disastrous consequences."
Wood's most important insight in 1911 was that this question could only
be answered by combining knowledge of geology, the built environment,
and human perceptions. Seismologists would have to study "the relations
between the places of origin and the distribution of the perceptible effects
of shocks; the relation between these effects and the geological character
of the ground where they occur; how the character of structures affects the
degree of the disaster,—in short, the interrelationships of all these things,
place of origin, phenomena, character of ground and of structures throughout the whole area in which the shock is felt perceptibly." Finally, true to
seismology's nineteenth-century European tradition, Wood insisted that

human observations could contribute to new scientific insights, even in the seismographic age: "Besides its practical bearing, the results of such correlations have much scientific value, no less than the purely instrumental studies."[114] Wood would argue this point with growing passion for the rest of his career: that the study of earthquakes as a geophysical problem must not be divorced from the study of earthquakes as an environmental hazard.

Again on the model of nineteenth-century seismology, Wood concluded that such a holistic investigation would depend on "the co-operation of large numbers of observers," for "even the keenest observer cannot compass all that is taking place about him." The greater the number of observers reporting on an earthquake, the "more complete its description." It remained only for Wood to furnish guidelines to potential witnesses—a full thirty pages of instructions. Thirteen years later, he would regret overburdening his readers.[115] Indeed, his guidelines were discursive, anecdotal, speculative, and open-ended. Like earlier European counterparts, they addressed the observer as both a register of geophysical effects and a naturalist in her own right. Wood encouraged observers to record their "Sensations and Emotions." These fell into two classes. In the first were those that were "part objective" and "part subjective," "undoubtedly effects of [the earthquake's] motion, subjectively modified." These included "faintness, dizziness, nausea, fear in varying degrees, and all analogous feelings." The second class included indirect results of the shaking and effects that preceded the earthquake by hours or even days—potential predictors, in other words. These included "nervous irritability, restlessness among brutes and birds, ill-defined dread, a sense of oppression, and the like," which possibly had "an objective basis in the physical conditions which prevail just before the shock." Since "effects of this sort are not susceptible of classification, observers should report them by giving brief descriptions." This held all the more for "Unclassified Phenomena," such as the appearance of lights or flames. Wood speculated that these might be of "still greater scientific importance, because uncommon and sporadic, and hence rarely subjected to observation, criticism, and interpretation," and he asked that "a full description" be given.[116] In this way, Wood preserved seismology's nineteenth-century character as a phenomenological, epistemically open field, dependent on and well suited to nonexpert contributions.

On the Margins

In 1913 Wood left his dead-end lectureship at Berkeley to become the first seismologist at the newly founded Hawaiian Volcano Observatory on the rim of the Kilauea crater. The setting was spectacular, and a hotel close by the volcano had been drawing tourists since 1866.[117] But the position was a marginal one, apparently one of the few open to Wood without a PhD. Finding himself in a wilderness 2,500 miles from California's booming cities, Wood began to ask new questions. He began to wonder, for instance, if the paucity of felt reports of earthquakes at Kilaeua was a function of something other than the low population density. It seemed that observers nearby were not sensing tremors that registered on his instruments as exceeding 1 cm/sec^2—the value of ground acceleration that Holden had postulated as the minimum perceptible to humans. Wood pointed out that this minimum unit had "never been determined by psychological experimentation."[118] The question he raised was a fundamental one: what exactly did humans feel when they felt an earthquake?

The answer, according to his analysis of twenty-nine earthquakes recorded in Kilauea, was that the relevant stimulus was not the acceleration of the shock alone, but rather the acceleration in combination with the amplitude of the ground movement. Ever since the work of Dutton, Holden, and Mendenhall on the mechanical interpretation of intensity in the 1880s, seismologists had assumed that the dependence of seismic intensity on the amplitude of ground motion was as simple as the dependence of sound intensity on the amplitude of sound waves. Wood was pointing to a more complex relationship: the amplitude-dependence of the intensity of an earthquake as judged by human observers actually varied with the acceleration of the shock. For weaker shocks, amplitude might even become more important than acceleration in determining perceptibility. Wood called for further study—"if possible, by experimentation."[119] A decade later, the British seismologist R. D. Oldham challenged Wood on this point. Oldham suspected that Wood's value for the "minimum unit of seismic perceptibility" was artificially low, due to the use of "skilled observers." He cited evidence that the sensitivity of observers rose with prior exposure to earthquakes or with training.[120] Wood objected (in an unpublished manuscript) that the observations had by no means come from "skilled observers, specially on the look-out, and living in lightly framed timber dwellings raised clear of the ground, especially suited, consequently, for the recognition of feeble shocks." On the contrary, the observations "were made and reported by *all sorts of people* and most of the structures."[121]

Wood was gaining a sense of the differences between studying nearby earthquakes and distant ones. It was a distinction he would often phrase in terms of "regional" versus "world" seismology, or "local" versus "teleseismic" studies. Thus it was at Kilauea that Wood determined the need for an instrument capable of recording weak local earthquakes. The seismographs in his care at Kilauea convinced him that existing instruments were unable to detect very weak shocks at distances of over one hundred kilometers. This worried Wood greatly, because weak tremors "should be telling harbingers of strong ones."[122] Close to its epicenter, a weak tremor produced faint, short-period oscillations, but sensitive seismographs attuned to distant quakes registered periods no shorter than six seconds.[123] Wood was convinced that seismologists must prioritize the study of the seismicity of their own regions, not the analysis of distant vibrations. To do so would require new methods.

In fact, much of the remainder of Wood's career was devoted to defining the aims and methods of "regional" as opposed to "world" seismology. Regional seismology comprised the study of the interrelations of various problems: "geological," "physical," "human," and "economic."[124] Thus a major article in the BSSA in 1916 called for research into earthquakes "in their importance to human life," which in turn "inevitably will increase immensely our knowledge of the physics of the earth, and of the specific, dynamical behavior of the earth's crust in this province." Wood looked to Europe and its many geophysical institutes as a model of "an enlightened public policy." He argued that the work of such an institute in California should consist of the close coordination of "laboratory" (seismographic) and "field" (geological and geodetic) studies, to be complemented by special collaborations with engineers (whose concerns would otherwise be outside the institute's "scientific" domain). Taking up this theme again in 1921, Wood described the "strictly scientific" and "practical" dimensions of earthquake research as "inextricably intertwined." By "scientific," Wood here meant questions of structural and tectonic geology; the "practical" dimension was "economic and humanitarian." There could be no strict division in the investigation of an earthquake: "the investigation of damage is required, its causes, from both the natural and the structural point of view, and its geographical distribution, all in relation to the underlying geologic structure, and in relation to surface developments and their bearing on living conditions (as sites for buildings, routes for transportation or conduits, and so forth), and the study of the problem from the point of view of engineering science, education, protective legislation and insurance."[125] Wood's "regional seismology" thus recognized the complex interplay of social and environmental factors in determining seismic risk.

The Seismo Lab

When the United States joined the First World War, Wood took the opportunity to move from the margins to the center of American science. He
found work at the Bureau of Standards in Washington, where his research
on the detection of distant artillery fire sharpened his expertise with seismic
registering devices. At the same time, he established contacts with scientists who would emerge from the war as key figures in the organization
of American science. These were men like George Ellery Hale, the founder
of the National Research Council, and John Merriam, the former Berkeley
geologist who became president of the Carnegie Institution of Washington
(CIW) in 1920. Naomi Oreskes has described the philosophy of the CIW at
this time as "science as service." Like Wood, the CIW viewed basic and applied research as "synergistic, not competing" pursuits.[126] Wood had found
the allies he needed.

In 1921, Wood became the director of the first seismological research
program in California, sponsored by the CIW and centered at Caltech. There
he got his first experience with public relations. He lectured to private clubs,
associations of businessmen, engineering and scientific societies, and civic
leaders, asking for sites to house seismographic stations and for help in
locating active faults.[127] He also got the names and addresses of all the cooperative earthquake observers in California. "It is my hope that I may be
able to enlist the services of some of these [observers], at a future time, in
the work I hope to develop in regional seismology here,—without prejudice
to their co-operation with the WB."[128] Wood was promised duplicates of
the observers' reports when they came in each month. Yet the shortcomings of the bureau's network soon became clear. These reports amounted to
no more than one per month, often fewer—not even keeping up with the
tremors reported in the press.

At the center of Wood's research program at Caltech was the detection of
weak shocks from nearby sources. As we have seen, weaker shocks could be
attributed with greater certainty to a given fault, because they were felt over
a smaller area—typically only directly over the causal fault.[129] Throughout
his years in Hawaii, Wood had focused on the development of instrumental techniques for measuring tremors close to the epicenter. In the early
1920s Wood and his collaborators worked, often frenetically, to devise a
suitable local seismograph. The results were mixed. The Wood-Anderson
seismograph introduced in 1922 succeeded in recording nearby shocks (as
well as the initial phases of distant ones). But it recorded only horizontal

movement, not vertical, and the construction of a regional network of these instruments was not completed until 1927. Even then, Wood and his California colleagues were repeatedly troubled by failures of the timekeeping mechanisms on these seismographs.[130]

By 1924, it was clear to Wood that tracking nearby earthquakes could not be done by instruments alone. He turned to the public with a new sense of urgency. From this point, Wood's outreach campaign was not merely a matter of raising awareness of seismic risk and attracting research money. It was an effort to recruit and train observers. This was not mere propaganda, but rather a two-way exchange.[131]

Wood realized that the observing guide he had published in 1911 had "failed to enlist the services of voluntary observers, probably in part because it did not present a concrete list of items or questions to be checked or answered."[132] So, having consulted with colleagues in Europe, and drawing on his knowledge of historical accounts of California earthquakes, he formulated a new questionnaire.[133] It was essentially a list of "a great many of the observations which commonly have been made in connection with the occurrence of earthquakes, great and small." He even provided readers with models of questionnaires completed by hand, which he described as "fictitious," but "*founded* on reports actually made by ordinary, untrained observers. They are made composite in order to bring out points not adequately emphasized in any single bona fide report at hand." Noteworthy on these reports were descriptions of the observer's state of mind. Sample replies included: "Dressing, still somewhat sleepy"; "Did not notice—too much confused and disturbed"; "Could not discriminate"; "Unknown." A report of a weak shock ended with the comment, "The observer was awakened with the impression of having experienced an earthquake. Some moments later, while still awake, two or three rather slow, undulatory movements were felt distinctly. The motion was slight and the shock would not have been felt if the observer had not been awake and attentive. If the observer was awakened by an earlier shock, as is believed, it must have been somewhat stronger."[134] In these examples, Wood was defining and modeling a standard of scientific observation. What he asked of the public was not unrealistic, but nor was it trivial. Beyond frank and thorough reporting, he demanded that observers calibrate their own degree of certainty. Wood's questionnaire thus codified his working knowledge of the psychology and sociology of earthquake response in Southern California, even as it was designed to elicit a more reflective response.

"Substantial Citizenship"

In renewing his call for volunteer observers, Wood may also have been inspired by a colleague—the Stanford geologist Bailey Willis, president of the SSA from 1921 to 1926. In his vigorous public activity, Willis hammered home the message that the work of the SSA could only be accomplished "in cooperation" with "laymen." Willis set the goal of expanding the society's membership from four hundred to one thousand, and by 1927 he had achieved it. He earned himself the popular nickname the "Earthquake Professor."[135]

Early one June morning in 1925 an earthquake struck the picturesque city of Santa Barbara, roughly a quarter of the way up the Pacific Coast from Los Angeles to San Francisco. The death toll of twelve was lower than it might have been if the city had been hit during business hours, but the damage was estimated at $5 million. The Santa Barbara Mission, which had been wrecked by earthquake in 1812 and rebuilt on the same spot, was again severely damaged. An earthquake with a similar distribution of intensities had struck Santa Barbara in 1883. Back then, the press had rejoiced that the shock had "waked her [the city] from her Rip Van Winkle sleep" and "stirred her pulses to activity." "Santa Barbara is not stagnant," the *San Francisco Times* had concluded in 1883; "It is in the bud now, but by and by it will open into the perfect blossom."[136] This style of California providentialism was repugnant to Willis. To him, the Santa Barbara earthquake of 1925 laid bare the folly of American-style development.

Willis painted a bucolic image of Santa Barbara back in Spanish colonial times: a place of "stateliness, license, piety, and poetic romance." The city's subsequent history was, on his telling, typically American. It had become a playground for the wealthy, where "wonderfully landscaped estates . . . bore forbidding 'No Trespass' signs." The town had lost a sense of community, of "substantial citizenship"—"where wealth is spent freely, lavishly, it is inevitably exploited, and the cohesion of society is weakened by the domination of self-interest." Its civic leaders were "thoroughly American, gifted with the American capacity for organization and engineering, but limited, as too many Americans are, in appreciation of history, art, and architecture." Such a society was apt to forget the lessons of past disasters: "Progress and common sense crowded tradition and romance to the wall, heedless of their charm, regardless of their permanent value in the life, yes, even in the prosperity of the community, forgetful also of the earthquake."[137] Willis suggested that Californians had even forgotten how to observe their landscape properly (see figure 10.2). The cliffs of Santa Barbara had become merely a

Fig. 10.2. Teaching Californians how to perceive seismic hazard: Willis's photographs of the Santa Barbara hills from his study of the 1925 earthquake. The climbers in the top photo highlight the beauty of the view; the dotted lines in the bottom photo delineate the fault. Bailey Willis, "A Study of the Santa Barbara Earthquake of June 29, 1925," *Bulletin of the Seismological Society* 15 (1925): 255–78, plate 28, after 256.

lure to a luxurious holiday. Willis made them speak of alternative values: of the romance of an undeveloped landscape, the wisdom of historical tradition, and the imperative of civic duty.

The wreckage of this charming town offered Willis an opportunity to demonstrate the value of felt reports. Writing in the *BSSA* for a nonspecialist audience, Willis called for "better methods of observing and recording those shocks which are perceptible to our senses."[138] He demonstrated what could be learned from such observations. For instance, reports of shaking from the mountains near the Santa Ynez fault indicated that the Mesa fault was not the sole source of the shocks. A fortuitous piece of evidence came from a civil engineer who happened to be driving down State Street (which runs northwest from the coast) when he "felt a blow from behind as though someone had run into the rear of his car." Willis took this observation as "definite evidence of an earthquake movement from the north or northwest."[139] Willis also offered his personal observations in great detail—perhaps, considering the circumstances, excessive detail:

> The writer was at the hotel Miramar, four and one-half miles east of Santa Barbara. Lying awake, he heard, as it seemed, a train approaching along the Southern Pacific tracks from the *east*, experienced such rapid vibrations as are produced by a train close at hand, and then felt the sharp jolt of the advancing wave of an earthquake, it came from the *west*. He was thrown sidewise in that direction. Recognizing the meaning of the shock, he noted the approximate time (6:44 a.m.) and began to count seconds. He had reached fifteen when the movement stopped. In the meantime the bed was rotating in an anti-clockwise direction with sufficient energy to cause him to put out his hand to steady himself. Had the motion continued or increased materially in violence it would have become alarming. As it was he and his friend dressed without haste, taking nineteen minutes, and in that interval there occurred six earthquake shocks including the first. Others followed, of course, but were not specifically noted.[140]

Willis did not note the gender of his "friend," and one can only imagine what Willis's wife thought upon reading this account. Whatever the circumstances were, however, Willis rose above them: he was the model of a cool, composed scientific observer.

Daring to challenge the American image of seismology as a matter of interest only to scientists, engineers, and businessmen, Willis set out to make the field genuinely popular. "The science of seismology has in the past excited but little interest among laymen, as compared for instance with astronomy, because it has offered too slight an appeal to the imagination, has not made itself known, and in relation to human affairs has been without practical significance."[141] Willis appealed to the "imagination" of Southern Californians by placing earthquakes in a cosmic context, in the tradition of Seneca. Like Eduard Suess and Albin Belar, he portrayed seismicity as a necessary feature of "the living globe," a developmental process continuous with the formation of the solar system. In the manner of Clarence King, he tried to teach his audiences to see for themselves the geological forces at work in familiar landscapes. He did not hide the romantic infatuation with mountains that had led him to landscape painting and the conservation movement. Indeed, he fed the public's fantasy with the "stupendous" image of "a mountain chain or an ocean deep, conceived as a growing thing."[142] He revealed "stories" of dramatic uplift and fracture: "He who would read the history of a mountain range cannot be guided by the age or nature of the rocks, since they are usually much older than the uplift, but must seek to read the story in the canyons, valleys, hills, and peaks. Their individual forms and their relations to one another are full of meaning, and he who rides [presumably on horseback] may read as his eye sweeps over them." In the end, his lesson was straightforward: "The universe is not an accomplished fact. It is a growing thing. It is evolving." The point was to enable the public to see this evolution with their own eyes: "With our own eyes we see that the rains wash away the soil and thus attack the hills, which in the course of ages must waste away, and thus we reason that the mountains themselves are but transient features of the landscape."[143] In this way, Willis worked to fuse the environmental perceptions of scientists and of ordinary Californians.

Willis presented the observation of earthquakes as an essential facet of the aesthetic experience of California's landscape. It was "well known to all *competent* observers that the mountains of California exhibit in their sculptured forms the evidence of having experienced uplift."[144] He asked his audience to train their senses on this subterranean process. His message was really quite simple: "We want to go to work and make [the earthquake record] complete so that we may be able to make deductions from it. That work also we shall ask for help on."[145]

Seeking Postmasters and Enthusiasts

Harry Wood agreed with Willis that seismology had suffered as a consequence of the decline of earthquake reporting in the California press. Weaker tremors, so revealing of seismic hazard, went unrecorded. "In early years," Wood wrote, "though places of residence in this part of the country were not numerous and were separated widely for the most part, yet the local earthquakes which were felt in those years were put on record about as well as could have been expected. In the last few decades, however, these happenings have not been recorded adequately. This is true especially of very weakly felt shocks which are, perhaps, more significant and interesting for serious seismologic studies than many of the shocks of greater energy. They tell us more definitely what they have to tell."[146] Felt reports could do more than supplement and corroborate other forms of data; they also held "practical value of their own which in some respects possesses *more immediate importance* than much of the more fundamental data of survey and measurement." Wood urged the formation of "permanent groups of cooperative observers, comprising from one person to a half-dozen persons in a given place . . . made up of people who will agree to make report upon *all earthquakes* which come to their notice in their vicinity, whether felt by themselves or by others." Of the utmost importance, Wood stressed, was that each and every shock be reported: "Those who make use of the assembled reports must feel confident that, on any given occasion, no earthquake came to notice in the locality of the observer, if no report is received from him."[147] What was needed was a permanent network of vigilant citizen-observers.

As Wood complained to his Berkeley colleague Perry Byerly in September of 1927, he had "at present no mechanism for obtaining reports efficiently."[148] At that time Wood was relying on the Coast Survey to distribute questionnaires following shocks. That November the survey's director Lester Jones telegraphed Wood: "Not ready to obtain reports on recent earthquakes. Hope to have plan organized soon." As Wood admitted to Byerly, "I have no machinery for gathering reports other than this."[149] It turned out that Byerly was already in the habit of sending out questionnaires after each shock in Northern California. He had "learned that speed is essential."[150] Wood's own stock of questionnaires from 1924 had unfortunately been "destroyed." He asked for Byerly's advice on designing a new form.[151] Wood thought it best that they share a standard questionnaire. Over the next three years, the two men frequently traded ideas for improving their reporting forms. Wood sent Byerly a draft in March of 1928 with the request to "criticize it as savagely as you please."[152] They agreed that brief postcards should be used for assessing intensities and epicenters. Full-length questionnaires, on the other hand, would ask for more detailed information, such as accompanying sounds and the direction of shaking (if felt outdoors, to avoid the distorting effects of buildings). They sought the aid of ordinary citizens and agreed to do "everything possible to prevent the card from having a formidable aspect."[153] In a bow to the legacy of the boosters, Byerly suggested that there be a check box for "damage: none," since "the loyal Californian is very anxious to tell when no damage has been done and he should be given the opportunity."[154] Wood and Byerly also considered how to target the best observers. Byerly insisted that postmasters were more reliable than schoolteachers, and he was not prepared to cede them to the Coast Survey.[155] In April of 1928 the pair agreed on a division of California for the purpose of earthquake investigations: the "Byerly-Wood line," as their colleagues called it, ran along the northern border of Santa Barbara County, then curved north of Inyo Valley. For events near the border, they agreed to "interchange telegrams before we decide who is to send out questionnaires and go into the field."[156]

Though trained as a physicist rather than a geologist, Byerly generally shared Wood's perspective on the field of seismology. Byerly complained to Wood that he resented a "slighting reference to seismologists" from the physicist Frank Wenner at the Bureau of Standards, who had lately been in California working on the design of seismographs. "The idea that we are ignorant of the elements of physics seems prevalent among a certain type of physicists." Byerly alluded to the ongoing dispute over the instrumental equivalent of seismic intensity—whether the "violence" of the shock corresponded to ground displacement, velocity, acceleration, or some combined function of these variables. "We are not wed to displacement meters as they seem to think. And our intellects are capable of comprehending even acceleration. After all there have been a few things accomplished in this world by men who did not bear the title of 'Physicist.'"[157] Wood tried to convince Byerly not to take the matter personally: "You must not forget that your own introduction to the subject [of seismology] was from the side of physics, whereas more than nine in ten have entered the field from meteorology, or geology, or astronomy, or scattering fields in which a thorough training in dynamics is rare."[158]

Despite his training in physics, Byerly was unimpressed by mathematical niceties. He felt his colleagues focused unduly on aspects of seismology that were highly mathematical and often poorly understood: "It impresses them far more than descriptions of what happened in the earthquake, which they themselves might have observed if they had taken the trouble to go into the field and look."[159] Byerly has been aptly described as a philosophical pragmatist. "Like William James," a colleague recalled, "it amused him to shock by insisting that truth must have a 'cash value.'"[160] Byerly's pragmatism led him to a healthy skepticism about claims to precision in seismology. As he was known to tell students, "an epicenter is a cross placed on a map by a seismologist."[161] Much as phenomenalism did for Mach (chapter 7), for Byerly pragmatism collapsed the Kantian tension between the quest for physical knowledge of the environment and the need for reflexive, critical evaluation. Byerly's suspicion of claims to absolute truth made him tolerant of the uncertainties of macroseismic evidence. All his life he was a firm believer in the value of felt reports.

Yet Byerly was discouraged by the results of his surveys. In 1930 he admitted to Wood that he was compelled to use an intensity scale with no more than ten degrees: "from the kind of data which I get I cannot make distinctions even as nice as the Rossi-Forel scale requires."[162] By 1934, he was forced to admit publicly that his analysis of a Northern California earthquake suffered from incomplete data. "The lists show that of the earthquakes recorded at more than one station 75 were not reported felt and 21 were reported felt. It seems certain that many of the 75 must have been felt but were not reported. Also the fact that only two shocks of maximum intensity I to III R-F were reported shows that observers are likely to allow slight shocks to pass unnoted."[163]

By contrast, Wood expressed satisfaction with the observations he was able to gather from the public in Southern California. "On the whole," he wrote to Byerly in 1930, "I think that this report form is working exceedingly well, especially in respect to the percentage of replies received. If we could receive reports from any adequate number of real enthusiasts, I would prefer my own old more complete form, but I recognize the impracticability of using this under present conditions or any likely to be encountered for some time to come."[164] Wood was learning what he could expect of the local population.

Wood's papers at Caltech include a sampling of felt reports from this period. Most observers simply crossed out or underlined the pertinent items from Wood's list of common effects. But some added further descriptions, along the lines of Wood's samples in his 1924 article. Thus a Miss Elizabeth Connor sought to describe the course of the shaking more precisely than the checklist allowed: "Following the first severe shock, when the observer was out of doors, the earth seemed sometimes to heave, at other times to jerk with a sideways motion. During the rest of the day, in the lighter shocks which came at intervals, the earth seemed to tremble." Another observer added an unusual analogy: "P.S. Earthquake similar to a large bird flying into aerial of radio fastened to house. The same kind of vibration and jerk." The most consistent reporter was Mrs. Muriel Sweet of Pasadena, who had been a member of the SSA since 1921.[165] She sent Wood a total of seven letters (those bearing dates were all marked the second half of 1925, following the Santa Barbara earthquake). One included a chart from the "Stormograph barometer" at her mother's home in Santa Barbara, which had been badly damaged in the earthquake. All her observations were made while lying down, all but one at night. In one undated letter she wrote that she had "not sent in the report of the last shake as I wanted to find out if you wanted more than one person checking up on them, you undoubtedly feel them all and note them down and basically mine are not necessary. I shall be very glad however to continue if you wish me to." Wood's response, from 29 October 1925, assured her: "While it is true, of course, that it would be possible to have so many earthquake reports for a given region that their use would be burdensome, in practice up to the present we have not had enough. A moderate number of reports from independent observers in the same small region usually present small differences which make them all useful. So, if you do not find it burdensome, we shall be glad to have you continue to send in reports on all occasions when you feel a shock, or learn with details that one has been felt by others. With thanks for your interest in this."[166]

This gracious response seems to have been typical of Wood's exchanges with the public in these years. In other cases, he sought to demonstrate to the observers the scientific value of their observations. In 1933 a resident of Santa Catalina Island wrote Wood noting the elapsed time between the radio announcement of the Long Beach earthquake and his own perception of it. Wood replied: "I thank you sincerely for your letter. . . . While of course the time arrival determined [sic] by direct observation is somewhat lacking in the degree of precision which we obtain from measurement on our records it nevertheless will prove very useful to us and I may say at once that it confirms roughly our latest estimates of distance from Pasadena and neighboring stations."[167] The following September a report arrived from a man in Los Angeles whose wife and two neighbors had felt what he termed an "earthquake" in scare quotes. Wood's note of thanks was dated two days later and offered thoughtful comments: "If this was actually a natural shock it must have been very small indeed. Our seismometers here register very small disturbances, but there is no trace of any record at the time referred to. A small artificial disturbance might be felt over a small area in your vicinity and yet not be large enough to register in Pasadena."[168] In 1933 the BSSA published a lengthy letter from a man who described visible surface waves; Wood appended a discussion of the possible physical, physiological, and psychological factors.[169] Through such exchanges, Wood was paving a broad avenue of communication with the public.

His efforts paid off. Evidence of success came with the investigation of the Santa Monica earthquake of 1930. The study was a joint effort by Wood, Richter, and Beno Gutenberg—Europe's premier geophysicist, newly arrived at Caltech. The researchers "obtained an unusually large number of observations on the action of the shock." Indeed, more than three hundred noninstrumental reports were supplied by "private individuals, postmasters and other public officials, and representatives of corporations, co-operating with the Survey." These reports were, in fact, "more numerous and more suitably and adequately distributed over the area of perceptibility than ever before for a shock in Southern California." Despite this auspicious start, a critical note soon crept into their article for the *BSSA*. The information available did not match what "has been available on some occasions in other parts of the world." This allusion to European seismology hints that the voice was now Gutenberg's. The critique continued: "little of the data is precise enough or explicit enough to permit critical study. On the whole very praiseworthy, necessarily in most instances the reports were not made by trained or expert investigators, and in many instances it is difficult to judge which of two grades of intensity is indicated." These lines depart radically from Wood's positive assessment of the felt reports in his letter to Byerly that February. They sound like the opinion of an observatory scientist with little previous experience in the use of nonexpert observations. The rest of the article vacillated over the significance of the noninstrumental data. Intensity values showed "inconsistencies" when plotted against a CGS map of the region that displayed relief and drainage. These were attributed in part to "the experience and psychology of the observers." It seemed fruitless to plot isoseismals, given the complexity of this intensity distribution. At the same time, however, the map was also said to demonstrate "important peculiarities." For instance, the "spotted" character of the intensity distribution was likely due to the brevity of the shaking, which may have subsided before producing stronger effects on humans or structures; thus the acceleration had perhaps been higher than the visible effects suggested. Meanwhile, "tentative isoseismals" seemed to display an interference pattern—was this a real effect or an artifact? The article struggled to account for it in physical terms and ultimately flagged it as a question for the future. In all, the article hinted at disagreements among its coauthors about what could and could not be learned from felt reports.[170]

Wood continued to press ahead with his recruitment of citizen-observers, even when others were ready to admit defeat. After a conversation with an acquaintance who owned a small country newspaper, he drew up a request to editors to print reports of each weak local tremor. Wood was confident that the California press had already shed the bad habit of seismic denial-ism.[171] In fact, he was favorably impressed by recent reporting: "I find that several of the newspapermen have in Imperial Valley had printed very good accounts of the recent shocks, and they appeared to be very interested in the whole subject."[172] If Wood saw a problem with the reporting, it was an "exaggerated tone." His goal in partnering with the press was not to sway public opinion. Rather, he was devising a research strategy, as became clear in the text that Wood proposed to send to the "rural" papers:

> A shock generally noticed in an out-of-the-way district may never come to the attention of earthquake students, nor find any place in any list of earth-quakes, unless some account of it is printed. Such earthquake lists or catalogs are the basis of practical or statistical studies, as well as more strictly scientific investigations, and the more complete the lists are the more value they have. By printing brief accounts of shocks whenever they occur the local papers can make a real and valuable contribution to the advancement of our knowledge of earthquakes.
>
> Knowledge of the occurrence of such weak earthquakes is of scientific, and *practical* value in various ways. For example, *mention*, and *non-mention*, of weak earth motion helps to determine the boundary between the region in which a shock is definitely felt and that in which it is definitely not felt. This aids in fixing the place of origin of the shock, in approximating its depth, and in estimating the amount of energy involved. Knowledge of the occurrence and place of origin of weak shocks help in determining geologic faults which are still active. The frequency with which such shocks originate on a given fault affords some index of its degree of activity and so the extent to which it may be a source of danger.
>
> Such information is valuable in many other ways which cannot be ex-plained here in detail.[173]

To his disappointment, Wood's colleagues were reluctant to help distrib-ute this letter to local newspapers. Maher, the chief of the CGS, refused to approach the press at all. Louderback at Berkeley thought that journalists were covering earthquakes well enough already. Even Byerly pointed out that "earthquakes are a ticklish business to mention to the public." Wood would not give up. In March he told Byerly that he hoped "that you and Dr. Louderback may reconsider your opinion about the desirability of ap-proaching the press. . . . By and large, day in and day out, the papers do not do anything like as good a job of reporting earthquakes as they should and I hope that you may find it feasible to make an approach in the name of the University along the line suggested. If the Carnegie Institution of Washington were well known to the rural press and had such a standing that its request would be heeded, I would take this action myself." In April Wood again urged Byerly to "reconsider the matter."[174] The initiative to enlist the help of local papers in seismological research seems to have died there.

The Language of Violence

Wood's growing experience with felt reports made him increasingly aware of the shortcomings of existing scales of seismic intensity. The Rossi-Forel scale was "ambiguous and badly sub-divided," while the Sieberg was "rather wordy." Neither was suited "to phenomena observed in American communities."[175] Wood hoped to develop a revised scale that would be adopted for use throughout California. Tellingly, however, he only pushed for the use of a standard intensity scale *in public.* For research purposes, he and his colleagues continued to use whatever scale best fit their data and interests. At heart, Wood shared Byerly's pragmatic attitude toward intensity measurement. As he put it to his Berkeley colleague, "I am under the impression that you, personally, attach less importance to this than I may appear to you to be doing. In reality, I think we make about the same appraisal of the situation. I am convinced that when it comes to publication it is desirable that we all use the same scale and that a better scale than the Rossi-Forel is now desirable and it is with this in mind that I have done this additional work. What scales we choose to use in our own offices is quite a different thing, but it is, I think, desirable to secure general adoption for professional use of a better scale than the old Rossi-Forel."[176] Wood's quest for an improved scale was not about producing a universal standard or an absolute measure. It was about reducing ambiguities in communication with the public.

Like Sieberg's 1912 scale, on which it was based, the "Modified Mercalli Scale" that Wood published with Frank Neumann in 1931 relied partly on human sensations to define its degrees. It simplified and streamlined Sieberg's descriptions, eliminating ambiguities. But it also inserted explicit statements about the mental states conducive to certain reported effects: a shock of degree 4 "awakened few, especially light sleepers; frightened no one, unless apprehensive from previous experience." These were among the social-psychological insights that Wood had gained from his experience with volunteer observers. Wood and Neumann also made clear that instruments would not soon replace the human seismograph. They insisted that "we do not know exactly what factors combine to constitute intensity as it is ordinarily understood. We are not yet in position to correlate destructive effects with instrumental data so as to establish an adequate measure of intensity. Though the importance of the factor of acceleration is recognized, we have as yet no satisfactory definition of intensity, no formula expressing earthquake violence in terms of ground movement."[177] They were not alone in judging that instruments had yet to be accurately correlated with felt reports. In 1915, C. F. Marvin had gone so far as to claim that "measurement of the intensity of earthquake movements on the acceleration basis is as yet chiefly a concept of the imagination of the seismologist" and reflected "grave ignorance of the imperfections of the seismograph."[178] Ten years later Richard Oldham still concurred: "For determining the acceleration and amplitude instrumental methods must be left out of count, as suitable instruments are still rare, sparsely scattered, and practically never happen to be where the records would be of use."[179]

This was the problem that Wood encouraged the young Charles Richter to take up: the determination of a mechanical equivalent of intensity. Richter had arrived at the Seismological Laboratory in 1927 with a PhD in theoretical physics. His thesis had dealt with the latest developments in quantum mechanics. "I never took courses in geology in my student years," he later admitted. He reasoned that working under Wood would allow him to remain in Pasadena and "keep in touch" with the physics community.[180] Richter was by all accounts an odd man—"introverted," "awkward," "unable to laugh at himself." Unbeknownst to most colleagues, among whom he seemed unable to make close friends, he was also a poet, womanizer, and, by the 1930s, a nudist. His biographer concludes that he suffered from Asperger's syndrome.[181] Richter was soon drawn into Wood's program for exploiting felt reports, down to the detail of seeking "a better word than the word 'furnishings' in the item 'moved small objects, furnishings.'"[182] At a personal level, though, Richter was hardly an ideal candidate to help maintain Wood's cooperative network.

Existing accounts of the development of the Richter scale take Richter's perspective, rather than Wood's.[183] They suggest misleadingly that the goal was an absolute geophysical measure. In a somewhat cryptic statement, Richter later explained: "We needed something which would not be subject to misinterpretation in terms of the size and importance of the events." This explanation only makes sense in light of the question that prompted it: whether the purpose of the scale had been above all a "public" one. Richter continued: "And also in the process of working with the scale it developed, which we had already suspected, that the statistics on earthquakes in general were in a very bad way because they had been too much influenced by accidental circumstances of local intensity. It seemed desirable to have some objective and instrumentally-founded means of comparing earthquakes with each other."[184] Thus the original purpose of Richter's scale was to avoid "misinterpretation" by the public of the comparative "importance" of earthquakes; only later did the goal emerge of achieving an "objective and instrumentally-founded" measure of relative intensity. Finally, Richter pointed to a further use of the scale, one which he implied was unanticipated: "Even within a limited region such as California it had advantages, and when it developed that it could be expanded to cover the entire world, the value of the thing was greatly increased." Read carefully, Richter's account contradicts the prevailing understanding of the scale. It did not originate in a quest for a universal, objective, and absolute measure of earthquake strength. To the contrary, the initial goal was a local and relative basis for classifying earthquakes—but one that the public would not "misinterpret."

Specifically, the goal was a basis for comparing actual damage with the damage to be expected based on the earthquake's intrinsic violence. It was, in other words, a matter of comparing the "is" and "ought" of an earthquake.[185] The same interpretation emerges from Wood's private correspondence at the time. He repeatedly searched for an adequate way to express the gap between an earthquake's intrinsic potential for violence and its actual effects. For instance, a shock that hit Santa Barbara in 1928 proved "too strong here for useful registration" by seismograph, forcing him to rely on felt reports. Yet his impression was that the damage was out of proportion to the quake's strength. "The close proximity of the origin, unsuitable building and unsuitable foundation ground—rather than great force (notice that I avoid intensity)—appear to be the factors which led to so much destruction."[186] Neither "intensity" (based on observed effects) nor "force" (proportional to acceleration) was the word Wood was looking for. Clearly, a new vocabulary was needed. Wood expressed a similar frustration toward the end of his work on the revised scale. He told Byerly, "Though I have worked hard on this scale, it still contains inconsistencies such as the one you point out respecting the chimney. I have never experienced an earthquake in which all works of construction were greatly damaged or destroyed, but such earthquakes have occurred. Such a fault slip as occurred in 1906 or in 1915 must be indicative of great *energy* whether or not productive of great *intensity*. There were anomalies in 1906 and in any experience there always are. . . . No scale is or can be perfect, but before we adopt this one or accept any other change we must do our best to adjust the definition and eliminate any absurdities, bearing in mind that some anomalies will always be found. Occasionally there will be things that indicate very high intensity when it is obvious that the intensity is low, and more rarely there will be effects which indicate low intensity when it is obvious that the intensity must have been high."[187] The purpose of an intensity scale was to rate earthquakes relative to each other in terms of destructive effects and human impacts. "Anomalies" were unavoidable because the thresholds for different forms of damage varied. Even after eliminating any "absurdities," the seismologist would still be left with the problem that no intensity scale could, in every case, express his own overall intuition of the violence of the shock. Still, a consistent rating scheme was necessary for seismology's *public* dimension.

This need became pressing in the aftermath of the devastation of Long Beach in 1933. Wood and his colleagues quickly got the sense that this earthquake had been far more destructive than it had any right to be: 120 deaths and over $40 million of damage for a shock that (as future commentators would note) rated only 6.2 on Richter's new scale.[188] It was clear to Wood that foolish building practices were to blame. The question in 1933 was how to express that perception to the public. Unusually, Wood chose to use the modified term "'apparent' intensity." Of course, he knew that intensity was always a measure of appearances. It was, however, essential at this moment to convey to the public that the damage due to this shock was out of all proportion to its intrinsic force—however that force might be measured. He struggled to convey this counterfactual claim to the press: "While the total amount of damage was large, there is good evidence from numerous well-distributed structures which survived the shock with little or no obvious injury, and from the sparseness of minor geological evidence of hard shaking, that the intensity or violence of the shaking was no greater than is usual in strong local shocks. . . . The nature and amount of the structural damage was out of proportion to the energy and violence of the shock. . . . 'Apparent' intensity of this value [grade 9, 1931 scale] was manifested in a number of places on exceptionally bad ground, where the intrinsic intensity as shown on good ground near by was substantially less."[189] Here we can see Wood grasping for words to convey his conviction that the earthquake had caused more damage than it should have. What was the telling variable: "energy," "violence," "intrinsic intensity"? At the time, rumors were circulating that Long Beach was one of the strongest earthquakes of all time. Wood was furious about these "wholly irresponsible and mischievous" reports. Different versions cited different forms of quantitative "evidence" to back their claim. In a bitter letter to the *Los Angeles Times*, Wood offered the following example:

"In order of severity the quakes have been: Tokio 5.8; Long Beach, 4.6; Santa
Barbara 3.4; San Francisco, 2.5."
Note—Anything man-built cannot withstand a velocity of 10.00.

Wood made clear that the paper's numbers were entirely meaningless. In
this context, it must have been a relief to Wood to be able to state, with the
support of Richter's new scale, that Long Beach was "not a great earthquake.
It was a fairly strong and moderately large shock in the local earthquake
class. It had about the same magnitude and violence as the Santa Barbara
shock in 1925, perhaps a little greater, perhaps a little less."[190]

Richter's scale originated, then, not in a quest for geophysical truth, but
within this pragmatic effort to classify earthquakes in terms of the damage they could be expected to cause. Richter's idea was straightforward: the
ratio of the amplitudes produced by a single shock at two stations should
be inversely related to the ratio of the distances of each station from the
epicenter. The scale defined a unit magnitude corresponding to the response
of a standard seismometer to a standard shock at a distance of one hundred
kilometers. This was not meant to solve the fundamental physical question
of what intensity measured—whether, for instance, acceleration, amplitude,
or duration of shaking. Indeed, Richter himself judged the scale according
to the pragmatic criteria of consistency. The only way to "demonstrate the
reality of the computed grades of magnitude" was to compare calculated
values of magnitude to the "observed effects of a number of representative
shocks."[191] Moreover, contrary to the air of precision with which earthquake
magnitudes are often cited, Richter himself had no expectation of producing an exact measure: "Precision in this matter was neither expected nor
required." Indeed, he stressed the method's many sources of inexactness,
including "inhomogeneity in the propagation of elastic waves, of varying
depth of focus, of difference in mechanism of shock production, of the
ground at the several stations, and of the instrumental constants." In other
words, instruments might respond nearly as contingently as people did.
Equally important, Richter did not aim from the start to make his scale universal. It emerged firmly within Wood's tradition of regional seismology. It
took decades to generalize the scale beyond Southern California because of
the need to develop corrections for different soil conditions and instrument
types.

Nevertheless, Richter's scale soon took on a life of its own, and its popularity edged out alternative perspectives. As Byerly would later write, "It has
been a struggle throughout the years to keep the record of felt earthquakes.
The public, both scientific and lay, have become so enamored of epicenters

and magnitudes, that there is danger of neglecting field observations."[192] Richter himself lost sight of the original meaning of his scale. In 1958, he argued that epicenters should be located only on the basis of instrumental data, not the distribution of macroseismic intensity: "The practice . . . of drawing isoseismals and then locating an 'epicentre' at the centre of the figure should be discontinued."[193] Even Richter had come to think of magnitude as a replacement for intensity, not a complement to it.

Conclusion

By 1935, California seismology had turned down a path quite different from the course Wood had charted in three ways. First, it had acquired a new leader. Beno Gutenberg had earned his fame using Göttingen's seismographic records to measure the depth of the earth's core. Caltech's president Robert Millikan pegged him as the future of seismology. To lure Gutenberg to Pasadena in 1930, Millikan offered him one of the highest salaries at the university.[194] When Wood contracted his debilitating illness in 1934, Gutenberg became the effective director of the Seismological Laboratory. There was no longer any question that the lab would prioritize geophysical studies of distant tremors over research on local seismicity.[195] Wood speculated that if he had not fallen sick, seismology might have developed differently: "Richter would have worked much more with me and less with Gutenberg, the final Long Beach earthquake paper would have been completed and published, and other local earthquake papers as well."[196] This points to a second divergence: as Goodstein puts it, "Technology now led the science." Field studies and felt reports were passé. Wood was informed by a colleague that "the most serious criticism I ever heard anyone make of your work in Pasadena, was substantially this, that you have the best instruments in the world but are doing nothing with them."[197] Third, California's seismologists had taken a new political tack. As Geschwind argues, they learned from the mistakes of Bailey Willis, who carried his public campaign too far. From 1925, Willis made increasingly bold predictions that a catastrophic earthquake would soon hit Southern California—indeed, within ten years. The evidence on which this claim rested was challenged by business interests and fellow scientists alike. Willis was publicly chastised, and his colleagues soured on public relations. The result was a turn away from public outreach, toward technocratic solutions. From then on, Wood's colleagues tended to focus on "quietly building consensus among technical experts and fostering alliances with government bureaucrats, who could bring the power of the state to bear."[198]

Wood, capable of working only in spurts after 1934, produced many pages of critical notes that never saw publication. In part, he was testing out various arguments for the value of felt reports in a high-tech age. Many of these ideas were incomplete, but they revealed the holes where further research was needed. "In more recent years there has come about a tendency to neglect [noninstrumental] studies, owing to the rapid development and utilization of seismometric instruments and methods." Wood's experience had convinced him that the public's observations were generally sound scientific evidence: "the great majority of persons whose occupation or temperament renders them observant are well qualified to aid in making observations and reports." The evaluation of felt reports required scientists' "critical judgment," of course, but the public could also be taught to be more critical: "Although the reporter or observer who cooperates with the professional seismologist is not immediately concerned with this critical evaluation of the data it is well that the situation be made clear so that the information submitted by reporters may be made as objective as possible." Wood also reflected on the scientific value of felt reports. Even in locations that were well covered by seismographic stations, noninstrumental studies "may indicate with greater certainty the place from which the maximum energy is radiated—as, for example, when the tract of origin of maximum vibration is not the same as that where the shaking begins." Even reports of nausea were not to be discounted, since they could be used to mark "the margin of perceptibility of a large shock, where other effects are few and uncertain," and where ground motion itself might not otherwise be perceived. Observational reports could also help guide field investigations and record transient effects. Local variations of intensity deserved close study, for they might shed light on the nature of surface waves; it was possible, for instance, "that these do not originate at the exact epicenter but rather at a small distance from it." Wood turned at this point to questions of basic geophysics; as the manuscript reflects, his hand could barely keep up with his train of thought. In these private notes Wood articulated better than anyone else the moral of seismology's history in Southern California: "Seismologists (and those who aid them by supplying reports of observations) are, or should be, interested in both apparent and 'intrinsic' values of intensity, and study of the distribution of these values geographically will serve more than one purpose, now and in future as well as in the past."[199]

Byerly, who lived another twenty years after Wood's death in 1958, continued to argue this point publicly. During the Cold War, while American seismology was swollen with defense funding and fixated on the problem of nuclear detection, Byerly dared to suggest that his colleagues were wrong to ignore low-tech research. In his pragmatic manner, he swiftly dismissed the charge that felt reports were "subjective." The search for an earthquake's "'true' measure" was futile, as he explained sardonically: "It has been popular, of recent years, to belittle ratings of intensity because the subject enters into the ratings. It is popular to talk of an as yet uninvented instrument, inexpensive and accurate, which will measure some as yet unselected physical quantity which will be a 'true measure' of the earthquake. Apparently this 'true' measure will enable one to conclude what damage occurred to buildings in the region without inspecting them." There was no good reason, he argued, to favor instruments over human observers: "instruments as we now know them are very selective in their response, and it would take batteries of them scattered all over a town to give us as much information regarding damage as one careful observer (if we can find him) will give us. The human being has a very wide range of sensitivity qualitatively, albeit his quantitative sense may not be all we would desire. We need intensity as well as 'true measure.'" Byerly's pragmatic reasoning brought him to the same conclusion as Wood: instruments might record the expected violence of an earthquake, but only human observers could survey the actual damage. The first could never replace the second. Wood and Byerly had realized what human geographers did not articulate until the 1970s: that "the common units of measurement employed for physical delimitation may be unsuited for assessment of social impact."[200] Byerly put the point memorably: "One of my colleagues in the days past used to harangue regarding 'apparent intensity' (a description of what happened), and 'true intensity' (what would have happened had the valley town been built on rock and all the buildings soundly constructed). It reminds me of Rudyard Kipling's reference to 'the god of things as they ought to be and the god of things as they are.'"[201]

Then Byerly took this argument one step further. "Every now and then throughout the years a man who makes the statement that we need quantitative measures for assessing the strength of an earthquake adds that the intensity evaluations now made should be suppressed. . . . It is the addition of the recommendation that we should suppress the history of what happened to the county court house, merely because it was poorly designed, shoddily constructed on filled ground, which has made me feel I should occasionally speak up. . . . Whatever the originators of intensity scales expected of it, its purpose today is to describe what happened. We must preserve the history, the record of what happened." Byerly was raising a provocative charge. Seismologists' search for a "true" measure of intensity had abetted the suppression of information about earthquake damage; it had allowed Californians to shirk responsibility for poor construction. Wood had distinguished "true" from "apparent" intensity precisely in order to reveal where damage had exceeded expectations. By the 1960s, however, it was widely assumed that an earthquake was fully described by its epicenter and Richter magnitude. Byerly was warning that such a description constituted willful ignorance. Macroseismology, the study of an earthquake's perceptible effects, was nothing short of an ethical obligation: to "preserve the history, the record of what happened."

Conclusion

By 1935, Wood's science of seismic disaster resembled an aging dam, barely able to contain the multiple currents that flowed into it. The publication of the Richter scale struck it like an earthquake, cracking its walls. Its waters rapidly diverged. Seismology's internationalist current was channeled into the high-tech observatory science of the International Seismological Summary. Its concern with the human impact of disaster was funneled into the nationalist enterprise of Cold War disaster studies. Its methods of measuring sensible earthquakes fed into the risk-management strategies of earthquake engineering. These new enterprises were generally managerial and technocratic, sharing little of seismology's earlier concern with environmental adaptation and public communication. Twentieth-century seismology replaced earthquakes with "microseisms," felt reports with seismographic traces, "violence" with so many meters-per-second-squared. The earthquake as scientific object became something unrecognizable to its victims.

Future research might compare the fates of other nineteenth-century sciences of disaster. Famines, droughts, and epidemics also became objects of scientific investigation in the late eighteenth century in frameworks that encompassed natural and social factors. In these early scientific accounts, victims would have seen their own experiences clearly reflected. Then, from the 1870s, scientific explanations of climate-related catastrophes grew increasingly reductive, focused on sunspot cycles and global atmospheric oscillations. Simultaneously, the hunt for microbes replaced the early nineteenth century's more multifaceted, socio-environmental explanations of disease. Human experiences of disaster no longer counted as scientific evidence. In this sense, the histories of climatology, epidemiology, and seismology since the 1870s all involve the construction of incommensurability between scientific expertise and common experience, with implications for public discussions of science today.[1]

At the same time, the built environment of the twentieth century was becoming increasingly vulnerable to devastating accidents triggered by natural hazards. By the 1970s, the realization was dawning that many complex feats of recent engineering—bridges, dams, skyscrapers, freeways, and, above all, nuclear reactors—had not been reinforced against earthquakes.[2] It was also becoming clear to a small group of engineers and seismologists that local seismic risk could not be evaluated on the basis of instrumental records alone. Seismographs have not been around long enough to give an accurate picture of a region's seismicity over time. Accelerometers can provide mechanical parameters that are easy to plug into engineers' equations, but they say nothing about future damage. Knowing the value of "peak ground acceleration" during past tremors won't tell you how many buildings are likely to collapse in the next one. The better measure of seismic risk is intensity, calculated from the observed effects of shaking. Intensity is a scientific measure of "actual earthquake experience"; it is therefore the best predictor of the human impact of future events.[3]

It was not easy, however, to revive the organization and habits of earthquake observing in the late twentieth century. The British were not alone in finding that the public had "other things on their mind" (chapter 2). Institutional memories of earthquake-observing networks survived in some places, but many of the lessons learned through their day-to-day operation had been forgotten. "One thing is sure," recalls the Ljubljana seismologist Ina Cecić, "the lack of written information about macroseismology in general was huge; this was something that was simply not discussed, and was considered to be simple and clear, but everybody had a different way how to deal with things. This goes both for data evaluation and data collection."

In May 1989, six months before the fall of the Berlin Wall, scientists from Yugoslavia, Austria, Hungary, and Italy gathered in a village in Slovenia to discuss—in English—macroseismic methodology. Since the collapse of the Austro-Hungarian monarchy in 1918, there had been little cooperation among these countries in macroseismological research. Cecić believes this was "the first known occasion that seismologists from several countries sat together and explained to everyone how the macroseismic studies were organized at their institutes."[4] One theme was the incongruity between the data obtained from citizen-observers and the analytical methods of the computer age. At one point, the Austrian representative admitted that, at his institute, aggregate intensities and isoseismals were generally "determined by experience." "That is not objective enough," responded an Italian colleague. This critic was a geophysicist familiar with theoretical models of macroseismic data, but with no experience collecting and evaluating observations. "Although the primary data are subjective," he continued, "it makes sense to use a more objective method of analysis, for example computer programs for the drawing of isoseismal lines." Those familiar with the intricacies of felt reports tried to convince the Italian that analysis could not be left to a computer. One colleague shot back, "Before we start working on any computer program for isoseismal lines, we must make a unique definition of what an isoseismal line is." Another noted that a computer "cannot take into account the credibility of the observations."[5] The bottom line was that interpreting macroseismic observations was something of an art. It often forced the geophysicist to think like a psychologist or sociologist. By the standards of other fields of twentieth-century physical science, the data and methods of macroseismology seemed inexcusably subjective. As Cecić puts it, "the rest of seismologists think we're strange (who wants to have physics without numbers? disgraceful!)."[6]

Standard measures of disaster continued to elude the international scientific community in the late twentieth century. When an earthquake occurred near an international border, it was not uncommon for seismologists in the affected countries to work independently of each other and come to conclusions that do not match up. The international seismological community has struggled in the past two decades to agree on uniform measures and methods. Since 2006, European governments and reinsurance companies have sponsored GEM, the Global Earthquake Model: "a uniform, independent standard to calculate and communicate earthquake risk worldwide." GEM plans to develop a comprehensive and universal model of seismic risk assessment (including a standard intensity scale and questionnaire), available through open-source software. It aims to serve a broad spectrum of users and prioritizes "community involvement" and "open debate." As of 2011, however, at least three different seismic intensity scales are in use internationally.[7]

This problem of international comparison is not confined to seismology, as the Fukushima catastrophe of 2011 demonstrated. The International Nuclear and Radiological Event Scale, introduced in 1990, has much in common with scales of seismic intensity. Its stated purpose is to give the public a clear sense of the severity of an accident, and thus "facilitate a common understanding between the technical community, the media and the public." Yet the scale seems to produce more confusion than clarity. Like seismic intensity scales, INES involves a baffling variety of parameters. When Japanese experts initially declared Fukushima a level 5 event, on par with Three Mile Island, the media expressed skepticism. When it was up-graded to the maximum of 7, matched only by Chernobyl, the reaction turned to exasperation. What meaning could such a comparison have, given that the radioactive material from Fukushima was estimated at one-tenth the amount from Chernobyl? Indeed, the INES scale was published with the unlikely caveat that it was meant to aid international communication, but *not* "international comparisons."[8]

Earthquake Observers in the 1970s

Seismic disaster abruptly recaptured the attention of American geophysicists in 1964, when Alaska experienced the most powerful shock ever recorded in North America. In 1965 Frank Press of Caltech began coordinating a national effort to address seismic hazards, with a notable emphasis on prediction.[9] While seismic risk assessment is about the future likelihood of earthquakes of a given severity, short-term prediction is about generat-ing timely warnings. Earthquake prediction briefly became part of main-stream geophysics in the United States in the 1960s and 1970s, but the use of untrained observers remained marginal. It was apparently not until the mid-1970s that efforts at prediction in California began to recruit local ob-servers. By then, it was clear that seismic observations could also serve the cause of California's antinuclear movement. A number of proposed nuclear reactors in California were canceled when residents argued against them on the basis of the local earthquake threat.[10] Then, in 1977, the question of earthquake prediction was taken up by the Center for Applied Intuition, a San Francisco–based organization founded by a Massachusetts Institute of Technology–trained electrical engineer, William Kautz. According to Kautz, the center's purpose was to employ "expert intuitives"—otherwise known as "psychics, channelers, clairvoyants, and healers"—to derive hypotheses that could be subjected to experimental testing.[11] Without revealing the source of his hypotheses, Kautz proposed four experiments to the USGS. Of the two that won funding, one was a statistical study of animal behavior pre-ceding earthquakes, Project Earthquake Watch. The experiment consisted of "an active network of 1500 volunteer observers of pets, farm animals, zoo animals and others—eventually a total of 200 species—in selected seismic areas throughout California." Kautz reported "positive results" for seven of the thirteen moderate earthquakes registered in the four-year period of the study.[12] Despite the support of the USGS, however, Project Earthquake Watch unfolded on the outer fringe of academic science, fueled by the coun-terculture of the 1970s.

A parallel experiment took place simultaneously in Europe. In 1976 Helmut Tributsch, a successful physical chemist, returned to his native village in the mountains of northern Italy to help his family rebuild after the devastating Friuli earthquake. "What left the most profound impression on me," he recalled, "was meeting and talking with the people living there, most of them peasants, whom I had known since childhood." Some of the neighbors mentioned that their animals had behaved strangely in the days before the earthquake. Tributsch took these comments seriously, because these were individuals "whom I trusted completely." Yet he recognized that "any scientist interested in studying this problem would risk not only his reputation but also any chance of getting support." The prospect of a fight strengthened his resolve. It seemed to him that twentieth-century science was failing to recognize its duty to these people: "After I had talked with the people of my village it became clear to me that injustice had been committed here and that science had failed." Tributsch pursued the hypothesis that animals responded to changes in atmospheric electricity preceding earthquakes. The observations of these peasants figured prominently in his research, as did nineteenth-century felt reports. "Against the scientific rules that demand that basic observations of nature must be as precise and reliable as possible, I have adopted these people's faith in what can be seen and experienced so that I might be better able to defend their case." Tributsch lamented that earthquake observing had been abandoned: "With the twentieth century came the seismograph, and with it the age of exact earthquake research. Stationmasters were no longer asked about their observations, and the uneducated people also lost their opportunity to pass on their observations to enlightened scholars." Appropriately, his account of this research for a general audience bore the dedication: "To the observers of nature without name, title, or career for their contributions to the progress of science."[13] Yet Tributsch's research on earthquakes remained marginal, even as he himself went on to a successful career in physical chemistry.

Only in one country did ordinary citizens become central to late twentieth-century research on earthquake prediction, as they had been to nineteenth-century seismology. Under Mao Tse-tung, the Chinese built a network of observers to report regularly on well-water variations, telluric currents, and animal behavior—all phenomena long suspected to be of value for predicting earthquakes. Fa-ti Fan argues that earthquake prediction "provided a perfect opportunity for mass science," where "mass science" meant the Maoist principle of the integration of technical and folk knowledge to utilitarian ends.[14] As Fan shows, this research drew admiring scientific visitors from around the world. Tributsch, for instance, remarked that "it is the Chinese earthquake researchers who are working more in the open-minded spirit of von Humboldt."[15]

Following the trail of the methods of observational seismology thus leads to a remarkable comparison. In states as different as republican Switzerland, imperial Austria, progressive-era California, and Communist China, a similar system of "citizen science" flourished. We can conclude that seismology's cultivation of amateurs made it ideologically useful to very different political regimes. As Fan points out, "citizen science" is an inherently political concept, an outgrowth of the "ideology, institutions, and functions of a state." As we have seen, seismic observing networks served as testing grounds for a remarkable range of political principles—populism in Comrie, anarchism in San Francisco, multinationalism in imperial Austria. Both the Swiss in the late nineteenth century and the Chinese in the 1960s and 1970s used seismological networks to inculcate scientific habits and a sense of national unity. Both states fostered a scientific epistemology that prioritized mass participation. As Fan observes, the mobilization of citizens to produce knowledge of environmental hazard has been a basic tactic of nation-building in very different political contexts.[16]

This analogy between direct democracy and totalitarianism might cast doubt on the alleged potential of citizen science to democratize the sciences. Yet we should recognize the limits of the comparison. Nineteenth-century seismology stands out for the modesty with which its practitioners approached the public. Maoist science was instead "mostly top-down despite its claim of the mass line."[17] Moreover, the Chinese demanded observations of phenomena that, by definition, preceded earthquakes. They relied on instruments, not observers, for records of ground movement. Nineteenth-century seismology, by contrast, asked observers to record the moment of potential catastrophe. It explicitly placed the conditions of knowledge under investigation, subverting Kant's division between the pragmatic and the critical. Whether in China, Europe, or the United States, seismology since the 1930s has avoided that radical move. As scientists canvassing survivors of the 1994 Northridge earthquake put it, "We emphasize that this questionnaire was constructed to study earthquake intensities and not the sociological and psychological aspects of human response to earthquake shaking."[18]

Translation

As I write, six seismologists face charges of manslaughter in Italy because of an alleged failure of public communication. These scientists and one government official are being held responsible for the deaths of over three hundred people in the earthquake that struck the city of L'Aquila in April 2009. International seismologists have responded with outrage but also concern. They admit to being hard-pressed to express their knowledge of seismic risk in terms that the public will understand and act on.[19]

Many of the scientists and engineers who began to worry about seismic hazard in the 1970s had little hope of explaining the risk to the public. Seismic risk was defined as a cost-benefit calculation to be made by seismologists and engineers.[20] "The risk communication expert conceived of the subject of environmental risk as needing to *know* conclusively, in order to act, in order to divert panic."[21] One of the first sociological studies of seismic risk perception, published in 1982, set out from the assumption that "public perceptions of seismic hazards rarely conform to *common sense.*" The author, today one of the leading sociologists of environmental risk, advocated "a vigorous public information and education campaign . . . that eliminates the biases introduced when the public interprets information."[22] But how can "common sense" be defined if not by means of public perceptions? In a similar vein, a Reaganite political scientist warned in 1978 that it was impossible to fully "translate" seismology into lay terms. "The problem of how to convert or translate scientific language into ordinary language is fundamental. The assumption of the undertaking is that popular speech is adequate for this translation. Yet, in making the translation, some of the meaning is bound to be lost." Seismologists were thus instructed to withhold information from the public in order to protect property values and business interests. The danger was too great that the public would "misunderstand it."[23] From this perspective, public communication is a form of marketing, best left to nonscientists. According to a recent article on "outreach" in *Seismological Research Letters,* "Scientific data must be *translated* into meaningful information to be used by potential users, that is, they must become *products.*"[24]

As the sociologist Brian Wynne cautions, talk of translating science for public consumption invites citizens to "sit back, and wait to be told what they must do, rather than go out and learn as well as take their share of responsibility for what could have been presented as a more complex, multidimensional and inherently indeterminate set of human problems, which citizens and their representatives can and should help define."[25] "Translation" had been an operative concept in nineteenth-century seismology, but it had nothing to do with diluting research for public consumption. Scientists patiently translated instructions, questionnaires, and research reports from one vernacular into another in order to enlist the participation of wider circles of the population. They drew on stores of local knowledge and the wisdom of oral and written traditions. They also took pains to translate earthquake effects between humans and instruments, correlating typical expressions and behaviors with mechanical traces. These moves were emphatically antireductive. They sought a comprehensive, multiperspectival description of the event: total knowledge, in Mach's sense. One might even say that in pursuing a scientific vernacular, nineteenth-century seismologists offered something akin to the "language therapy" called for by central European philosophers like Mach and Wittgenstein. Like Mach, they assumed that science was only of value to the extent that its conclusions could be *translated* back into the observations of ordinary people.[26] In this way, nineteenth-century seismology cultivated a scientific vernacular, a language that mediated between expert analysis and common experience. It was not that science had "contradicted and corrected" ordinary sensibilities.[27] Rather, the scientific description of the environment merged with the public's own perceptions. When scientists announced that "the notion that the ground is naturally steadfast is an error,"[28] the public was already developing the capacity to feel that instability. A common language shaped a common vision: a living earth meant a quaking earth.

As Karl Kraus observed over a century ago, disasters create a knowledge vacuum. Immediate official information tends to be uncertain and inconsistent, and the public quickly grows suspicious of scientific expertise.[29] In the past few years, Internet tools like Twitter and Google Maps have enabled citizens to self-organize and produce knowledge to fill that void—tracking wildfires in California or reporting personal dosimeter readings in Japan. Scientists in turn are thinking creatively about how to use data like this. In the early 1990s, the USGS launched a website to collect data on seismic events in real time. Called Did You Feel It?, the site mimics the questionnaires developed by Swiss scientists in the 1870s. As a former USGS geologist explains, "Another indicator for the future is that thousands of ordinary people, just by using the CCI [Community Internet Intensity] system, are becoming better earthquake observers. As their skill grows, their awareness of earthquake issues also rises. That in turn promises to make each of them—*each of you*—more effective members of society. And that is one of the great benefits of this kind of citizen science." Like the nineteenth-century science of seismology, Did You Feel It? seeks to "provide an important human perspective on earthquakes, providing documentation of the way people behave and respond, and how they perceive risk." Like nineteenth-century observing networks, the site is meant to be a "two-way street" between experts and nonexperts. It is "fundamentally a *citizen science* endeavor."[30] Yet Internet-based questionnaires have radically constricted such exchanges. Nineteenth-century observers were free to modify the terms of a survey or replace it with a free-form letter. Today, data that do not fit the expected format never reach scientists at all.

Fear

In 1908, three days after Christmas, an earthquake and tsunami killed more than eighty thousand people in southern Italy. Four weeks earlier, the Swiss psychoanalyst Carl Jung had welcomed his second child into the world. Jung's four-year-old daughter had been peppering her parents with questions about the baby, refusing to believe it had been brought by a stork. Then she overheard the adults discussing the Italian disaster, and her curiosity took a new turn:

> She repeatedly asked her grandmother to tell her how the earth shook, how the houses fell in and many people lost their lives. After this she had nocturnal fears, she could not be alone, her mother had to go to her and stay with her; otherwise she feared that an earthquake would happen, that the house would fall and kill her. . . . Many means of calming her were tried, thus she was told, for example, that earthquakes only occur where there are volcanoes. But then she had to be satisfied that the mountains surrounding the city were not volcanoes. This reasoning led the child by degrees to a desire for learning, as strong as it was unnatural at her age, which showed itself in a demand that all the geological atlases and text-books should be brought to her from her father's library. For hours she rummaged through these works looking for pictures of volcanoes and earthquakes, and asking questions continually.

At last, the girl was given a more plausible account of human reproduction (this time with a botanical theme). With that, her fear of earthquakes "entirely vanished." Jung even tested her with illustrations of disaster scenes: "Anna remained unaffected, she examined the pictures with indifference, remarking, 'These people are dead; I have already seen that quite often.' The picture of a volcanic eruption no longer had any attraction for her. Thus all her scientific interest collapsed and vanished as suddenly as it came."[31]

Jung wrote the incident up as the case of "Little Anna," and it figures in the history of psychoanalysis as a counterpart to Freud's "Little Hans," revealing the neurotic effects of sublimation in children. Jung used it to argue against "prudishness" with respect to children's interest in sexual matters. But consider the story from another angle. Perhaps Little Anna's curiosity about earthquakes was no less natural than her interest in sex. Perhaps she was just as clever to doubt her parents when they claimed that earthquakes were caused by volcanoes as when they claimed that her brother was brought by a stork. "Fear is the expression of converted libido," wrote Jung, implying that the child could not genuinely be afraid of earthquakes.[32] Perhaps, in Jung's denial of this possibility, we have met with a resistance that requires its own interpretation.

The tendency of modernist thought in the human sciences is to peer behind allusions to natural disasters for what is presumed to be the hidden source of anxiety, whether it be sexual, socio-economic, or political. Fear of an unmasterable, external threat is recast as a diffuse "anxiety," requiring therapy to identify its true origin.[33] As in the case of Little Anna, the disaster itself is quickly made to "vanish" from the story. Too quickly. It seems that we have stumbled on a mighty defense mechanism—and behind it, a profound anxiety about nature out of control.[34] It falls to the historian to probe the background to stories like Jung's. The pat answers that adults offered his daughter about the earthquake masked their own ignorance. This book has tried to reveal the processes by which earthquakes were alternately confronted and displaced in the modernist age, a process much like the rapid awakening and extinction of Little Anna's "scientific interest" in earthquakes. It falls to the historian to follow the clues back to the scene of knowledge in the making.

Enlightenment

On 11 May 1910, a light tremor shook Vienna. Among the eyewitness reports published in the papers the next day was that of a bank clerk: "At the start of the earthquake, before I could form a judgment of the nature of the phenomenon, I already saw how my wife, with jerky motions and a terrified expression, threw her head to the side and gripped at her heart. . . . She whispered just one word: earthquake. Only with this utterance did I find enlightenment [*Erst durch diese Äußerung bin ich aufgeklärt worden*]." The sharp eyes of Karl Kraus seized on the word *aufgeklärt*—literally, "enlightened." He mocked the clerk's scientific ambitions: "In the event of the world's destruction, the learned professions must immediately form a judgment of the nature of the phenomenon, this they have learned from the newspapers." Kraus feigned wonder at the clerk's observation that the earthquake had "begun with an explosive sound, during which I felt a vertical shock from below, which was followed by a horizontal shock, probably in the direction northwest-southeast." "That is the marvelous thing," Kraus remarked, "and it has always amazed me about these learned professions: that they at the instant of an earthquake immediately look at the clock and are even clearly oriented with respect to the direction of the shock. The likes of us are not capable to distinguish west from east when all is calm, certainly not in the manner of an Englishman and that of a reader of *Die Zeit*. But the readers of *Die Zeit*, as soon as their wives whisper the word earthquake, they are already *enlightened*, whisper northwest-southeast, and moreover have the presence of mind to sit down immediately and send the shiksa from the country, who is *not enlightened*, and is instead howling and clanging plates in the kitchen expecting the end of the world, to the editorial office, so that the letter will be in the morning paper for sure."[35]

Kraus's insight was worthy of his nemesis Freud. All around him, Kraus saw denial—denial of the power of nature to escape human control. It had become impossible to perceive disasters as disasters. All one saw were representations of disaster through the lens of the press, recast in the "scientific intonation" that stripped them of the horrific. Well before Jünger, Kraus perceived that firsthand accounts of disaster were being transformed into a predictable genre. He thus recognized the tendency of modernism to repress natural disasters beneath a veneer of stylization and metaphor. Kraus portrayed the reports of earthquake observers as a narcissistic, self-indulgent genre akin to the *feuilleton:* a jargon-ridden and vacuous form of "cosmic prattle." He decried the very goal of training the public to perceive a threatening natural world through the lens of modern science. It was a lens of false security, transforming disasters into geophysical specimens. "A lowly genre," as Kraus once described satire—"as far below the dignity of a historian as an earthquake."[36]

Kraus saw as clearly as anyone what was at issue when the newspapers talked earthquakes. Since the eighteenth century, a society's response to earthquakes had been a measure of its enlightenment. Yet, in Kraus's view, what a poor bank clerk took for "Aufklärung" in 1910 could not have been further from the real thing. In these terms, only the "unenlightened" would be so crass as to display fear at such a moment, to "expect the end of the world." Kraus, the messenger of apocalypse, cast his lot with the unenlightened. Once again, Kant's ghost was haunting Europeans' encounters with seismic instability. The question was plain: what was enlightenment?

What Kraus failed to see was that this bank clerk was engaged in a dialogue. What did Kraus really know of the project that Eduard Suess had launched in 1873 with his appeal to the public for seismic observations?

To Kraus, Suess embodied the shared hubris of science and liberalism. Kraus failed to recognize the modesty with which nineteenth-century seismologists approached citizens. Far from tolerating "cosmic prattle," these scientists took pains to express themselves in the vernaculars of their lay observers and to preserve even the original idioms of their testimony. Kraus took no notice when scientists admitted how little they knew. He did not read the instructions in which they urged observers to gauge their own uncertainty. To be sure, this was a project of "enlightenment," yet with only a superficial resemblance to Kraus's caricatures. For scientists like Suess, Albert Heim, Charles Davison, Harry Wood, and Perry Byerly, enlightenment required a frank and accurate assessment of the limits that nature posed to human ambitions. Disasters—floods, earthquakes, avalanches—were central to this worldview, as to Kraus's. Suess was, after all, the man credited with having reintroduced disasters to geohistory. For Suess, as for Kraus, the need to adjust humanity's sense of scale was an urgent ethical imperative. It was the task of natural science "to ascertain the place of humanity in the universe."[37]

The earthquake observers of the nineteenth century succeeded in reframing Kant's question for a scientific age. In the face of the "concept-quake" that was tearing through the foundations of traditional beliefs, the dialogue between scientists and citizens offered an intellectual compass. It furnished no easy answers, but it showed how to pose the questions. Exactly how much fear of untamed nature did enlightenment call for? Precisely how much skepticism toward intellectual authority did enlightenment warrant? In the midst of disaster, who was the model of enlightenment: The European who slipped into "hysteria"? The native of the "savage regions" who kept her "presence of mind"? The distant scientist counting oscillations on a seismograph? The bank clerk who reported a northwest-southeast shock? How, in short, might the experience of enlightenment *feel* when the ground itself was adrift?

ACKNOWLEDGMENTS

The seeds for this book were planted by two generous colleagues and nurtured by many others. Raine Daston first piqued my interest in the Lisbon earthquake eight years ago, with a casual note on a draft of my dissertation. She has quietly continued to drop crumbs of genius ever since. But I knew nothing of human seismographs until I had the good fortune to meet Andrea Westermann in 2007. From Andrea I learned about the curious project of the Swiss Earthquake Commission, and she has offered excellent advice and kind support ever since. I have been lucky to share my interest in nineteenth-century earthquakes with Conevery Valencius, with whom I have happily indulged in conversations about newspaper clippings, fallen chimneys, and much else. Fa-ti Fan and Jeremy Vetter have been generous with their expertise about Chinese seismology and American meteorology, respectively. Carl-Henry Geschwind shared his incomparable knowledge of archival collections on American seismology. Ina Cecić, Monika Gisler, Christa Hammerl, Roger Musson, and Leonardo Seeber have patiently discussed seismology's recent history with me. For very helpful conversations or comments on drafts I warmly thank Marwa Elshakry, Peter Galison, Michael Gordin, Christopher Harwood, Matthew Jones, Alexandre Métraux, Mark Mazower, Nara Milanich, Ted Porter, Jonathan Schorsch, Pamela Smith, Jan Surman, Lisa Tiersten, and Conevery Valencius; and thanks to Stefan Andriopoulos for advice on German translation. I owe a particular debt to Andre Wakefield, who provided indispensable feedback first as a conference commentator and then as a reviewer for the University of Chicago Press. My editor, Karen Darling, has been an absolute pleasure to work with. Valuable suggestions have also come from audiences at the University of British Columbia, Yale, and Columbia's Earth Institute; from commentators, panelists, and audiences at meetings of the History of Science Society and the American Historical Association; and from the organizers and participants of the workshop "Earth Science, Global Science" at York University. Special thanks to those who attended the workshop on comparative histories of earthquake science and response held at Barnard in 2008, especially Peder Anker, Paola Bertucci, Gregory Clancey, Jan Kozák, Stuart McCook, Matthew Stanley, Charles Walker, and Paul White.

I gratefully acknowledge the help of the research assistants who gathered archival documents for me: Cathrin Hermann in Vienna, Valérie Graf in Zurich, Jaime Raba in Berkeley and Palo Alto, and Lan Li in Pasadena; and many thanks to Manuela Krebser for transcribing several of these documents for me. At Barnard, I had the pleasure of working with two excellent undergraduate research assistants, Sally Davis and Amelia Steinman.

My research was generously supported by a grant from the University of Chicago's Defining Wisdom Project and the John Templeton Foundation. I am grateful to the project's directors, staff, and participants for the assistance and intellectual stimulation they provided. I also thank the provost's office at Barnard for research funds and a leave from teaching in 2010–11.

Portions of the book have appeared in *Science in Context, Modern Intellectual History,* and *The Nationalization of Scientific Knowledge in the Habsburg Empire, 1848–1918,* edited by Mitchell Ash and Jan Surman (Palgrave, 2012).

Most importantly, I thank my family for making this book possible. Without the constant support of my husband, parents, and in-laws, I would not know how to begin to combine motherhood with scholarship. The sensitivity and good humor of my children, Amalia and Adam, make motherhood a delight and a privilege. I dedicate this book with love and appreciation to my husband, Paul.

NOTES

INTRODUCTION

1. Kraus, "Das Erdbeben" (1908). Unless otherwise noted, all translations are my own.
2. To my chagrin, a week after completing this manuscript (23 August 2011), I mistook a rare East Coast tremor for the subway passing underground.
3. Nietzsche, "On the Uses and Disadvantages of History for Life," 120. In fact, "The earthquake metaphor is a veritable topos of the reception of Nietzsche" (Meyer, "Nietzsche und die klassische Moderne," 13).
4. Exner, Der Schlichten Astronomia, 138.
5. Popper quoted in Dawson, "The Reception of Gödel's Incompleteness Theorems," 84; Menger, "The New Logic," 336.
6. Dawson, "Reception of Gödel's Incompleteness Theorems."
7. Schütz, Der Grubenhund, 37.
8. Voss, Symbolische Formen, 13.
9. De la Beche, How to Observe, 142. On representations of earthquakes, see Dombois, Über Erdbeben. Jamie Rae Bluestone's excellent dissertation on pre-instrumental seismology, "Why the Earth Shakes," came to my attention as I was completing this project.
10. Neumayr, Erdgeschichte, 305.
11. Isaac Esterbrock to Charles Rockwood, 10 August 1884, book 3, Rockwood Papers, Princeton University Archives.
12. Heim, Die Erdbeben und deren Beobachtung, 24.
13. Favre, "Tremblement de terre à Fleurier," 132. Visual representations of earthquakes are another crucial source of information for historical seismologists today; see Kozák and Čermák, Illustrated History of Natural Disasters. On the divergence between scientific illustrations of earthquakes and artistic conventions in the late eighteenth century, see Keller, "Sections and Views."
14. Fan, "'Collective Monitoring'"; Valencius, "Accounts of the New Madrid Earthquakes."
15. Palm and Carroll, Illusions of Safety.
16. Quervain, "Erdbeben der Schweiz 1910."
17. Steinberg, Acts of God; Hoffman and Oliver-Smith, "Anthropology and the Angry Earth." "Risk" is defined as "hazard" multiplied by "vulnerability," where hazard is a

natural condition and vulnerability is the probability of damage, given an event of a certain magnitude.

18. Stein, "Continental Intraplate Earthquake Issues," 2; Fréchet et al., *Historical Seismology*; Valencius, "New Madrid Earthquakes."
19. Hilhorst and Bankoff, introduction, 1.
20. Sarton, "Secret History," 187.
21. Mauch, *Natural Disasters, Cultural Responses*, 7; Davids, "River Control"; Poliwoda, *Aus Katastrophen Lernen*.
22. Suess, *Antlitz der Erde*, 1:25.
23. Latour, *Never Been Modern*.
24. There was a delay of four weeks before the quake was reported in Hamburg and Berlin papers (Wilke, "Das Erdbeben von Lissabon als Medienereignis").
25. Fischer, *Geschichte der neueren Philosophie*, 151.
26. Kant, "Naturbeschreibung des Erdbebens," 434.
27. Gerland, "Immanuel Kant."
28. Fisher, Geschichte der neueren Philosophie, 152; Benjamin, "Lisbon Earthquake," 538. For a more sober assessment of the scientific significance of the Lisbon earthquake and of Kant's research, see Oldroyd et al., "The Study of Earthquakes."
29. Wilson, *Kant's Pragmatic Anthropology*, 11.
30. Kant, *Physische Geographie*, vol. 1, part 1, 14.
31. Until the past few years, this tension tended to hold the field of history of science aloof from environmental history (Anker, "Environmental History versus the History of Science"). Doing environmental history seemed to preclude critical analysis of the construction of environmental knowledge. Recent work—in particular, by Gregg Mitman, Michelle Murphy, and Linda Nash—has shown how to integrate the strong points of each field, and this book seeks to follow their examples.
32. Neumayr, *Erdgeschichte*, 305. On the Portuguese use of questionnaires to investigate the quake of 1755, see Oliveira, "1755 Lisbon Earthquake."
33. Herbert, *Victorian Relativity*, 147.
34. Gerland, "Immanuel Kant," 449.
35. White, "Darwin, Concepción." On the naturalization of disaster, see Steinberg, *Acts of God*, and Davis, *Late Victorian Holocausts*.
36. Richter, "Instrumental Magnitude Scale"; Hough, *Richter's Scale*; Goodstein, "Waves in the Earth"; Geschwind, *California Earthquakes*.
37. Barth, "Politics of Seismology."
38. Other romantics pursued natural knowledge by scaling mountain peaks, cultivating somnambulism, sticking pins in their eyes, or even electrocuting themselves. See Strickland, "Ideology of Self-Knowledge"; Felsch, *Laborlandschaften*; Dettelbach, "Face of Nature."
39. On nonexpert observers, see Vetter, ed., "Lay Participation"; Charvolin et al., *Des sciences citoyennes?*; on scientific observation more broadly, Daston and Lunbeck, *Histories of Scientific Observation*.
40. The reference is to Willy Hellpach's "geopsyche" theory; Hennig, "Abhängigkeit vom Wetter," 780.
41. Golinski, *British Weather*, chapter 5; on sensitivity to air pollution in particular, see Thorsheim, *Inventing Pollution*; on the regimentation of meteorological observation in the eighteenth century, see Daston, "Empire of Observation." On amateur weather observers in the nineteenth century, see Locher, *Le savant et la tempête*, chapter 3. On the pathologization of weather, see Janković, *Confronting the Climate*. On the

development of meteorological instruments to mimic lay "weather wisdom," see Anderson, *Predicting the Weather*, chapter 5. On climatology and colonization, see Osborne, "Acclimatizing the World," and Jennings, *Curing the Colonizers*.

42. Whitman, "Kosmos" (1860), quoted in Walls, *Passage to Cosmos*, 280.

43. Bennett, *Vibrant Matter*, 120.

44. Katharine Anderson hints that such a "reflexive turn" is appropriate for the history of meteorology, in which "the meaning of global science needs to be investigated through exactly these shifting contemporary characterizations of meteorology as global, national, or local science" (*Predicting the Weather*, 290).

45. Clark, *Academic Charisma*.

46. Kafka, "Public Misconceptions."

47. Kuhn, *Structure*, 20.

48. Koerner, *Linnaeus*, 40. On Galileo's scientific vernacular, see Redondi, "Galilée et Comte."

49. Porter, "How Science Became Technical"; Cohen and Wakefield, "Introduction"; Valencius, *Health of the Country*; Hamblyn, *Invention of Clouds*; Anderson, *Predicting the Weather*; Daston, "Empire of Observation"; Huler, *Defining the Wind*; for examples of the transformation of vernacular sciences into technical sciences, see Barrow, *Passion for Birds*, and Endersby, *Imperial Nature*; on recent initiatives to vernacularize science, see Irwin, *Citizen Science*; on induced seismicity and the citizen science movement, see Schneider and Snieder, "Putting Partnership First."

50. Carl Schmitt identified the judicial response to the Messina earthquake of 1908 as a prototype for the "state of exception" (Orihara and Clancey, "The Nature of Emergency"). See too Walker, *Shaky Colonialism*.

51. Baudrillard, "Paroxysm."

52. Daston, "Life Chances," 14.

53. Robin, *Fear*.

54. Matthias Dörries shows how fear has been invoked in recent climate discourse both to motivate action and inaction (Dörries, "Climate Catastrophes and Fear").

55. Quoted in Gardner, *Science of Fear*, front matter.

56. Smith, "Fear," 19.

57. Hume, *Treatise of Human Nature*, 438–48. Adela Pinch explains that "empiricism allows emotion to be a way of knowing" (*Strange Fits*, 18).

58. Luhmann, *Ecological Communication*.

59. Murphy, *Sick-Building Syndrome*.

60. Wilkinson, *Anxiety*, 19.

61. Smith, "Preface," 1, cited in Bourke, *Fear*, 283.

62. Beck, *Risk Society*.

63. Bourke, *Fear*; Orr, *Panic Diaries*.

64. In addition, a Norwegian earthquake service was founded in 1887; by 1900 it had collected felt reports on 269 earthquakes, revealing a higher seismicity than anticipated for the territory. See Muir Wood et al., "Earthquakes in the Northern North Sea." Serbia organized a permanent observing network in 1906, relying both on private citizens and civil servants; see Radovanović and Mihailović, "Die Erdbeben in Serbien."

65. Kozák and Plešinger, "Regular Seismic Service"; Früh, "30-jährige Tätigkeit," 61.

66. On the paucity of macroseismic sources for the late Ottoman Empire, see Ambraseys and Finkel, "Seismicity of Turkey." In the late Russian Empire, scientists investigated individual strong quakes macroseismically and compiled a catalog, but there was no

attempt to organize an observational network; see Tatevossian, "Earthquake Studies in Russia." The French Académie des Sciences began collecting macroseismic observations in the 1760s, but systematic macroseismic investigations were not undertaken in France until the founding of a seismological service in 1908; Quenet, *Les tremblements de terre;* Fréchet, "Historical Seismicity."

67. On technocratic trends, see Alder, *Engineering the Revolution,* and Porter, *Trust in Numbers.*

68. Steinberg, *Acts of God,* 20; Hecht, "Introduction" (thanks to Gabrielle Hecht for sharing this in advance of publication).

CHAPTER ONE

1. The OED cites Mallet, *Earthquake Catalogue,* as the first use of "seismology"; it was adopted into French and German by the early 1860s. William Whewell coined "scientist" in 1833: Yeo, *Defining Science,* 110.

2. Montaigne, "Of Cannibals," 152.

3. Gisler, *Göttliche Natur.*

4. Valencius, "New Madrid Earthquakes."

5. Gisler, *Göttliche Natur.*

6. Cited in Thouvenot and Bouchon, "Lowest Magnitude Threshold," 314.

7. Keller, "Sections and Views," 158; see too Oldroyd et al., "Study of Earthquakes."

8. Dean, "Robert Mallet."

9. Ferrari and McConnell, "Robert Mallet," 53.

10. Mallet, *Observational Seismology,* 1:7.

11. Mallet, "First Report," 2.

12. Davison, *Founders of Seismology,* 99–100.

13. Mazzotti, "Jesuit on the Roof," 76–77.

14. Davison, *Founders of Seismology,* 100.

15. Muir Wood, "Earthquakes Incorporated," 124.

16. Davison, "Scales of Seismic Intensity," 49.

17. Milne, "Earthquakes in Great Britain," 346. Davison remarked that Milne "during his whole career investigated only one earthquake, that of February 22, 1880 in Japan" and "received only 120 replies to his circular, a number far too small for drawing the isoseismal lines of a strong earthquake." Davison, "Scales of Seismic Intensity," 121. Milne's textbook, *Seismology,* does not discuss the use of human observers at all. As Montessus de Ballore observed, "It is a noteworthy fact that John Milne almost never made any use of intensity scales, nor did he delineate isoseismal curves in his many seismological papers" (quoted in Cecić et al., "Epicentre Determination from Macroseismic Data," 1014).

18. Clancey, *Earthquake Nation,* 157.

19. Rockwood, "Japanese Seismology," 469.

20. Mallet, *Observational Seismology,* 2: 2878.

21. Fort, *Book of the Damned,* 7, 15.

22. This summary is based on Dewey and Byerly, "Early History of Seismometry," and Howell, Jr., *Introduction to Seismological Research.*

23. Quervain, "Erdbeben des Wallis," 85.

24. Thovenot and Bouchon, "Lowest Magnitude Threshold," 313.

25. Brush, "Discovery of the Earth's Core."

26. Oreskes, *Rejection of Continental Drift.*

27. Wegener, *Entstehung der Kontinente,* 20, 37.

28. On Suess, see Hamann, *Eduard Suess*.
29. Suess, *Boden der Stadt Wien*, 255–57.
30. Tollmann, "Eduard Suess," 48, 69.
31. Suess, *Erinnerungen*, 238.
32. Suess, *Erdbeben Nieder-Österreichs*, 2.
33. Hoefer, "Erdbeben Kärntens," 2.
34. Schmidt, "Erdbeben am 15. Jänner 1858," 131. On the significance of the 1858 investigation, see Vaněk and Kozák, "First Macroseismic Map."
35. Suess, *Erdbeben Nieder-Österreichs*, 1.
36. Greene, *Geology in the Nineteenth Century*.
37. Strehlau, "Eduard Sueß' Study of Earthquakes."
38. Oreskes, *Rejection of Continental Drift*.
39. Greene, *Geology in the Nineteenth Century*, 171.
40. Greene, *Geology in the Nineteenth Century*, 178.

CHAPTER TWO

1. Quoted in Gray, "Highland Potato Famine," 358.
2. Carment, *Scenes and Legends*.
3. Quoted in Carment, *Scenes and Legends*, 6, 7.
4. Carment, *Scenes and Legends*, 29.
5. Davison, "Earthquakes of Comrie."
6. "Experiences of Earthquakes in Great Britain."
7. Musson, "Comrie."
8. Davison, *British Earthquakes*, 65.
9. Carment, *Scenes and Legends*, 21.
10. Morrell and Thackray, *Gentlemen of Science*.
11. He added his wife's last name when she inherited a set of estates in 1852, becoming known as David Milne Home. See Milne Home, *Biographical Sketch*.
12. Milne, "Earthquake-Shocks Felt in Great Britain," 93.
13. Cited in Milne Home, *Biographical Sketch*, 65.
14. Milne Home, *Biographical Sketch*, 72.
15. Milne, "Earthquake-Shocks Felt in Great Britain," 260.
16. Davison, *British Earthquakes*, 64.
17. The failures were blamed on friction between the writing stylus and the surface; Dewey and Byerly, "Early History of Seismometry," 188.
18. *Times* (London), 13 December 1824; for the 1957 report, see Muir Wood and Melville, "Who Killed the British Earthquake?," 170.
19. Darwin to Milne, 20 February 1840, Letter 560, and Milne to Darwin, 28 March 1840, Letter 562f, Darwin Correspondence Project.
20. Darwin to Milne, 20 February 1840, Letter 560, Darwin Correspondence Project.
21. Drummond, *Comrie Earthquakes*, 12.
22. Drummond, *Comrie Earthquakes*, 3, 4, 5, 12.
23. Drummond, "Table of Shocks," 240.
24. Drummond, *Comrie Earthquakes*, 7.
25. Drummond, *Comrie Earthquakes*, 8.
26. Hunt, *Maxwellians*, 169.
27. Varley reported to the Society of Telegraph Engineers in 1873 that the line between London and Valentia was more strongly disturbed by earth currents on the day preceding an earthquake in India. See John Graves, "On Earth Currents," 115–16.

28. Drummond, *Comrie Earthquakes*, 8.
29. Carment, *Scenes and Legends*, 112.
30. Davison, "Earthquakes in Great Britain (1889–1914)," 358.
31. Musson, "Curious Seismological Monument."
32. Buckle, *Civilization in England*, 87; see chapter 5, below.
33. *Times* (London), 8 October 1863.
34. Davison, *British Earthquakes*, 245.
35. Lowe, "Earthquake of 1863," 60.
36. "The Earthquake," *Times* (London), 8 October 1863, 8.
37. "Earthquakes," *All the Year Round* 3 (1860): 197–201, on 201.
38. Editorial, *Times* (London), 8 October 1863, 6.
39. Musson, "Seismicity: Fatalities in British Earthquakes," 15.
40. "Earthquake in England," *Times* (London), 23 April 1884, 7.
41. Davison, *The Hereford Earthquake*, 224.
42. Milne, Letter to the *Times* (London), 28 September 1881, 8; "The Earth in Its Vigour," *Times* (London), 8 September 1883 (this article was reprinted in the Glasgow and Bristol papers).
43. Milne, "The Recent Earthquake in Essex," 9.
44. Quoted in Orange, "The British Association for the Advancement of Science," 329.
45. Meldola and White, *East Anglian Earthquake*. Meldola was professor of chemistry at Finsbury Technical College, and White was a member of the Geologists' Association.
46. Meldola and White, *East Anglia Earthquake*, vi.
47. Davison, *Hereford Earthquake*, 3.
48. Davison, "British Earthquakes of 1889."
49. H. J., "Charles Davison"; Richards, "Dr. C. Davison," 805.
50. Davison, *Hereford Earthquake*, 4.
51. Musson, "A Critical History of British Earthquakes."
52. Davison, "Robinson Crusoe's Earthquake."
53. Davison, "Effects of Earthquakes on Human Beings." See too Davison, "Effects of an Observer's Conditions."
54. Roger Musson, conversation with author, 29 July 2011; Lovell, *The Dr ATJ Dollar Papers*; Musson, "A Critical History of British Earthquakes."
55. Milne, Letter to the *Times* (London), 29 December 1896, 9.
56. "The Belt of Earthquakes . . . ," *Aberdeen Journal*, 10 September 1886, 1.
57. "People Must Not Run Away . . . ," *Aberdeen Journal*, 14 July 1894, 1.
58. "Comrie as an Earthquake Centre: The Reputation of the Village at Stake," *Pall Mall Gazette*, 25 August 1898, 6.
59. "Comrie as an Earthquake Centre," *Pall Mall Gazette*, 25 August 1898, 6.
60. *Times* (London), 18 December 1896.
61. "Objects of the Society," *Proceedings of the Society for Psychical Research* 1 (1882–83): 3–6, on 4; Oppenheim, *The Other World*; Owen, *The Darkened Room*.
62. "On the Trail of a Ghost," *Times* (London), 12 June 1897, 11.
63. Davison, "On the Comrie Earthquake of July 12, 1895," 77; Davison, "The Earthquakes of Comrie."
64. "On the Trail of a Ghost," *Times* (London), 15 June 1897, 12.
65. "On the Trail of a Ghost," *Times* (London), 21 June 1897, 4.
66. Noakes, "Telegraphy Is an Occult Art."
67. Lang, *The Book of Dreams and Ghosts*, xiv. The reference is to Byron, *Childe Harold's*

Pilgrimage: "and now, the glee Of the loud hills shakes with its mountain-mirth, As if they did rejoice o'er a young earthquake's birth."

68. "On the Trail of a Ghost," *Times* (London), 24 June 1897, 10.

69. "On the Trail of a Ghost," *Times* (London), 6 November 1897, 9. In a final bit of hairsplitting, "A Cynical Observer" responded to "An Earnest Inquirer": "Were he really to inquire earnestly he would find that the noises said to have been heard in Ballechin were not accounted for as the results of 'seismic disturbance.' It was merely suggested that, if there were any noises which could not be referred to ordinary physical causes, then it was more probable that they were due to seismic disturbances than that they were caused by a 'bogie.' There is much virtue in an 'if.'" "On the Trail of a Ghost," *Times* (London), 8 November 1897, 6.

70. Dunning, "The Story of the Earth," 101.

71. Quoted from an unpublished manuscript by Dunning, in Dombois, *Über Erdbeben,* 178.

72. Roger Musson, e-mail to author, 28 October 2010.

73. Roger Musson, e-mail to author, 28 October 2010.

74. Cecić and Musson, "Macroseismic Surveys," 52.

75. Musson, "From Questionnaires to Intensities," 512.

76. Musson and Cecić, "Intensity and Intensity Scales," 12.

77. *New Scientist,* 20 October 1983; "Shaken to the Core," *Times* (London), 27 August 1983.

78. Muir-Wood and Melville, "Who Killed the British Earthquake?"

79. Musson, "Comrie: A Historical Scottish Earthquake Swarm," 479.

80. "Quake Missed Due to Paper Shortage," *Times* (London), 3 March 2008, 27.

CHAPTER THREE

1. Conboy, *Journalism,* 110.

2. Conboy, *Journalism,* 125.

3. Quoted in Conboy, *Journalism,* 122.

4. Le Bon, *La psychologie des foules;* Vyleta, *Crime, Jews, and News,* 72; for media influences on public perceptions of hazards, see Wilkinson, *Anxiety in a Risk Society,* chapter 5.

5. Anderson, *Predicting the Weather,* chapter 2.

6. Hoernes, "Erdbeben in Steiermark 1880," 66. See too Belar, "Praktische Bedeutung."

7. Keilhack, *Lehrbuch der praktischen geologie,* 295.

8. Sieberg, "Makroseismische Bestimmung der Erdbebenstärke," 238. For example, the 1902 earthquake in Salonika was reported to have caused heavy damages and casualties, but Hoernes found evidence of only four deaths. Hoernes, "Erdbeben von Saloniki."

9. "Vom Tage," 727.

10. *Neue Freie Presse,* 7 January 1873, 18 July 1876, 22 and 23 September 1885; Kraus, "Das Erdbeben." Likewise, the newspaper accounts of the 1812 earthquakes in New Madrid, Missouri, were "generally agreed to be intelligible only when many individual accounts from many different places could be collected and compared" (Valencius, "New Madrid Earthquakes").

11. For examples of "participatory journalism" with a meteorological theme, see Markovits, "Rushing into Print."

12. Wood, "Earthquake Reports," 60.

13. Martin, "Earthquake of December, 1874," 191.

14. On the iconography of Victorian weather maps, see Anderson, "Mapping Meteorology." On macroseismic intensity maps, see chapter 4, below.

15. Daniel Martin to Charles Rockwood, 11 May 1875, box 1, book 1, Rockwood Papers, Princeton University Archives.

16. Kafka et al., "Earthquake Activity in the Greater New York City Area"; Kafka, "Public Misperceptions."

17. Burton, "Volcanic Eruptions," 51–52.

18. Martin to Rockwood, 18 June 1875, box 1, book 1, Rockwood Papers, Princeton University Archives.

19. Milne, *Earthquakes and Other Earth Movements*, 276

20. On their sons' stamp collections, see Alexis Perrey to Charles Rockwood, 2 August 1879 and 23 February 1882, box 1, book 1, Rockwood Papers.

21. "Prof. Charles G. Rockwood."

22. Fuchs to Rockwood, 4 April 1880, box 1, book 2, Rockwood Papers.

23. Fuchs to Rockwood, 1 June 1880, box 1, book 2, Rockwood Papers.

24. J. Goodrich (*Boston Post*) to Rockwood, 14 November 1877, box 1, book 1, Rockwood Papers.

25. E. H. Little to Rockwood, 6 May 1876, box 1, book 1, Rockwood Papers.

26. Illegible name in Caracas to Rockwood, 13 August 1883, box 1, book 2, Rockwood Papers.

27. Rockwood, "Notices of Recent American Earthquakes II" (1878), 21.

28. "Shock of an Earthquake," *Times* (London), 13 December 1824.

29. "From Notices in Our Papers . . . ," *Times* (London), 11 December 1839.

30. "Les Tremblements de terre italiens," *Journal de Genève*, 30 December 1908.

31. Cited in Belar, *Laibacher Erdbebenstudien*, 7.

32. Davison, *Founders of Seismology*, 49.

33. Cited in Milne, "Seismology at the British Association," 124.

34. Belar, "Die jüngsten amerikanischen Katastrophen," 13 (originally published in *Neues Wiener Tagblatt*).

35. Davison, *Founders of Seismology*, 88.

36. Te Heesen, *Cut and Paste*.

37. Perry Byerly to Harry O. Wood, 19 February 1930, Wood Papers, Caltech.

38. Te Heesen, *Cut and Paste*, 326, 327.

39. Falb, "Aus dem Leben Rudolf Falbs."

40. Falb, *Umwälzungen im Weltall*.

41. *Neue Freie Presse*, 16 March 1885.

42. "The Falb Earthquake Theory," *New York Times*, 26 June 1887.

43. Hoernes, *Erdbeben-Theorie Rudolf Falbs*, 88.

44. *Neue Freie Presse*, 28 December 1887; the second is cited in Makowsky, "Ueber die Erdbeben-Theorie Rudolf Falb's."

45. Cited in Hammerl and Lenhardt, *Erdbeben in Österreich*, 131.

46. In 1877 Hoernes called Falb's claim "ingenious" (*geistreich*). Hoernes, "Das Erdbeben von Belluno," 45.

47. "Zur Falbschen Theorie," 362; Fuchs, "Vulkanische Ereignisse 1871," 718–19; Schmidt, *Vulkane und Erdbeben*, 1.

48. Falb, *Land der Inca*, xi.

49. Hoernes, *Erdbeben-Theorie Rudolf Falbs*.

50. "Ursache der Erdbeben."

51. Makowsky, "Erdbeben-Theorie Rudolf Falb's," 54.

52. Belar, "Über die praktische Bedeutung der modernen Erdbebenforschung," 30.
53. Fleming, *The Pleasures of Abandonment,* 47.
54. *New York Herald,* 11 August 1884, box 2, book 3, Rockwood Papers.
55. *Puget Sound Daily Courier,* 16 December 1872; *Olympia Weekly Pacific Tribune,* 21 December 1872, box 1, book 1, Rockwood Papers.
56. "Wiener Erdbeben Chronik," *Die Bombe,* 14 November 1880, and *Neue Freie Presse,* 11 November 1880, late edition, 2.
57. "Das Erdbeben," *Der Floh,* 14 November 1880.
58. Twain, *Early Tales,* 2:289. On seismic denialism in nineteenth-century California, see Geschwind, *California Earthquakes,* and chapter 9, below.
59. Twain, *Roughing It,* 1:140.
60. Twain, "The Great Earthquake in San Francisco," 304.
61. Twain, "A Page from a Californian Almanac," 132–34.
62. Salpeter, "Das Erdbeben," 339.
63. "Die Erde will nicht mehr. Es war bloß ein nervöses Zucken,—und der Jammer ist unendlich. Wenn ihr aber wirklich einmal die Geduld reißt?"; Kraus, "Erdbeben" (1908).
64. Reitter, *The Anti-Journalist.*
65. Benjamin, "Karl Kraus," 433, 437.
66. Edward Timms identifies the 1908 essay "Apocalypse: An Open Letter to the Public" as the moment when Kraus first adopted his "apocalyptic" role. Timms, *Karl Kraus,* 206.
67. Kraus, "The Discovery of the North Pole," 56.
68. Beck, *Risk Society.*
69. Kraus, "Das Erdbeben" (1908), 19.
70. Kraus, "Das Erdbeben" (1908), 23.
71. Kraus, "Das Erdbeben" (1908), 21–22.
72. Kraus, "Das Erdbeben" (1908), 22.
73. Reitter, *Anti-Journalist,* 86.
74. Kraus, "Das Erdbeben" (1908), 23.
75. Kraus, "Antworten des Herausgebers: Geolog," 20.
76. Belar, "Praktische Bedeutung der modernen Erdbebenforschung," 28.
77. Kraus, "Antworten des Herausgebers: Geolog," 20.
78. Kraus, "Messina," 1.
79. "Das, was Eduard Suess so geistvoll den Pulsschlag des Erdballs genannt hat, wird mit wissenschaftlicher Genauigkeit bekannt sein . . . Das wird aber den Pulsschlag der Erde nicht weiter genieren. Und ihre Bonmots sind überraschender." Kraus, "Erdbeben" (1909), 46.
80. Kraus, "Erdbeben" (1909), 46.
81. Kraus, "Nach dem Erdbeben," 22.
82. Schütz, *Der Grubenhund,* 38.
83. "Indem sie den Journalismus hineingelegt hat, hat sie ihre Identität bewiesen und sich selbst dazugelegt." Kraus, "Nach dem Erdbeben," 21.
84. Kraus, "Nach dem Erdbeben," 24.
85. "Kraus, "Das Erdbeben" (1908), 20.

CHAPTER FOUR

1. Forel, "Les tremblements de terre 1879–1880," 463.
2. From the 1890s, articles in the *Schweizerische Bauzeitung* took earthquakes into account as factors in the design of tunnels and buildings; e.g., Locher, "Zum

Jungfraubahnproject von Oberst Locher"; Koch, "Ursachen des Verfalles der Hoch-bauten"; "Die Chemie der hohen Temperaturen," *Schweizerische Bauzeitung* 27 (1896): 133–35; "Ueber die Geschwindigkeit von Erdbebenstössen," *Schweizerische Bauzeitung* 32 (1898): 100; plus several articles on the restoration of buildings in Basel that were damaged by the great earthquake of 1356.

3. Früh, "Ergebnisse fünfundzwanzigjähriger Erdbebenbeobachtungen," 147.

4. Quervain, *Erdbeben der Schweiz 1909*, 1.

5. Gisler et al., *Nachbeben*.

6. Früh, "Ergebnisse fünfundzwanzigjähriger Erdbebenbeobachtungen," 146.

7. Forel, "Tremblement de terre du 30 décembre 1879," 14.

8. Favre, "Tremblement de terre à Fleurier," 134.

9. Früh, "Ergebnisse fünfundzwanzigjähriger Erdbebenbeobachtungen," 146.

10. Westermann, "Disciplining the Earth."

11. Volger, *Erdbeben in der Schweiz*, 2:67.

12. Gisler et al., "The 1855 Visp Earthquake."

13. Daum, *Wissenschaftspopularisierung*, 168.

14. Volger, *Erdbeben in der Schweiz*, 67.

15. Volger, *Erde und Ewigkeit*, 266.

16. Daum, *Wissenschaftspopularisierung*, 170, 253.

17. Quoted in Daum, *Wissenschaftspopularisierung*, 406.

18. Volger, *Erdbeben in der Schweiz*, 363, 380, 81.

19. Volger, *Erdbeben in der Schweiz*, 84.

20. Oeser, "Historical Earthquake Theories."

21. Vaněk and Kozák, "First Macroseismic Map," 595, 597. On the popularization of weather maps, see Anderson, *Predicting the Weather*, chapter 5.

22. Gisler et al., "The 1855 Visp Earthquake"; Heim, *Erdbeben und deren Beobachtung*, 10. The first generalizable seismic intensity scale was devised in Prussia by P. N. C. Egen in 1828, with six degrees ranging from "slight traces sensed" to "significant damage." Davison, "Scales of Seismic Intensity."

23. Quervain, "Erdbeben von Wallis," 75.

24. Früh, "Ergebnisse fünfundzwanzigjähriger Erdbebenbeobachtungen," 148.

25. Heim, "Über den Stand der Erdbebenforschung."

26. Heim to Charles Rockwood, n.d., box 2, book 3, Rockwood Papers, Princeton University Archives.

27. Mousson, "Organisation meteorologischer Beobachtungen," 197.

28. Mousson, "Organisation meteorologischer Beobachtungen," 104.

29. Heim to Charles Rockwood, 5 March 1885, box 2, book 3, Rockwood Papers, Princeton University Archives.

30. Heim, *Erdbeben und deren Beobachtung*, 27.

31. Heim, "Zur Prophezeiung der Erdbeben," 141.

32. Heim, *Erdbeben und deren Beobachtung*, 5, 22.

33. Seneca, *Dialogues and Letters*, 112; Williams, "Greco-Roman Seismology," 146.

34. Heim, "Bergsturz und Menschenleben," 128; Heim, "Bergsturz von Elm."

35. Speich, "Draining the Marshlands," 437; Pfister, *Am Tag danach*.

36. The *Helvetii* were the Celtic tribe that inhabited the Swiss Plateau prior to the Roman conquest. Zimmer, *Contested Nation*, 200, 206.

37. Forster, "Bericht der Erdbeben-Commission, 1879–1880," 102.

38. Porter, "How Science Became Technical," 299.

39. Bensaude-Vincent, "Science and Its 'Others.'"

40. "Bergschläge im Simplontunnel"; Sulzer-Ziegler, "Der Bau des Simplon-Tunnels"; Heim, "Ueber die geologische Voraussicht beim Simplon-Tunnel"; Heim et al., "Étude géologique sur le nouveau projet."

41. Brockmann-Jerosch et al., *Albert Heim*, 206.

42. Heim et al., "Étude géologique sur le nouveau projet," 20, 24.

43. Brockmann-Jerosch et al., *Albert Heim*, 181, emphasis in original.

44. For the figure of the cautious and modest engineer, see Christopher Harwood's *Human Soul of an Engineer*. Harwood brilliantly analyzes the Russian novelist Andrei Platonov's development of a philosophically nuanced and literarily dexterous critique of the Stalinist cult of technology at its height. The ideal to which Heim held himself can be compared to the Platonovian hero, the "humane engineer."

45. Quervain, "Erdbeben als Folge von Tunnelbau."

46. Heim had a hand in other progressive causes as well, including temperance, abstinence, and cremation. Brockmann-Jerosch et al., *Albert Heim*.

47. Aubert, "La protection des blocs erratiques"; Bachmann, *Zwischen Patriotismus und Wissenschaft*.

48. Brockmann-Jerosch et al., *Albert Heim*, 24.

49. Heim, *Sehen und Zeichnen*, 13, 21.

50. Heim, *Erdbeben und deren Beobachtung*, 23.

51. Taussig, *Mimesis and Alterity*, 81.

52. Tributsch, *When the Snakes Awake*.

53. Daston and Galison, "The Image of Objectivity."

54. Heim, "Über den Stand der Erdbebenforschung," 147.

55. Forel, *Le Léman*, 1:x.

56. Forel, *Handbuch der Seenkunde*, 668–69.

57. Forel, *Le Léman*, 1:xi.

58. Forel, *Handbuch der Seenkunde*, 171.

59. Forel, "Tremblement de terre du 30 décembre 1879," 1, 2; Forel, "Les tremblements de terre de 1879 à 1880," 463, 481.

60. Forel, "Tremblements de terre de 1879 à 1880," 463, 464.

61. Forel, "Tremblements de terre de 1879 à 1880," 465.

62. Forel, "Tremblement de terre du 30 décembre 1879," 8.

63. Forel, "Tremblements de terre de 1879 à 1880," 490.

64. Quervain, *Erdbeben der Schweiz 1909*, 9.

65. Früh, "Bericht der Erdbebenkommission 1907/08," 68.

66. Quervain, *Erdbeben der Schweiz 1912*, 12.

67. Forel, "Tremblements de terre de 1879 à 1880," 490.

68. Quervain, *Erdbeben der Schweiz 1909*, 2.

69. Forel, "Tremblements de terre de 1879 à 1880," 465.

70. Forel, "Tremblements de terre de 1879 à 1880," 465.

71. Rossi, "Programma dell'osservatorio," 67–68; Davison, "Scales of Seismic Intensity," 47.

72. Anderson, *Predicting the Weather*; Huler, *Defining the Wind*; Daston, "Empire of Observation"; Hamblin, *Invention of Clouds*.

73. Musson et al., "Comparison of Macroseismic Intensity Scales."

74. Forel, "Tremblements de terre de 1879 à 1880," 4.

75. Quervain, *Erdbeben der Schweiz 1909*, 1.

76. Daston and Galison, *Objectivity*; Daston and Lunbeck, eds., *Scientific Observation*.

77. Forel, "Tremblements de terre de 1879 à 1880," 7.

78. Mercalli, "Scala sismica De Rossi-Forel," 189.

79. Früh, "Ueber die 30-jährige Tätigkeit der Schweizerischen Erdbebenkommission," 65.

80. Früh, "Ergebnisse fünfundzwanzigjähriger Erdbebenbeobachtungen," 146; Davison, "Effects of an Observer's Conditions," 69.

81. Forel, "Tremblements de terre de 1879 à 1880," 8.

82. Cooper, *Inventing the Indigenous;* Mendelsohn, "The World on a Page"; Golinski, *British Weather;* Bowker, *Memory Practices;* Edwards, *A Vast Machine;* Strasser, "Laboratories, Museums, and the Comparative Perspective."

83. Forster, "Bericht der Erdbeben-Commission, 1878–1879," 115.

84. Forel, "Tremblements de terre de 1879 à 1880," 464.

85. Billwiller, "Bericht der Erdbeben-Commission, 1899–1900," 167.

86. Schröter, "Prof. Christian Brügger," x.

87. Schröter, "Prof. Christian Brügger," xviii.

88. Brügger Nachlass, Signatur B564/4, Staatsarchiv Graubünden, Chur.

89. Nyhart, *Modern Nature,* 253–56.

90. Neumayr, *Erdgeschichte,* 263.

91. Forster, "Die Schweizerischen Erdbeben in den Jahren 1884 und 1885," 17.

92. Vidal, *Piaget before Piaget.*

93. Suter, "Prof. Dr. Hans Schardt."

94. Leuba, "Le professeur Hans Schardt," 106.

95. Hs 389:1077.10 and 1077.8, Nachlass Schardt, ETH, Zurich.

96. Quervain, *Erdbeben der Schweiz Jahre 1909,* 3–4.

97. Hs 389:1080.5, Hs 389:1080.2, Hs 389:1077.15, Hs 389:1077.8, Hs 389:1077.10, Nachlass Schardt, ETH, Zurich.

98. Hs 389:1080.5, Nachlass Schardt, ETH, Zurich.

99. Hs 389:1077.2.7, Nachlass Schardt, ETH, Zurich.

100. Hs 389:1080.3, 1080.4, Nachlass Schardt, ETH, Zurich.

101. Brügger Nachlass B 564/4, Staatsarchiv Graubünden, Chur.

102. Forster, "Bericht der Erdbebenkommission" (1888), 128.

103. Mousson, "Organisation meteorologischer Beobachtungen," 223.

104. Vetter, "Regional Development of Science."

105. Phillips, "Friends of Nature."

106. Hs 389:1078.7, Nachlass Schardt, ETH, Zurich.

107. Bel-Perrin to Schardt, 20 February 1909, Hs 389:1077.3, Nachlass Schardt, ETH, Zurich.

108. Hs 389:1077.1.3, Nachlass Schardt, ETH, Zurich.

109. Hs 389:1077.17, Nachlass Schardt, ETH, Zurich.

110. Hs 389:1079.3.2, Nachlass Schardt, ETH, Zurich.

111. Josef Pernter to education ministry, 19 March 1903, ÖStA/AVA/folder 678, 4A/document 9523.

112. Josef Pernter to education ministry, 3 February 1909, ÖStA/AVA/folder 678, 4A/document 4511. Cf. Clark, *Academic Charisma.*

113. Schardt, "Tremblement du terre du 29 mars 1907," 314.

114. "Das Erdbeben vom 26. Mai," *Intelligenzblatt,* 28 May 1910.

115. Forel, "Tremblements de terre dans les cantons de Vaud et de Neuchâtel."

116. Forster, "Bericht der Erdbebenkommission" (1888), 127.

117. Tribelot, "Tremblements de terre ressentis dans le canton de Neuchâtel," 373.

118. Früh, "Ueber die 30-jährige Tätigkeit der Schweizerischen Erdbebenkommission."

119. "Les Tremblements de terre italiens," *Journal de Genève*, 30 December 1908, 3.

120. Schardt, "Les causes du tremblement de terre de Messine," 111.

121. Hs 389:1076.6, Schardt Nachlass, ETH, Zurich.

122. Forster, "Die Schweizerischen Erdbeben in den Jahren 1884 und 1885," 19, 29, 30.

123. Früh, "Bericht der Erdbeben-Kommission für das Jahr 1913/14," 121.

124. Quervain, "Erdbeben von Wallis," 79.

125. Quervain, "Erdbeben von Wallis," 85.

126. Phillips, "Friends of Nature."

CHAPTER FIVE

1. Milne, "Earthquake Effects," 91.

2. For example: "Like cyclones and other electric disturbances in the atmosphere, earthquakes and volcanoes prevail chiefly in the tropical regions" ("New Theory of Earthquakes and Volcanoes," 396).

3. Buckle, *History of Civilization in England*, 1:87–89, 90. Darwin was often quoted to this effect: "Earthquakes alone are sufficient to destroy the prosperity of any country." Yet unlike Buckle, Darwin emphasized the absence of innate differences between Europeans and natives of the New World. Darwin went on to imagine that if England were prone to earthquakes, its civilization would be reduced to bankruptcy, anarchy, and death; "the hand of violence and rapine would go uncontrolled." In private, he added portentously: "Who can say how soon such will happen?" Darwin, *Voyage of the Beagle*, 228, 233; White, "Darwin, Concepción, and the Geological Sublime."

4. Branco, *Wirkungen und Ursachen der Erdbeben*, 54; Middendorf, *Peru*, 138; compare Charles Kingsley on the "idle" colonials plagued by earthquakes in Peru (*Madame How and Lady Why*, 38–39).

5. On nineteenth-century scientific internationalism, see Mazower, *Governing the World*. On the construction of natural hazards as a "tropical" problem and its political consequences, see Bankoff, "The Historical Geography of Disaster: 'Vulnerability' and 'Local Knowledge' in Western Discourse." When the interwar German field of *Geopolitik* became concerned with earthquakes as a cause of economic instability, its evidence came largely from nineteenth-century travelers' accounts of panic-stricken natives. One scholar devised a simplistic "instability factor" to produce, on a "purely quantitative" basis, an international map of economic risk from earthquakes. Severit, "Anthropogeographische Bedeutung der Erdbeben," 99.

6. Coit, *A Memoir*, 110; Forel quoted in Davis, "Swiss Earth-Quake Commission," 197. On the modern desire for "riskless risk," see Rozario, *Culture of Calamity*.

7. Gilbert, "California Earthquake of 1906," 215.

8. Wallace, *Malay Archipelago*, 2:238.

9. James, "Mental Effects of the Earthquake," 209.

10. Sachs, *Humboldt Current*, 57, 138, 211, 225.

11. *New York Times*, 12 August 1881.

12. Muir, *Our National Parks*, 262–67.

13. *New York Times*, 26 July 1872.

14. Rozario, *Culture of Calamity*, 101.

15. Dauncey, *The Philippines*, 286.

16. Mallet, "Earthquakes," 234.

17. Seebach, "Erdbebenkunde."

18. On "Humboldtian science": Cannon, *Science in Culture*; Tresch, "Even the Tools Will

Be Free"; Sachs, *Humboldt Current*; Dettelbach, "Global Physics and Aesthetic Empire." For critical views of Humboldt's relationship to imperialism, see Pratt, *Imperial Eyes*, and Cañizares-Esguerra, "How Derivative Was Humboldt?"

19. Brown, "The 'Demonic' Earthquake."
20. On 1755 as "an iconographic moment," see Lauer and Unger, "Angesichts der Katastrophe: Das Erdbeben von Lissabon und der Katastrophendiskurs im 18. Jahrhundert," 43.
21. Walls, *Passage to Cosmos.*
22. Humboldt, *Personal Narrative* (1995), 126–29
23. Humboldt, *Personal Narrative* (1995), 130.
24. Janz et al., "Einleitung," 22.
25. Humboldt, *Personal Narrative* (1995), 130.
26. Humboldt, *Personal Narrative* (1995), 129.
27. Dettelbach, "Humboldt between Enlightenment and Romanticism," 20.
28. In *A Natural History of Revolution*, Mary Ashburn Miller argues that metaphors of geophysical cataclysm made the Revolution seem like the product of unstoppable forces rather than individual agency.
29. Walls, *Passage to Cosmos*, 129.
30. Quoted in Dettelbach, "Global Physics and Aesthetic Empire," 283.
31. Humboldt, *Cosmos*, 1:212–13.
32. Humboldt, *Personal Narrative* (1995), 93.
33. Humboldt, *Personal Narrative* (1853), 3:5.
34. Eliot, *Middlemarch*, 189.
35. Struck, "Kenntnis afrikanischer Erdbebenvorstellungen," 90.
36. Severit, "Anthropogeographische Bedeutung der Erdbeben."
37. Tylor, *Primitive Culture*, 1:285.
38. Ratzel, *Völkerkunde*, 1:178.
39. Lasch, "Erdbeben im Volksglauben," 237.
40. White, "Darwin, Concepcíon, and the Geological Sublime."
41. Darwin, *Voyage of Beagle*, 228.
42. John McPhee tells a more recent story of seismic epiphany: the geologist Anita Harris was instantly convinced of the significance of catastrophes to geohistory by her own experience of an earthquake. McPhee, *Annals of the Former World*, 168–71.
43. Fuchs to Rockwood, 1 June 1880, box 1, book 2, Rockwood Papers.
44. Davison, *Founders of Seismology*, 39.
45. Davison, *Founders of Seismology*, 48
46. Quoted in Davison, *Founders of Seismology*, 73.
47. Mallet and Mallet, *Earthquake Catalogue*, 46.
48. Mallet and Mallet, *Earthquake Catalogue*, 61.
49. Mallet and Mallet, *Earthquake Catalogue*, 58.
50. Davison, *Founders of Seismology*, 72.
51. "The Earthquakes," *Times* (London), 13 January 1885.
52. Cisternas, "Montessus de Ballore."
53. Montessus de Ballore, *Ethnographie sismique*, 30.
54. Montessus de Ballore, "La théorie sismico-cyclonique du déluge"; see too Montessus de Ballore, "La Sismologia en la Biblia"; Girard, *La théorie sismique du deluge.*
55. Milne, "Notes," *Nature* 26 (4 May 1882): 16–18, on 17.
56. Milne to Rockwood, 17 December 1881, box 1, book 2, Rockwood Papers.
57. Clancey, *Earthquake Nation*; Fan, "Redrawing the Map."

58. Fan, "Redrawing the Map," 533.
59. Milne, "Prehistoric Remains," 62.
60. Anonymous (John Milne), "The Earthquake," *Japan Gazette*, 13 December 1879; box 1, book 2, Rockwood Papers.
61. Milne, "Earthquake Effects," 109–10.
62. Milne, "Earthquake Effects," 112.
63. Clancey, *Earthquake Nation*, 83.
64. Mallet, "Earthquakes," 225.
65. Herschel, *Manual of Scientific Inquiry*, 333.
66. Neumayer, *Anleitung zu wissenschaftlichen Beobachtungen*, 1st ed., 1:310.
67. Neumayer, *Anleitung zu wissenschaftlichen Beobachtungen*, 2nd ed., 1:xi.
68. Richtofen, *Führer für Forschungsreisende*, iv.
69. Ella, "Physical Phenomena of the South Pacific Island," 563.
70. Nunn, "Fished Up or Thrown Down." On myths as geohistorical evidence, see Piccardi and Masse, *Myth and Geology*.
71. E.g., Friedländer, "Beiträge zur Geologie der Samoa-Inseln" (1910) cites Samoan names for geological formations and refers to their oral tradition as at least "halfway reliable" as a source of historical information for the past three hundred years (on 513).
72. Pyenson, *Cultural Imperialism and the Exact Sciences*, 33–138. On the use of native technicians in volcanology on Montserrat (and its link to deskilling in twentieth-century experimental physics), see Galison, *Image and Logic*, 164.
73. Wagner, "Das Samoa Observatorium," 11.
74. Wegener, "Die seismischen Registrierungen am Samoa-Observatorium 1909–1910."
75. Tetens, "Bericht über die Arbeiten des Samoa-Observatoriums in den Jahren 1904 und 1904," 40.
76. Linke, "Bericht über die Arbeiten des Samoa-Observatoriums in den Jahren 1905 und 1906," 65.
77. Linke, "Bericht über die Arbeiten des Samoa-Observatoriums in den Jahren 1905 und 1906," 65.
78. Wegener, "Die seismischen Registrierungen am Samoa-Observatorium 1909–1910," 325.
79. Tetens, "Bericht über die Arbeiten des Samoa-Observatoriums in den Jahren 1904 und 1904," 41.

CHAPTER SIX
1. Palmer, "California Earthquakes during 1916," 17.
2. Dror, "Seeing the Blush."
3. Hagner, "Psychophysiologie und Selbsterfahrung."
4. Janz et al., *Schwindelerfahrungen*.
5. Review of *De effectibus terraemotus in corpore humano*, by Vincenzo Domenico Mignani, *Monthly Review* 72 (1785): 526.
6. Valencius, "Histories of Medical Geography"; Nash, *Inescapable Ecologies*, 53–54.
7. Mayer-Ahrens, "Beziehungen des Vulkanismus," 294, 332.
8. Schivelbusch, *The Railway Journey*.
9. E. B. [Edgar Bérillon], "Les névroses," 98.
10. Charcot, *Leçons du Mardi*, 36.
11. Micale, *Hysterical Men*, 143.

12. Guinon, *Les agents provocateurs de l'hystérie*, 52.
13. Review of Atlasoff, "Ueber den Einfluss der Erderschütterungen auf den geistigen Zustand," *Zentralblatt für Nervenheilkunde* 7 (1889): 146.
14. Quoted in Neumayr, *Erdgeschichte*, 286; repeated by Hoernes, *Erdbebenkunde*, 133, and Phleps, "Psychosen nach Erdbeben," 384.
15. Phleps, "Psychosen nach Erdbeben," 383.
16. Phleps, "Psychosen nach Erdbeben," 402.
17. Phleps, "Psychosen nach Erdbeben," 404.
18. Phleps, "Psychosen nach Erdbeben," 405.
19. Phleps, "Psychosen nach Erdbeben," 404, emphasis added.
20. Hentig, "Über die Wirkung von Erdbeben auf Menschen," 559.
21. Hentig, "Über die Wirkung von Erdbeben auf Menschen," 562.
22. Matus, *Shock, Memory, and the Unconscious*, 56.
23. Adams, *The Education of Henry Adams*, 495.
24. This was the case even in fields as disparate as British ecology in Africa and Soviet ethnography in Central Asia; e.g., Anker, *Imperial Ecology*; Hirsch, *Empire of Nations*.
25. Lombroso, *Crime, Its Causes and Remedies*, chapters 1–2.
26. Agamben, *State of Exception*, 17; Orihara and Clancey, "The Nature of Emergency."
27. Stierlin, "Störungen nach Katastrophen," 2029.
28. Lombroso and Lombroso, "La Psicologia dei Terremotati," 126, 128.
29. "Homo nobilis u. bête humaine," *Archiv für Kriminologie* 27 (1906): 362–64, on 362.
30. Komine and Maki, "Psychiatric Observations." An American physician noted that "irritable heart," a typical earthquake symptom, had been a frequent complaint during the US Civil War "and again, when the stress and strain in the struggle for existence was markedly intensified during the World War, many observers recorded the same phenomena" (Marshall, "Biological Reactions to Earthquakes").
31. Simson, "Psychische und psychotische Reaktionen," 138.
32. Hentig, *Kosmische, biologische und soziale Krisen*; on Hentig, see Geyer, *Verkehrte Welt*, 101.
33. Hentig, "Wirkung von Erdbeben auf Menschen": quote on 568; on Jewish earthquake sensitivity, 547.
34. Hentig, *Kosmische, biologische und soziale Krisen*, 104–5.
35. Brussilowski, "Beeinflussung der neuropsychischen Sphäre," 462.
36. Simson, "Psychische und psychotische Reaktionen," 134.
37. Brussilowski, "Beeinflussung der neuropsychischen Sphäre," 463–64.
38. Brussilowski, "Beeinflussung der neuropsychischen Sphäre," 444.
39. Brussilowski, "Beeinflussung der neuropsychischen Sphäre," 470.
40. Simson, "Psychische und psychotische Reaktionen," 134.
41. Menand, *The Metaphysical Club*, 138.
42. Quoted in Palm and Carroll, *Illusions of Safety*, 14.
43. James, "Mental Effects of the Earthquake," 209, 211–14. On James's evolutionary thought, see Richards, *Emergence of Evolutionary Theories*, 409–50; on James and the 1906 earthquake, see Solnit, *Paradise Built in Hell*, 49–57.
44. Baelz, "Ueber Emotionslähmung," 718, emphasis added.
45. "Bericht über die Jahresversammlung des Vereins der deutschen Irrenärzte in Berlin," *Psychiatrische Wochenschrift* 1901: 65–68, on 66.
46. Stierlin, "Nervöse und psychische Störungen nach Katastrophen," 2032.
47. Jaspers, *Allgemeine Psychopathologie*, 305–7.
48. Cannon, *The Wisdom of the Body*.

49. Heim, "Notizen über den Tod durch Absturz," 335.

50. Matus, *Shock, Memory, and the Unconscious*, 7.

51. Baelz, "Über Besessenheit."

52. Compare Samuel Prince in what is typically cited as the founding work of disaster studies: "Catastrophe opposes the tendency to eliminate from life everything that requires a calling forth of unusual energies." Prince, *Catastrophe and Social Change*, 53.

53. "Wie ein Seismograph hatten seine empfindlichen Nerven die unterirdischen Erschütterungen schon dann verzeichnet, als andere sie noch völlig überhörten." Heise, *Erinnerungen an Aby Warburg*, 56.

54. "Beide sind sehr empfindliche Seismographen, die in ihren Grundfesten beben, wenn sie die Wellen empfangen und weitergeben müssen." Quoted in Krummel, *Nietzsche und der deutsche Geist*, 2:55.

55. "Man hält die Feder hin, wie eine Nadel in der Erdbebenwarte, und eigentlich sind nicht wir es, die schreiben; sondern wir werden geschrieben." Frisch, *Tagebuch 1946–49*, 19.

56. "Nach dem Erdbeben schlägt man auf die Seismographen ein. Man kann jedoch die Barometer nicht für die Taifune büßen lassen, falls man nicht zu den Primitiven zählen will." Jünger, *Strahlungen*, 9.

57. "Den Ablauf der Geschehnisse zeichnen die 'Stahlgewitter' mit der ganzen Macht der Frontjahre am stärksten, ohne jedes Pathos geben sie das verbissene Heldentum des Soldaten wieder, aufgezeichnet von einem Menschen, der wie ein Seimograph alle Schwingungen der Schlacht auffängt." Quoted in Esposito, *Mythische Moderne*, 320.

58. "Die Feder ist aber nur ein seismographischer Griffel des Herzens. Erdbeben lassen sich damit festhalten, aber nicht vorhersagen." Kafka quoted in Janouch, *Conversations with Kafka*, 47.

59. "Ich . . . denke mir den Verleger—wie soll ich sagen—etwa als Seismograph, der bemüht sein soll, Erdbeben sachlich zu registrieren. Ich will Aeusserungen der Zeit, die ich vernehme, soweit sie mir irgendwie wertvoll erscheinen, überhaupt gehört zu werden, notieren und für die Oeffentlichkeit zur Diskussion stellen. (Seismograph nicht Seismologe sein.)" *Karl Kraus und Kurt Wolff Briefwechsel*, 47.

60. "[Der Dichter] gleicht dem Seismographen, den jedes beben, und wäre es auf Tausende von Meilen, in Vibrationen versetzt. Es ist nicht, daß er unaufhörlich an alle Dinge der Welt dächte. Aber sie denken an ihn. Sie sind in ihm, so beherrschen sie ihn. Seine dumpfen Stunden selbst, seine Depressionen, seine Verworrenheiten sind unpersönliche Zustände, sie gleichen den Zuckungen des Seismographen, und ein Blick, der tief genug wäre, könnte in ihnen Geheimnisvolleres lesen als in seinen Gedichten." Hofmannsthal, "Der Dichter und diese Zeit," 34.

61. "Der Dichter erscheint als Zeiger, als Seismograph, von dem der Gewissenszustand seiner Umwelt abgelesen werden kann." Hesse, *Briefe*, 178–79.

62. Huyssen, "Fortifying the Heart"; Herf, *Reactionary Modernism*.

63. Quoted in Huyssen, "Fortifying the Heart," 16, emphasis added.

64. Jünger, *Copse 125*, 1.

65. Treitel, *Science for the Soul*, 141.

66. Quoted in Taussig, *Mimesis and Alterity*, 254.

67. Adams, *Education of Henry Adams*, 495; Shelley, *Frankenstein*, 47, 216; Nordau, *Degeneration*, 7.

68. Spector, *Prague Territories*, 110.

69. Taussig, *Mimesis and Alterity*, 255.

70. Bucholtz, "Vorwort," n.p.

71. Jünger, "Über die Gefahr," 16.
72. Long, "Ernst Jünger, Photography, Autobiography, and Modernity."

CHAPTER SEVEN

1. Mach, *Analysis of Sensations*, 155.
2. Between 1867 and 1918, the two halves of the dual monarchy were bound by a joint army, common foreign policy and currency, and economic agreements, as well as by allegiance to the emperor. Within each half of the monarchy, burgeoning national movements competed for privileged positions, with the Czechs and Poles in Austria and the Croatians in Hungary winning limited concessions to independence. For historiographical discussion, see Coen, "Rise, Grubenhund"; for climatology as exemplary of Habsburg science, see Coen, "Imperial Climatographies."
3. Kozák and Plešinger, "Regular Seismic Service," 105; Lehner, *Geschichte der Naturkatastrophen*, 143.
4. *Für Laibach.*
5. Wo gab's Slovenen da, wo Deutsche,
Wo Sprachenzwist und Kampf um Macht?
Nur Menschen gab es, bange, bleiche,
Nur Menschen, zitternd, angstverwirrt,
Nur Brüder, Einige und Gleiche—
Die nun das Unglück—coalirt.

"Naturlehre," *Der Floh*, 21 April 1895.
6. Neumayr, *Erdgeschichte*, 1:305–6.
7. Reprinted in Hochstetter, "Ueber Erdbeben," 11.
8. Neumayr, *Erdgeschichte*, 306.
9. Hantken, *Erdbeben von Agram*.
10. Skoko and Mokrović, *Andrija Mohorovičić*, 94–95.
11. Neumayr, *Erdgeschichte*, 306.
12. Wähner, *Erdbeben von Agram*, 5.
13. Wähner, *Erdbeben von Agram*, 17.
14. Wähner, *Erdbeben von Agram*, 149–52.
15. "Das fürchterliche Getöse [*tutnjava*] begann mit der Erde und den Häuserern immer stärker und stärker zu schütteln" (Wähner, *Erdbeben von Agram*, 153). *Getöse* is commonly translated into English as "roar," *tutnjava* as "boom."
16. Wähner, *Erdbeben von Agram*, 141.
17. Wähner, *Erdbeben von Agram*, 289.
18. Spengler, "Franz Wähner," 310.
19. Ash, "Wissenschaftspopularisierung."
20. Wähner, *Erdbeben von Agram*, 288.
21. Wähner, *Erdbeben von Agram*, 288.
22. Wähner, *Erdbeben von Agram*, 4.
23. Namely, the Styrian Naturwissenschaftlicher Verein and the Carinthian Naturhistorisches Landesmuseum.
24. Hoernes, "Erdbeben und Stoßlinien Steiermarks," 1–2.
25. Albini, "Earthquakes in the Eastern Adriatic."
26. Quoted in Skoko and Mokrović, *Andrija Mohorovičić*, 96.
27. Canaval, "Erdbeben von Gmünd"; Hoernes, "Erdbeben in Steiermark 1880."
28. Hoernes, *Erdbebenkunde*, 168.
29. Hoernes, "Erdbeben in Steiermark," 114.

30. E.g., Suess, *Erdbeben Nieder-Österreichs;* Wähner, *Erdbeben von Agram;* Faidiga, "Erdbeben von Sinj."
31. Canaval, "Erdbeben von Gmünd," 354, emphasis added.
32. Davison, *Hereford Earthquake,* 11.
33. Hoernes, *Erdbeben-Theorie Falbs,* 3.
34. Hochstetter, "Ueber Erdbeben." The two lectures together raised nine hundred gulden for the victims.
35. Hochstetter, "Ueber Erdbeben," 13. See too the popular lecture by geologist Franz Toula, "Ueber den gegenwärtigen Stand der Erdbebenfrage."
36. Wähner, *Erdbeben von Agram,* 4.
37. Günther Hamann observes that although Suess took a "German-Austrian standpoint," he advised tolerance and equality, and warned against "disdain for one's neighbors" (Hamann, "Eduard Suess als liberaler Politiker," 88–89).
38. A Buddhist proverb quoted by Suess in his 1888 rectorial address, "The Progress of the Human Race"; Hamann, "Eduard Suess als liberaler Politiker," 93.
39. *Neue Freie Presse,* 21 July 1898
40. "Die Erde lebt und bebt—und jeder Stillstand bedeutet Weltentod." "Erdbebenkatastrophen und ihre Ursachen," 21. Reprinted in *Die Erdbebenwarte* 1908–9: 17–21, on 21.
41. Clerke, "Earthquakes and the New Seismology," 312.
42. Quoted in Hamann, "Eduard Suess als liberaler Politiker," 93.
43. Westermann, "Overcoming the Division of Labor."
44. Suess, *Über den naturgeschichtlichen Unterricht,* 12.
45. Hoernes, "Erdbeben in Steiermark," 73.
46. Suess, *Erdbeben von Laibach,* 7.
47. Suess, *Erdbeben von Laibach,* 203.
48. Herscher, "Städtebau as Imperial Culture."
49. Vidrih and Mihelič, *Albin Belar.*
50. Bettelheim, ed., *Biographisches Jahrbuch und deutscher Nekrolog,* 7:114.
51. "Zur Geschichte der Gründung der Erdbebenwarte in Laibach," reprinted in Hammerl et al., *Die Zentralanstalt für Meteorologie und Geodynamik,* 87.
52. See Belar's flattering obituary for Joseph Suppan in *Die Erdbebenwarte* 2 (1902–3): 99–100.
53. Vidrih and Mihelič, *Albin Belar,* 127.
54. On the emergence of German and Slovene identities through 1848, see Hösler, *Von Krain zu Slowenien.*
55. Vidrih and Mihelič, *Albin Belar,* 135.
56. See, for instance, his chart comparing three years of human and instrumental records in *Laibacher Erdbebenstudien,* Tafel II.
57. Belar, "Was erzählen uns die Erdbebenmesser," 101.
58. Vidrih and Mihelič, *Albin Belar.*
59. Kozák and Plešinger, "Seismic Service."
60. Commenda, "Aufruf zur Einsendung von Nachrichten über Erdbeben und andere seltene Naturereignisse," 3.
61. Láska, "Die Erdbeben Polens," 1–2.
62. Mojsisovics, "Allgemeiner Bericht und Chronik 1900," 6.
63. Mojsisovics, "Allgemeiner Bericht und Chronik 1900," 71.
64. Minutes of the Meeting of the Earthquake Commission, 1 October 1901, folder 2, carton 1, Archive of the Earthquake Commission, Austrian Academy of Sciences.

65. Minutes of the Meeting of the Earthquake Commission, 9 November 1900, folder 2, carton 1, Archive of the Earthquake Commission, Austrian Academy of Sciences.

66. For details see Coen, "Faultlines and Borderlands." For the parallel case of imperial climatology see Coen, "Climate and Circulation in Imperial Austria."

67. Kozák and Plešinger, "Seismic Service."

68. Jan Surman ("Figurationen der Akademia") refers to a "philosophy of science of small nations": the principle that small nations contribute to world science not in spite of but by virtue of their unique histories and languages.

69. Vidrih and Mihelič, *Albin Belar*, 87.

70. Vidrih and Mihelič, *Albin Belar*, 149.

71. Mach, *Analysis of Sensations*, 155. Mach had previously cited earthquakes as events that disrupt the normal experience of gravity, producing a "sensation of constant ascent" and rendering useless a terrestrial system of coordinates. In the *Analysis of Sensations* Mach was concerned with earthquakes as a problem not of mechanical explanation but of observation.

72. Mach was a very active member at the time; *Lotos* 11 (November 1869): 165.

73. Mach, *Analysis of Sensations*, 19.

74. On other possible reasons for Mach's resignation, see Blackmore, *Ernst Mach*, 80. Blackmore points out that such resignations were not common.

75. Daston and Galison, *Objectivity*.

76. Klotz, "Seismological Work in the Pacific," 300.

77. Steve Fuller notes, "If Planck's principle of unity was *reduction*, Mach's was *translation*" (*Thomas Kuhn*, 118).

78. E.g., Woldřich, "Zemětřesení v Pošumaví."

79. Montgomery, *Science in Translation*.

80. See Coen, "Rise, Grubenhund."

CHAPTER EIGHT

1. The official founding date of the ISA was 1904, but the association's work began with the First International Seismological Conference held in Strasbourg, Germany, in April 1901. Far from representing a cross-section of the world or even of active scientists, most members of the ISA came from German-speaking central Europe. Britain and France had both declined to send official delegates and, along with Austria and the United States, refused their support until 1906–8. Moreover, the selection of Strasbourg as the site of an international scientific organization could not but be perceived in geopolitical terms. Ever since Prussia had won Alsace from the French in 1871, Strasbourg had served as a theater for demonstrations of German cultural power. The premier English seismologist John Milne regarded Gerland's bid to control global seismology as an insult to his own priority and expertise. He attempted to shame the British into backing his own network against Strasbourg's, accusing his government of having "accepted shelter from a Continental aegis" by subscribing to the ISA (Milne, "Recent Earthquakes," 593). At the insistence of the Russian and Japanese delegates to the 1901 meeting, the ISA was formed as an association of states rather than of scholars; it was thus an international organization in the literal sense. Marielle Cremer's dissertation, "Seismik zu Beginn des 20. Jahrhunderts," gives the diplomatic and institutional history of the ISA; see too Kozák, "100-Year Anniversary" and Schweitzer, "Birth of Modern Seismology." I will use the French spelling of Strasbourg since it is the familiar one today, but until 1918 the city was officially known by the German spelling, Straßburg or Strassburg.

2. Gerland, "Die moderne seismische Forschung," 148, 157.
3. In other branches of environmental science in this period, international cooperation seemed to function smoothly. Crawford, *Nationalism and Internationalism;* Crawford et al., *Denationalizing Science;* Rozwadowski, "Internationalism, Environmental Necessity"; Anderson, *Predicting the Weather;* Edwards, *A Vast Machine.*
4. Vetter, introduction to *Knowing Global Environments,* 13.
5. Aubin et al., introduction to *The Heavens on Earth.* According to John Tresch, Arago envisioned a Humboldtian, democratic future for the observatory sciences ("The Daguerreotype's First Frame").
6. Gerland, foreword to the first issue of *Beiträge zur Geophysik* (1887), vi.
7. Sieberg, "Methoden der Erdbebenforschung," 295.
8. Greene, *Geology in the Nineteenth Century,* esp. chapters 6 and 7.
9. Gerland, "Die Kaiserliche Hauptstation für Erdbebenforschung," 465. For Gerland's own perfunctory version of a macroseismic questionnaire, see Sieberg, *Erdbebenkunde* (1904), 259.
10. Montessus de Ballore, "Mémorandum pour la conférence sismologiques internationale," 134.
11. Cited in Bernard, "Erdbebenstudien des Grafen de Montessus de Ballore," 133.
12. Montessus de Ballore, "Les visées de la sismologie moderne," 20.
13. For inevitable constraints on the internationalization of disaster management, see Burton et al, *Environment as Hazard,* chapter 7.
14. Gerland, "Die moderne seismische Forschung," 151.
15. Gerland, "Die kaiserliche Hauptstation für Erdbebenforschung," 433.
16. Gerland, foreword to the first issue of *Beiträge zur Geophysik,* xv.
17. Gerland, foreword to the first issue of *Beiträge zur Geophysik* , xxix.
18. Gerland, "Die kaiserliche Hauptstation für Erdbebenforschung," 431.
19. Gerland, "Die moderne seismische Forschung," 152.
20. Gerland, "Die moderne seismische Forschung," 152.
21. Quoted in Sieberg, *Handbuch,* 313.
22. Aubin et al., introduction to *The Heavens on Earth,* 7; see too the essays in that volume by Theresa Levitt and Ole Molvig. On industrial and military applications of seismology in the First World War, see Cremer, "Seismik zu Beginn des 20. Jahrhunderts," 193–208.
23. Jaehnike, "Das Gebäude der Kaiserlichen Hauptstation."
24. Gerland, "Die kaiserliche Hauptstation für Erdbebenforschung," 427.
25. Gerland, "Moderne Erdbebenforschung," 448.
26. Gerland, "Moderne Erdbebenforschung," 449.
27. Montessus de Ballore, "Visées de la sismologie modern," 12
28. Gould, *Time's Arrow,* 67–70. Thanks to Andre Wakefield for this reference.
29. Montessus de Ballore, "Introduction à un essai de description sismique," 352.
30. Milne, "Seismological Observations and Earth Physics," 2.
31. Quoted in Rudolph, *Verhandlungen der Zweiten Internationalen Seismologischen Konferenz,* 48.
32. Montessus de Ballore, "Géosynclinaux et Régions à Tremblements de Terre," 244.
33. Montessus de Ballore, "Physique du globe," 609.
34. Westermann, "Eduard Suess' The Face of the Earth."
35. Austrian Academy of Sciences to Imperial Ministry for Religion and Education, 13 November 1902, folder 678_4A, document 36277, Meteorological Institute Papers.
36. Suess, preface to Hobbs, *On Some Principles of Seismic Geology,* 220.

37. Leighly, "Methodologic Controversy in Nineteenth-Century German Geography," 254.

38. Darwin, *Descent of Man*, 90, 230.

39. Boas, *The Mind of Primitive Man*, 13.

40. Gerland, *Aussterben der Naturvölker*, 123, 141.

41. Zimmerman, *Anthropology and Antihumanism*, 11.

42. Gerland, "Immanuel Kant."

43. Gerland, *Anthropologische Beiträge*, 1:iv.

44. Gerland, *Anthropologische Beiträge*, 424.

45. Sarton, "Secret History," 191–92.

46. Montessus de Ballore, "Introduction à un essai de description sismique," 333.

47. Montessus de Ballore, "Introduction à un essai de description sismique," 333, 334.

48. Fréchet, "Past and Future of Historical Seismicity Studies," 135.

49. Rayward, introduction to *European Modernism and the Information Society*, 12. See too Krajewski, *Zettelwirtschaft*.

50. *Verhandlungen des Siebenten Internationalen Geographen Kongress*, 202.

51. Von Kövesligethy, ed., *Procès-Verbaux de l'Association Internationale de Sismologie 1907*, 23.

52. Von Kövesligethy, ed., *Procès-Verbaux de l'Association Internationale de Sismologie 1907*, 22, 195.

53. Forel, "A la Commission du Catalogue de l'Association sismologique internationale," 190–95.

54. Sieberg, "Methoden der Erdbebenforschung," 287.

55. "Réponse à la note 'Denkschrift, betreffend die Herstellung der dem Zentralbureau übertragenen makroseismischen Kataloge,'" 198.

56. Palazzo, "Projet de réforme du catalogue macrosismique," 202–3.

57. "Katalog der im Jahre 1903 bekannt gewordenen Erdbeben," *Gerlands Beiträge zur Geophysik*, Ergänzungsband 3, Vorwort, xi.

58. Montessus de Ballore, "Réponse à la note . . .," 197, 198–99.

59. Forster, "Bericht der Erdbebencommission für 1882/83," 93.

60. Davison, "Scales of Seismic Intensity," 104.

61. Apparently, ISA meetings were conducted primarily in French. The Hungarian representative provided translations into French of remarks in German or English. One attendee wryly noted that language posed a problem, since "becoming a seismological expert does not immediately make one a polyglot." Berloty, "Bulletin Scientifique: Sismologie," 872.

62. Sieberg, "Über die makroseismische Bestimmung der Erdbebenstärke," 230.

63. Sieberg, "Über die makroseismische Bestimmung der Erdbebenstärke," 231–35.

64. Wood and Neumann, "Modified Mercalli Scale"; Musson, "The Comparison of Macroseismic Intensity Scales."

65. Montessus de Ballore, "Earthquake Intensity Scales," 230.

66. Rockwood, "Notes on American Earthquakes No. 15," 7.

67. Townley and Allen, "Descriptive Catalog," 1.

68. Montessus de Ballore, "Earthquake Intensity Scales," 229.

69. Davison, "Scales of Seismic Intensity."

70. Davison, "On Scales of Seismic Intensity" (1921), 129.

71. Author's conversation with Roger Musson, 29 July 2011; Davison, "On Scales of Seismic Intensity" (1921), 129.

72. *Actes de la Conference internationale de bibliographie et de documentation* (1908), 87.

73. Lecointe, "Motion sur la publication annuelle de la bibliographie sismologique," in

Verhandlungen der Intentionationalen Seismologischen Association 1906, 181–83, and discussion at the ISA transcribed in *Verhandlungen 1907*, 21–23.

74. Quoted and translated in van den Heuvel, "Building Society, Constructing Knowledge," 132.

75. Rayward, introduction to *European Modernism*.

76. Quoted and translated in Rayward, introduction to *European Modernism*, 15.

77. Quoted and translated in Rayward, introduction to *European Modernism*, 16.

78. Bigourdan, "Classification bibliographique de la Sismologie actuelle."

79. Berloty, "Bulletin Scientifique: Sismologie," 876.

80. This discussion draws on Bowker, *Memory Practices*, and Bowker, "Biodiversity Data Diversity."

81. Guidoboni and Eben, *Earthquakes and Tsunamis in the Past*, chapter 10.

82. Rohr, "The Danube Floods and Their Human Response and Perception (14th to 17th C)."

83. Discussion of Gerland, "Die moderne seismische Forschung," 196.

84. Von Kövesligethy, ed., *Procès-Verbaux des séances de la deuxième conférence de la Commission Permanente et de la première assemblée générale de l'Association Internationale de Sismologie réunie a la Haye du 21 au 25 septembre 1907*, 192.

85. Montessus de Ballore, "Les visées de la sismologie moderne," 21.

86. Calhoun, "The Imperative to Reduce Suffering."

87. Montessus de Ballore, "Réponse à la note," 199.

88. Pyenson, *Cultural Imperialism*, 39

89. Wiechert, "Entwurf einer Denkschrift über seismologische Beobachtungen in den Deutschen Kolonien," 314.

90. Schweitzer, "Birth of Modern Seismology"; see the list of stations in Schweitzer, "Old Seismic Bulletins to 1920," table 1.

91. Pyenson, *Cultural Imperialism*, 126–35.

92. Kaddoura, "A Tremendous Privilege," 489; Barth, "Politics of Seismology."

93. Tilley, *Africa as a Living Laboratory*; Cremer, "Seismik zu Beginn des 20. Jahrhunderts," 118; Crawford, *Nationalism and Internationalism*.

CHAPTER NINE

1. Willis, "A Study of the Santa Barbara Earthquake of June 29, 1925," 256.

2. Buckle, *History of Civilization in England*, 71.

3. Congressman Abram S. Hewitt, quoted in Sachs, *Humboldt Current*, 254.

4. Shaler, "The Stability of the Earth," 270. For the many sides of Shaler's thought—utilitarian, romantic, conservationist, racist—see Livingstone, *Nathanial Southgate Shaler and the Culture of American Science*; see too Kingsland, *Evolution of American Ecology*, and Hones, "Distant Disasters, Local Fears." On disasters in California, see Davis, *Ecology of Fear*.

5. Shaler, *The Story of Our Continent*, 246.

6. Shaler, "The Stability of the Earth," 269, 267, 268.

7. Shaler, "California Earthquakes," 351–60, on 359.

8. Hones, "Distant Disasters, Local Fears," 185; Rozario, *Culture of Calamity*.

9. Nash, *Inescapable Ecologies*, 53–54.

10. Geschwind, *California Seismology*; Meltsner, "Public Communication"; Steinberg, *Acts of God*, chapter 2; Proctor and Schiebinger, eds., *Agnatology*.

11. Meltsner, "Communication of Scientific Information"; Geschwind, *California Earthquakes*, 12–18.

12. Geschwind, *California Earthquakes,* 17.
13. Soulé et al., *The Annals of San Francisco,* 165.
14. Kevles, *The Physicists,* 49.
15. Şengör, *The Large-Wavelength Deformations of the Lithosphere,* 225.
16. Sachs, *Humboldt Current,* chapter 7.
17. Whitney, "Earthquakes," 608.
18. Whitney, "Owen's Valley Earthquake.".
19. Reprint from *San Francisco Bulletin* in *Boston Advertiser,* 5 June 1872, box 1, book 1, Rockwood Papers (concerning the 17 May earthquake).
20. Stearns, "Dr. John B. Trask, a Pioneer of Science on the West Coast," 243.
21. Trask, *Report on the Geology of the Coast Mountains, and Part of the Sierra Nevada: Embracing Their Industrial Resources in Agriculture and Mining,* 8.
22. Trask, "Earthquakes in California, 1812–1855."
23. Trask, "Earthquakes in California in 1856," 342.
24. See Aldrich et al., "The 'Report' of the 1868 Haywards Earthquake."
25. Twain, *Roughing It,* 423.
26. Huber, "The San Francisco Earthquakes of 1865 and 1868," 270.
27. Huber, "The San Francisco Earthquakes of 1865 and 1868," 266.
28. Rowlandson, *Treatise on Earthquake Dangers,* 58.
29. Rowlandson, *Treatise on Earthquake Dangers,* 95.
30. Rowlandson, *Treatise on Earthquake Dangers,* 57.
31. Fleming, *Meteorology in America, 1800–1870,* 81–93.
32. Abbe to Rockwood, 19 August 1884, box 2, book 3, Rockwood Papers.
33. Unknown to Rockwood, 17 February 1887, box 2, book 4, Rockwood Papers.
34. Kafka et al., "Earthquake Activity in the Greater New York City Area: Magnitudes, Seismicity, and Geologic Structures."
35. Contained in box 3, Rockwood Papers; see Rockwood, "Notes on American Earthquakes, No. 14."
36. "Miscellaneous Scientific Intelligence," *American Journal of Science* 28, no. 165 (September 1884): 242.
37. *New York Herald,* 12 August 1884, 3.
38. *New York Herald,* 12 August 1884, 4.
39. *New York Herald,* 12 August 1884, 4.
40. *New York Herald,* 12 August 1884, 4.
41. Abbe to Rockwood, 19 August 1884, box 2, book 3, Rockwood Papers.
42. Powell to Rockwood, 12 November 1884, box 2, book 3, Rockwood Papers.
43. "Washington Letter," *Science* 6 (1885): 491–92, on 492.
44. *New York Herald,* 14 November 1884, 12.
45. Davis, "The Work of the Swiss Earth-Quake Commission," on 197.
46. Heim to Rockwood, n.d., received January 1885; Heim to Rockwood, 5 March 1885, box 2, book 3, Rockwood Papers.
47. Vetter, "Regional Development of Science"; biographical sketches of cooperative observers are included in Hinrichs, *Report of the Iowa Weather Service 1886.*
48. Hinrichs, *Report of the Iowa Weather Service 1886,* 14.
49. Hinrichs, *First Annual Report of the Iowa Weather Stations;* Alter, "The Cooperative Weather Bureau Observers of Utah." Only observers involved with telegraphic networks were paid a small sum (Jeremy Vetter, e-mail to author, 20 July 2011).
50. Alter, "The Cooperative Weather Bureau Observers of Utah," 273.
51. Compare Vetter, "Lay Observers, Telegraph Lines, and Kansas Weather," 268.

52. Hinrichs, *Report of the Iowa Weather Service 1886*, 16.
53. Vetter, "Lay Observers, Telegraph Lines, and Kansas Weather: The Field Network as a Mode of Knowledge Production," 264.
54. Pietruska, "Looking Forward." See too Goldstein, "'Yours for Science.'"
55. Bauer, "Monthly Meteorological Reports of Volunteer Observers," 174.
56. Hinrichs, *First Annual Report of the Iowa Weather Stations*.
57. Hinrichs, *Fifth Biennial Report of the Central Station of the Iowa Weather Service*, 13.
58. Francis Nipher to Charles Rockwood, 9 August 1880, box 1, book 2, Rockwood Papers.
59. Alter, "The Cooperative Weather Bureau Observers of Utah," 273.
60. H. Dow to Rockwood, 21 January 1881, box 1, book 2, Rockwood Papers.
61. Fleming, *Meteorology in America*, 162.
62. The Signal Service had been "unfriendly to State Service work" in the 1870s (Nipher, "The Missouri Weather Service: Shall It Be Sustained as a State Service?," 340); in the 1880s, however, under H. A. Hazen and Horace Greeley, the state services enjoyed "liberal and hearty co-operation" (*Annual Report of the Iowa Weather and Crop Service for the Year 1890* [Des Moines: Ragsdale, 1891], 6). The decentralization of the weather service continued when it passed to civilian hands at the Department of Agriculture in 1891.
63. Dutton and Hayden, "Abstract of the Results of the Investigation of the Charleston Earthquake."
64. Steinberg, *Acts of God*, 6.
65. Taber, "Seismic Activity in the Atlantic Coastal Plain," 118.
66. McGee, "Some Features of the Recent Earthquake."
67. Dutton, *Earthquakes in the Light of the New Seismology*, iii.
68. "Notes," *Nature*, 3 May 1888, 16.
69. Rockwood, "Notes on American Earthquakes No. 14," 426.
70. Dutton and Hayden, "Abstract of the Results of the Investigation of the Charleston Earthquake," 489, emphasis added.
71. Oldham, "The New Seismology: Two New Text-Books," 321, emphasis added.
72. Holden, "Earthquakes in California (1888)," 392.
73. Holden, *List of Recorded Earthquakes*, 3.
74. Holden, *List of Recorded Earthquakes*, 18. Holden had already published a proposed local warning system for tornadoes: Holden, "A System of Local Warnings against Tornadoes."
75. Osterbrock, "The Rise and Fall of Edward S. Holden: Part 2," 172.
76. Rothenberg, "Organization and Control: Professionals and Amateurs in American Astronomy, 1899–1918," 308.
77. Quoted in Neubauer, "A Short History of the Lick Observatory," 369.
78. Holden, "The Work of an Astronomical Society," 13.
79. Holden, "Earthquakes and How to Measure Them," 759.
80. Holden, "Earthquake Observations," 73.
81. Holden, "Earthquakes and How to Measure Them," 749.
82. Holden, "Earthquakes and How to Measure Them," 750.
83. Holden, "Earthquakes and How to Measure Them," 758.
84. Maximum velocity $V = \dfrac{2\pi a}{t}$, Maximum acceleration $= \dfrac{V^2}{a}$ (a = maximum displacement or amplitude). Holden, "Note on Earthquake-Intensity in San Francisco."
85. Holden, *A Catalogue of Earthquakes on the Pacific Coast, 1769–1897*, 22.

86. As, for instance, Harry O. Wood did in 1911: "By acceleration is meant the rate of change of speed in the motion of the earth particles. It will be obvious after a little thought that it is not either a rapid motion of the earth particles nor a slow motion which produces the oversetting of objects and the wrecking of structures on the earth's surface—because, for example, there is no effect of this sort in a rapidly but uniformly moving train nor in a vehicle moving with a slow, uniform speed—but it is a sudden, rapid change from a very slow speed or from absolute rest at one instant to a relatively rapid velocity at the instant following, or vice versa. It is *change of speed*, bringing inertia into action, which produces earthquake damage, and the *rate of change of speed* is the acceleration." Wood, "The Observation of Earthquakes: A Guide for the General Observer," 61.
87. See Mendenhall, "On the Intensity of Earthquakes, with Approximate Calculations of the Energy Involved."
88. Mendenhall, "On the Intensity of Earthquakes," 190–95.
89. Neubauer, "A Short History of the Lick Observatory," 370.
90. Neubauer, "A Short History of the Lick Observatory," 370.
91. Osterbrock, "The Rise and Fall of Edward S. Holden: Part 2," 171.
92. Reid, "Records of Seismographs in North America and the Hawaiian Islands."
93. Tarr and Martin, "Recent Change of Level in Alaska."
94. Martin and Tarr, *The Earthquakes at Yakutat Bay*, 15.
95. Martin, "Recent Changes of Level in the Yakutat Bay Region, Alaska," 45.
96. Martin and Tarr, *The Earthquakes at Yakutat Bay*, 26, 62.
97. Martin and Tarr, *The Earthquakes at Yakutat Bay*, 27, 15, 16.
98. Martin and Tarr, *The Earthquakes at Yakutat Bay*, 17.
99. Martin and Tarr, *The Earthquakes at Yakutat Bay*, 64.
100. Gilbert, preface to Martin and Tarr, *The Earthquakes at Yakutat Bay*, 9.
101. Martin and Tarr, *The Earthquakes at Yakutat Bay*, 63.
102. Martin and Tarr, *The Earthquakes at Yakutat Bay*, 63.
103. Hobbs, review of Martin, "Alaskan Earthquake of 1899," 96.
104. Martin, "Alaskan Earthquake of 1899," 405.
105. Tarr and Martin, "Recent Change of Level in Alaska," 42.

CHAPTER TEN

1. http://earthquake.usgs.gov/regional/nca/1906/18april/revolution.php, accessed 5 June 2010.
2. Meltsner, "Public Communication"; Geschwind, *California Earthquakes*; Rozario, *Culture of Calamity*; Steinberg, *Acts of God*.
3. This is one lesson of Peter Galison's *Image and Logic*, which argues, against Thomas Kuhn's concept of paradigm shifts, that theory, experiment, and instruments rarely change at once.
4. Wood to Byerly, 12 March 1928, box-folder 14.6, Wood Papers.
5. Box-folder 24.13, Wood Papers.
6. Hough, *Richter's Scale*, 60.
7. Goodstein, "Seismology Comes to Southern California," 206.
8. Quoted in Hough, *Richter's Scale*, 51.
9. Wood, "Earthquakes in Southern California, Part 1."
10. Wood to Buwalda, 1938, quoted in Goodstein, "Waves in the Earth," 230.
11. Solnit, *A Paradise Built in Hell*, 36–37. On the 1906 earthquake's role in shaping

twentieth-century forms of international aid, see Hutchinson, "Disasters and the International Order."

12. Lawson, *California Earthquake of April 18, 1906*, vol. 1, pt. 2, 374.
13. Gilbert, "San Francisco Earthquake," 97.
14. Lawson, *California Earthquake of April 18, 1906*, 2:4.
15. Winchester, *A Crack in the Edge of the World*, 270.
16. Lawson, *California Earthquake of April 18, 1906*, 2:3.
17. McAdie, "Muir of the Mountains," 20.
18. Quoted in Rozario, *Culture of Calamity*, 113.
19. Norris, *Noon*, 24
20. Atherton's advice is quoted in Rozario, *Culture of Calamity*, 101; see too Solnit, *Paradise Built in Hell*, 13–70; Rozario, *Culture of Calamity*, 101–33. Atherton, Sisters-in-Law, 8.
21. Genthe, *As I Remember* , 88.
22. Lawson, *California Earthquake of April 18, 1906*, vol. 1, pt. 2, 374.
23. Branner, "Geology and the Earthquake," 66.
24. Lawson, *California Earthquake of April 18, 1906*, vol. 1, pt. 2, 380.
25. Lawson, *California Earthquake of April 18, 1906*, 2:48.
26. Richter and McAdie, "Phenomena Connected with the San Francisco Earthquake," 505.
27. Lawson, *California Earthquake of April 18, 1906*, vol. 1, pt. 2, 377.
28. Omori, "Seismic Experiments."
29. Lawson, *California Earthquake of April 18, 1906*, 2:22.
30. Lawson, *California Earthquake of April 18, 1906*, vol. 1, pt. 2, 321.
31. Lawson, *California Earthquake of April 18, 1906*, vol. 1, pt. 1, 4, 162, emphasis added.
32. Lawson, *California Earthquake of April 18, 1906*, 2:49–56.
33. Byerly, review of the 1969 edition of the Lawson Report, 2089.
34. Rozario traces the effects of the earthquake on urban planning in San Francisco (*Culture of Calamity*, 75–100).
35. Austin, "The Temblor: A Personal Narration," 359.
36. Quoted in Sachs, *Humboldt Current*, 340.
37. Kroeber, "Earthquakes," 322–23; Barrett, "Indian Opinions of the Earthquake of April, 1906," 324–25.
38. Chamberlin, "Letter to George Darwin, June 30, 1906," 66; quoted in part in Oreskes, *Rejection of Continental Drift*, 132.
39. Geschwind, *California Earthquakes*, 36.
40. Geschwind, *California Earthquakes*, 7.
41. Geschwind, *California Earthquakes*, 7–8.
42. See Jeremy Vetter's insightful review of Geschwind's book: *Environmental History 7* (2002): 701–2.
43. *Publications of the Astronomical Society of the Pacific; Sierra Club Bulletin.*
44. Townley, "John Casper Branner," 1.
45. Geschwind, *California Earthquakes*, 40.
46. Geschwind, *California Earthquakes*, 41–42.
47. "The Seismological Society of America," *Science* 25 (1907): 437.
48. Byerly, "History of the Seismological Society of America."
49. Koelsch, "Alexander McAdie."
50. Branner, "Earthquakes and Structural Engineering," 3.
51. Branner to W. H. Hobbs, 15 February 1911, box 9, 30:453, Branner Papers. In 1909,

fifteen Jesuit colleges across the United States formed the Jesuit Seismological Service, but they showed little interest in enlisting the "laity"—in either the religious or scientific sense. See Geschwind, "Embracing Science and Research."

52. Branner to Taber, 27 March 1911, box 9, 31:53, Branner Papers.
53. Branner, *How and Why Stories*, vi.
54. Kevles, *The Physicists*, 104.
55. Branner, "Survey of the Coal Fields of Arkansas," 537.
56. Branner, "The Policy of the U.S. Geological Survey," 728.
57. Townley, "J. C. Branner."
58. Branner, "Seismologic Work on the Pacific Coast," 6.
59. Branner to Willis Moore, 23 February 1910, box 8, 28:264–66, Branner Papers.
60. Branner to Frank Briggs, 5 March 1910, box 8, 29:88, Branner Papers.
61. "To Foretell 'Quakes,'" *Washington Post*, 15 May 1910. Walcott hinted at this intention in 1909, and he had consulted on the matter with Harry Reid in 1910 (Walcott, "Progress of the Smithsonian Institution," 526; Branner to Andrew Palmer, 27 August 1912, box 10, 33:216; on Walcott, see too Branner to Taber, 13 March 1914, box 11, 36:181, Branner Papers).
62. "Topics of the Times," *New York Times*, 28 August 1911, 6.
63. Kevles, *The Physicists*, 53.
64. Branner to Marvin, 13 May 1914, box 11, 36:310, Branner Papers.
65. Palmer, "Inauguration of Seismological Work."
66. Palmer, "California Earthquakes during 1918."
67. Palmer, "California Earthquakes during 1916," 1.
68. Palmer, "California Earthquakes during 1915," 16.
69. Palmer, "California Earthquakes during 1919," 4.
70. Humphreys, "Seismology," 687.
71. Geschwind, *California Earthquakes*, 46.
72. Branner to Chas. N. Gould, 18 September 1911, box 9, 31:289, Branner Papers.
73. Branner to J. S. Rossiter, 5 September 1911, box 9, 31:239, Branner Papers.
74. Branner to J. S. Rossiter, 13 September 1911, box 9, 31:272, Branner Papers.
75. Branner to Gordon Surr, 2 October 1911, box 9, 31:340, Branner Papers.
76. Templeton, "Central California Earthquake."
77. Branner, "Earthquakes and Structural Engineering," 2.
78. Branner, "Earthquakes and Structural Engineering," 5.
79. Carl S. Clemans to J. C. Branner, 26 February 1911, box 43, folder 203.
80. Branner, "Earthquakes and Structural Engineering," 4.
81. Beal, "The Earthquake in the Santa Cruz Mountains," 215.
82. Beal, "The Earthquake in the Santa Cruz Mountains," 215, 216.
83. Beal, "The Earthquake at Los Alamos, Santa Barbara County, California, January 11, 1915," 22.
84. Beal, "The Earthquake in the Imperial Valley, California, June 22, 1915," 130, 148.
85. Beal, "The Earthquake in the Imperial Valley," 132.
86. Taber, "The Los Angeles Earthquakes of July, 1920," 66; Geschwind, *California Earthquakes*, 50–51.
87. Taber et al., "The Earthquake Problem in Southern California," 285.
88. Taber, "Los Angeles Earthquakes of July, 1920," 67, 75.
89. Taber, "Los Angeles Earthquakes of July, 1920"; Taber et al., "Earthquake Problem in Southern California," 283, 287. He mentioned Chile, Japan, Russia, and England.

90. Distefano, "Disasters, Railway Workers, and the Law in Avalanche Country, 1880–1910."

91. On reinforced concrete, and on boosterism after 1906, see Geschwind, *California Earthquakes*, 28–31.

92. Quoted in Mulholland, *William Mulholland and the Rise of Los Angeles*, 131.

93. Taber et al., "Earthquake Problem in Southern California," 297.

94. Taber et al., "Earthquake Problem in Southern California," 289–93.

95. Hamlin to Lawson, 10 July 1906, Bancroft Library, University of California, Berkeley, http://bancroft.berkeley.edu/.

96. Hamlin, "Earthquakes in Southern California," 20.

97. Townley, "The San Jacinto Earthquake of April 21, 1918," 56.

98. "Aftershocks of the San Jacinto Earthquake, from Data Collected by Homer Hamlin of Los Angeles, California," 132, 133.

99. Townley, "San Jacinto Earthquake"; Rolfe and Strong, "The Earthquake of April 21, 1918 in the San Jacinto Mountains"; Arnold, "Topography and Fault System of the Region of the San Jacinto Earthquake."

100. Rolfe, "The Southwest Section of the Seismological Society of America," 5.

101. Taber et al., "Earthquake Problem in Southern California," 298.

102. See Friedman, *Appropriating the Weather*, 97–137.

103. Geschwind, *California Earthquakes*, 55.

104. Kevles, *The Physicists*, 110.

105. "The American Meteorological Society," *Bulletin of the American Meteorological Society* 1 (1920): 1–6, on 1, 5.

106. Marvin, "Foreword: Co-Operative Observers' Department."

107. Shepard, "Dear Fellow Co-Ops."

108. Donald R. Whitnah states incorrectly that responsibility for collecting felt reports from cooperative observers remained with the Weather Bureau. Whitnah, *A History of the United States Weather Bureau*, 164.

109. http://www.lib.noaa.gov/noaainfo/heritage/coastandgeodeticsurvey/Joneschapter .pdf.

110. Jones, *Earthquake Investigation in the United States*, 5.

111. Maher, "The United States Coast and Geodetic Survey—Its Work in Collecting Earthquake Reports in the State of California," 78

112. Maher, "The United States Coast and Geodetic Survey," 79.

113. Wood, "California Earthquakes," 62.

114. Harry O. Wood, "The Observation of Earthquakes: A Guide for the General Observer," 49, 51, 52, 62.

115. Wood, "Earthquake Reports."

116. Wood, "Observation of Earthquakes."

117. Apple, "Thomas A. Jaggar, Jr., and the Hawaiian Volcano Observatory."

118. Wood, "Concerning the Perceptibility of Weak Earthquakes and Their Dynamical Measurement."

119. Wood, "Concerning the Perceptibility of Weak Earthquakes," 38.

120. Oldham, "The Depth of Origin of Earthquakes," 74.

121. Wood, "The Observation of Earthquakes: Second Paper," ms., n.d. (post-1942), Wood Papers, emphasis added.

122. Wood, "A Further Note on Seismometric Bookkeeping"; Wood, "'Regional' vs. 'World' Seismology in Relation to the Pacific Basin"; Wood, "The Earthquake Problem in the Western United States," 209.

123. Goodstein, *Millikan's School*, 134.
124. Wood, "Earthquake Problem in Western United States," 201.
125. Wood, "'Regional' vs. 'World' Seismology," 382.
126. Goodstein, *Millikan's School*, 131. Oreskes, "Weighing the Earth from a Submarine: The Gravity Measuring Cruise of the U.S.S. S-21," 65 and 66.
127. Geschwind, *California Earthquakes*, 59, 64.
128. Wood to E. A. Beals, 20 June 1921, Wood Papers.
129. Wood, "California Earthquakes," 64.
130. Byerly to Wood, 7 March 1928; Byerly to Wood, 14 December 1929; Byerly to Wood, 4 December 1931; Wood to Byerly, 7 December 1931.
131. Note that noninstrumental observations had not been on the agenda for the Carnegie's Advisory Committee on Seismology; nor had Wood systematically collected felt reports on Hawaii. Wood, "The Study of Earthquakes in Southern California"; Wood, "The Earthquake Problem in the Western United States"; Wood, "The Seismic Prelude to the 1914 Eruption of Mauna Loa," 40. Geschwind (*California Earthquakes*, 98–99) treats Wood's outreach as simple propaganda.
132. Wood, "Earthquake Reports," 62.
133. Wood to Byerly, 23 May 1927, box-folder 14.7
134. Wood, "Earthquake Reports," 65–68.
135. Waters, "Memorial to Bailey Willis (1857–1949)"; Geschwind, *California Earthquakes*, 67–96.
136. "Santa Barbara's 'Quake,'" *Los Angeles Times*, 8 September 1883, 3.
137. Willis, "A Study of the Santa Barbara Earthquake of June 29, 1925," 256.
138. Willis, "A Study of the Santa Barbara Earthquake," 275; Geschwind, *California Earthquakes*, 76–7.
139. Willis, "A Study of the Santa Barbara Earthquake," 264
140. Willis, "A Study of the Santa Barbara Earthquake," 263.
141. Willis, "Coöperation," 120.
142. Willis, "Coöperation," 120. On Willis's involvement in the conservation movement, see Geschwind, *California Earthquakes*, 67.
143. Willis, "Essays on Earthquakes," 35–36.
144. Willis, "A Study of the Santa Barbara Earthquake," 261, emphasis added.
145. Willis, "The Next Earthquake," 379.
146. Wood, "Earthquake Reports," 60.
147. Wood, "Earthquake Reports," 61, 62, emphasis in original.
148. Wood to Byerly, 26 September 1927, Wood Papers, box-folder 14.7.
149. Wood to Byerly, 12 November 1927, box-folder 14.7.
150. Byerly to Wood, 14 November 1927, box-folder 14.7.
151. Wood to Byerly, 23 May 1927, box-folder 14.7.
152. Wood to Byerly, 5 March 1928, box-folder 14.6.
153. Wood to Byerly, 12 March 28, box-folder 14.6.
154. Byerly to Wood, 22 March 1930, box-folder 14.6
155. Byerly to Wood, 13 April 1928, box-folder 14.6
156. Byerly to Wood, 13 April 1928, box-folder 14.6; see too Bolt, "Memorial: Perry Byerly (1897–1978)," *BSSA* 69 (1979): 928–45, on 932.
157. Byerly to Wood, 5 July 27, box-folder 14.7
158. Wood to Byerly, 7 July 1927, box-folder 14.7
159. Louderback, in Bolt, "Memorial: Perry Byerly," 934.

160. Bolt, "Memorial: Perry Byerly," 934.
161. Bolt, "Memorial: Perry Byerly," 935.
162. Byerly to Wood, 19 November 1930, box-folder 14.6.
163. Byerly, "Northern California Earthquakes, April 1932–1933," 117.
164. Wood to Byerly, 20 February 1930, box-folder 14.6.
165. "Members of the Seismological Society of America: December 20, 1921," *BSSA* 11 (1921): 205–17.
166. Wood to Mrs. M. Sweet, 29 October 1925, and letters from Sweet to Wood, box-folder 3.1.
167. Wood to Lawrence F. Cook, 24 December 1932, box-folder 14.6.
168. Wood to R. W. Lohman, 26 September 1933, box-folder 14.6.
169. Warne, "Earthquake Waves."
170. Gutenberg et al., "The Earthquake in Santa Monica Bay, California, on August 30, 1930."
171. Wood, "Earthquake Reports."
172. Wood to Byerly, 10 March 1930, box-folder 14.6.
173. A copy of the text is included in Wood to Byerly, 6 February 1930, box-folder 14.6.
174. Byerly to Wood, 19 March 1930, box-folder 14.6; Wood to Byerly, 22 April 1930, box-folder 14.6.
175. Wood to Byerly, 25 March 1928, box-folder 14.6.
176. Wood to Byerly, 20 February 1930, box-folder 14.6.
177. Harry O. Wood and Frank Neumann, "Modified Mercalli Scale of 1931."
178. "Discussion of President McAdie's Address." McAdie responded that Marvin had "an unnecessarily gloomy outlook regarding instruments," 189.
179. Oldham, "The Depth of Origin of Earthquakes," 73, 74.
180. Charles Richter, interview by Ann Scheid, Pasadena, CA, 15 February–1 September 1978, Oral History Project, California Institute of Technology Archives, http://resolver.caltech.edu/CaltechOH:OH_Richter_C, 19, accessed 10 June 2010.
181. Hough, *Richter's Scale*, 212–40.
182. Wood to Byerly, 20 February 1930, box-folder 14.6.
183. The canonical interpretation of the origins of the Richter scale is Goodstein, "Waves in the Earth: Seismology Comes to Southern California," which is based on Richter's publications and on Scheid's interview.
184. Richter interview, 33.
185. See Byerly, "Seismic Intensity Scales."
186. Wood to Byerly, 23 July 1928, box-folder 14.6.
187. Wood to Byerly, 26 March 1930, box-folder 14.6.
188. Geschwind, *California Earthquakes*, 105.
189. Wood, "Preliminary Report on the Long Beach Earthquake," 51.
190. Wood to *Los Angeles Times*, 4 May 1933, A4
191. Richter, "An Instrumental Magnitude Scale," 14.
192. Byerly, "History of the Seismological Society of America."
193. Quoted in Cecić et al., "Do Seismologists Agree upon Epicentre Determination from Macroseismic Data? A Survey of ESC Working Group 'Macroseismology,'" 1015.
194. Goodstein, *Millikan's School*, chapter 7.
195. Knopoff, "Beno Gutenberg," 131.
196. Quoted Goodstein, "Waves in the Earth," 230.
197. Quoted Goodstein, "Waves in the Earth," 213.

198. Geschwind, *California Earthquakes*, 118.
199. Wood, "The Observation of Earthquakes: Second Paper," ms., n.d. (post-1942).
200. Burton et al., *Environment as Hazard*, 2nd ed., 34.
201. Byerly, "Seismic Intensity Scales."

CONCLUSION

1. Davis, *Late Victorian Holocausts*, chapter 7; Nash, *Inescapable Ecologies*; Abbasi, *Americans and Climate Change*, 24.
2. For example, a 1973 article on seismic risk aimed at an engineering audience opened with the observation, "The technological age in which we live is marked among other things by large structures, construction of which would have been impossible just half a century ago." Wohnlich, "Erdbebenprognose und seismisches Risiko," 1139.
3. Musson and Henni, "Probabilistic Seismic Hazard Mapping," 385; Lapajne, "MSK Intensity Scale and Seismic Risk."
4. Ina Cecić, e-mail to author, 18 November 2010.
5. "First AB Workshop on Macroseismic Methods," Polyče, Slovenia, 9–11 May 1989; thanks to Ina Cecić for providing a scan of the workshop proceedings.
6. Ina Cecić, e-mails to author, 18 November 2010 and 2 February 2011. An important question for macroseismological research today is how best to use an observed intensity distribution to locate the fault that caused the tremor. See Cecić et al., "Do Seismologists Agree?"
7. Tertulliani et al., "Unification of Macroseismic Data Collection Procedures"; Musson, "Intensity and Intensity Scales"; author's conversation with Roger Musson, 29 July 2011.
8. The International Nuclear and Radiological Event Scale User's Manual, 1. On Fukushima, see, e.g., "Fukushima Crisis Raised to Level 7, Still No Chernobyl," http://www.newscientist.com/blogs/shortsharpscience/2011/04/fukushima-crisis-raised-to-lev.html, accessed 19 July 2011.
9. Geschwind, *California Earthquakes*, 142.
10. Gyorgy, *No Nukes*, 120–21.
11. Kautz, *Opening the Inner Eye*, 3.
12. Kautz, *Opening the Inner Eye*, 182.
13. Tributsch, *When the Snakes Awake*.
14. Fan, "'Collective Monitoring, Collective Defense.'"
15. Tributsch, *When the Snakes Awake*, 131.
16. Fan, "'Collective Monitoring, Collective Defense;'" I thank Fa-ti Fan for clarifying this point in conversation.
17. Fan, "'Collective Monitoring, Collective Defense.'"
18. Dengler and Dewey, "Households Affected by the Northridge Earthquake," 444.
19. "Trial over Earthquake in Italy Puts Focus on Probability and Panic," *New York Times*, 4 October 2011.
20. Cornell, "Engineering Seismic Risk Analysis."
21. Fortun, "From Bhopal to the Informating of Environmentalism," 285.
22. Mileti, "Public Perceptions of Seismic Hazards," 18, emphasis added.
23. Meltsner, "Communication of Scientific Information," 343, 345, 331.
24. Goltz, "Science *Can* Save Us," 492.
25. Wynne, "Strange Weather," 300.

26. "Language therapy" is a concept associated with Wittgenstein, but Steve Fuller argues convincingly that it also describes Mach's goals, for instance, his engagement with forms of "folk science." Fuller, *Thomas Kuhn*, 119.
27. Canguilhem, "The Living and Its Milieu," 27.
28. Shaler, "The Stability of the Earth," 259.
29. Button, *Disaster Culture*.
30. http://geology.about.com/od/quakemags/a/didyoufeelit.htm, accessed 14 February 2012; http://earthquake.usgs.gov/earthquakes/dyfi/background.php, accessed 24 January 2011. See too Allen, "Transforming Earthquake Detection."
31. Jung, "Experiences Concerning the Psychic Life of the Child," 140, 144. For the biographical context, see Bair, *Jung*.
32. Jung, "Experiences Concerning the Psychic Life of the Child," 141.
33. This recalls the Austrian painter Alfred Kubin's response to a proposed treatment for his depressive delusions: "They want to take away my fear, but it is my only asset!" (Dittmar, "Mein Kapital: Die Angst!")
34. Only in its second generation did psychoanalysis begin to acknowledge the sensitivity of psychic health to the natural environment. As the Hungarian analyst Michael Balint observed, "without water it is impossible to swim, without earth impossible to move on." Quoted in Sedgwick, "The Weather in Proust."
35. Kraus, "Erdbeben" (1910), 5, emphasis added.
36. Kraus, "Nestroy und die Nachwelt. Zum 50. Todestag," 16.
37. Quoted in Hamann, "Eduard Suess."

BIBLIOGRAPHY

ARCHIVAL COLLECTIONS

Charles Greene Rockwood Collection on Earthquakes, Princeton University Library, Princeton, NJ

Christian Brügger Papers, Graubünden Canton Federal Archive, Chur, Switzerland

Darwin Correspondence Project, www.darwinproject.ac.uk

Earthquake Commission Papers, 1895–1919, Austrian Academy of Sciences, Vienna

John Casper Branner Papers, Stanford University Archives, Stanford, CA

Hans Schardt Papers, Library of the Swiss Federal Institute of Technology (ETH), Zurich

Harry O. Wood Papers, California Institute of Technology Archives, Pasadena, CA

Meteorological Institute Papers, 1849–1914, General Administrative Archives, Austrian National Archives, Vienna

ABBREVIATIONS

Bulletin of the Seismological Society of America: BSSA

Verhandlungen der Schweizerischen Naturforschenden Gesellschaft: VSNG

PRINTED SOURCES

Abbasi, Daniel R. 2006. *Americans and Climate Change: Closing the Gap between Science and Action.* http://environment.yale.edu/climate/americans_and_climate_change.pdf. Accessed 17 November 2011.

Adams, Henry. *The Education of Henry Adams.* Boston: Houghton Mifflin, 1918.

"Aftershocks of the San Jacinto Earthquake, from Data Collected by Homer Hamlin of Los Angeles, California." *BSSA* 8 (1918): 131–34.

Agamben, Giorgio. 2005. *State of Exception.* Translated by Kevin Attell. Chicago: University of Chicago Press.

Albini, Paola. "A Survey of Past Earthquakes in the Eastern Adriatic." *Annals of Geophysics* 47 (2004): 675–703.

Alder, Ken. *Engineering the Revolution: Arms and Enlightenment in France, 1763–1815.* Princeton, NJ: Princeton University Press, 1997.

Aldrich, Michele L., Bruce A. Bolt, Alan E. Leviton, and Peter U. Rodda. "The 'Report' of the 1868 Haywards Earthquake." *BSSA* 76 (1986): 71–76.

Allen, Richard M. "Transforming Earthquake Detection?" *Science* 335 (2012): 297–98.

Alter, J. Cecil. "The Cooperative Weather Bureau Observers of Utah." *Monthly Weather Review* 40 (1912): 272–74.

Ambraseys, N. N., and C. F. Finkel. "Seismicity of Turkey and Neighbouring Regions, 1899–1915." *Annales Geophysicae* 5B, no. 6 (1987): 701–26.

"The American Meteorological Society." *Bulletin of the American Meteorological Society* 1 (1920): 1–6.

Anderson, Katharine. "Mapping Meteorology." In *Intimate Universality: Local and Global Themes in the History of Weather and Climate,* edited by James R. Fleming, Vladimir Jankovic, and Deborah R. Coen, 69–92. Sagamore Beach, MA: Science History Publications, 2006.

———. *Predicting the Weather: Victorians and the Science of Meteorology.* Chicago: University Chicago Press, 2005.

Anker, Peder. "Environmental History versus History of Science." *Reviews in Anthropology* 31 (2002): 309–22.

———. *From Bauhaus to Ecohouse: A History of Ecological Design.* Baton Rouge: Louisiana State University Press, 2010.

———. *Imperial Ecology: Environmental Order in the British Empire, 1895–1945.* Cambridge, MA: Harvard University Press, 2001.

Apple, Russell A. "Thomas A. Jaggar, Jr., and the Hawaiian Volcano Observatory." http://hvo.wr.usgs.gov/observatory/hvo_history.html. Accessed 28 June 2010.

Arnold, Ralph. "Topography and Fault System of the Region of the San Jacinto Earthquake." *BSSA* 8 (1918): 68–73.

Ash, Mitchell. "Wissenschaftspopularisierung und bürgerliche Kultur im 19. Jahrhundert. Essay-Rezension," *Geschichte und Gesellschaft* 28 (2002), 322–34.

Atherton, Gertrude. *The Sisters-in-Law.* New York: Frederick A. Stokes, 1921.

Aubert, Daniel. "La protection des blocs erratiques dans le canton de Vaud." *Bulletin de la Société Vaudoise des Sciences Naturelles* 79 (1989): 185–207.

Aubin, David, Charlotte Bigg, and H. Otto Sibum, eds. Introduction to *The Heavens on Earth: Observatories and Astronomy in Nineteenth-Century Science and Culture.* Durham, NC: Duke University Press, 2010.

Austin, Mary. "The Temblor: A Personal Narration." In *The California Earthquake of 1906,* edited by David Starr Jordan, 339–61. San Francisco: A.M. Robertson, 1907.

B., E. [Edgar Bérillon]. "Les névroses provoquées par les tremblements de terre." *Revue de l'Hypnotisme* 4 (1905): 97–98.

Bachmann, Stefan. *Zwischen Patriotismus und Wissenschaft: Die schweizerischen Naturschutzpioniere (1900–1938).* Zurich: Chronos, 1999.

Baelz, Erwin. "Über Besessenheit und verwandte Zustände." *Verhandlungen der Gesellschaft Deutscher Naturforscher und Ärzte* 78 (1906): 120–38.

———. "Ueber Emotionslähmung." *Allgemeine Zeitschrift für Psychiatrie* 58 (1901): 717–21.

Bair, Deirdre. *Jung: A Biography.* New York: Little, Brown, 2003.

Bankoff, Greg. "The Historical Geography of Disaster: 'Vulnerability' and 'Local Knowledge' in Western Discourse." In *Mapping Vulnerability,* edited by Greg Bankoff, Georg Frerks, and Dorothea Hilhorst, 25–36.

Barrett, S. A. "Indian Opinions of the Earthquake of April, 1906." *Journal of American Folklore* 19 (1906): 324–25.

Barrow, Mark. *A Passion for Birds: American Ornithology after Audubon.* Princeton, NJ: Princeton University Press, 1998.

Baudrillard, Jean. "Paroxysm: The Seismic Order." European Graduate School. http://www.egs.edu/faculty/jean-baudrillard/articles/paroxysm-the-seismic-order/ Accessed 1 May 2008.

Barth, Kai-Henrik. "The Politics of Seismology: Nuclear Testing, Arms Control, and the Transformation of a Discipline." *Social Studies of Science* 33 (2003): 743–81.

Bauer, J. W. "The Examination of Monthly Meteorological Reports of Volunteer Observers: Is It Desirable to Report Back to the Volunteer Observer the Errors and Irregularities Discovered in His Report?" In *Proceedings of the Second Convention of Weather Bureau Officials (1901)*, edited by James Berry and W. F. R. Phillips, 172–75. Washington, DC: Government Printing Office, 1902.

Beal, Carl H. "The Earthquake at Los Alamos, Santa Barbara County, California, January 11, 1915." *BSSA* 5 (1915): 14–25.

———. "The Earthquake in the Imperial Valley, California, June 22, 1915." *BSSA* 5 (1915): 130–49.

———. "The Earthquake in the Santa Cruz Mountains, California, November 8, 1914." *BSSA* 4 (1914): 215–19.

Beck, Ulrich. *Risk Society: Towards a New Modernity*. London: Sage, 1992.

Belar, Albin. "Die jungsten amerikanischen Katastrophen im Lichte der modernen Erdbebenforschung." *Die Erdbebenwarte* 6 (1907): 72–75.

———. "Erdbebenkatastrophen und ihre Ursachen." *Die Erdbebenwarte* 8 (1909): 16–21.

———. *Laibacher Erdbebenstudien*. Ljubljana: Kleinmayr & Bamberg, 1899.

———. Obituary for Joseph Suppan. *Die Erdbebenwarte* 2 (1902–3): 99–100.

———. "Über die praktische Bedeutung der modernen Erdbebenforschung." *Erdbebenwarte* 8 (1908–9): 21–30.

———. "Was erzählen uns die Erdbebenmesser von den Erdbeben." *Die Erdbebenwarte* 6 (1906–7): 101–10.

Benjamin, Walter. "Der Erzähler: Betrachtungen zum Werk Nikolai Lesskows." In *Illuminationen*, 385–410. Frankfurt: Suhrkamp, 1977.

———. "Karl Kraus." In *Selected Writings: 1931–1934*, vol. 2, edited by Michael W. Jennings, Howard Eiland, and Gary Smith, 433–58. Cambridge, MA: Harvard University Press, 1999.

———. "The Lisbon Earthquake." In *Selected Writings*, vol. 2, part 2, edited by M. W. Jennings et al., 536–42. Cambridge, MA: Harvard University Press, 1999.

Bennett, Jane. *Vibrant Matter: A Political Ecology of Things*. Durham, NC: Duke University Press, 2010.

Bensaude-Vincent, Bernadette. "A Historical Perspective on Science and Its 'Others.'" *Isis* 100 (2009): 359–68.

"Bergschläge im Simplontunnel." *Schweizerische Bauzeitung* 64 (1914): 68–70.

"Bericht über die Jahresversammlung des Vereins der deutschen Irrenärzte in Berlin." *Psychiatrische Wochenschrift* 1901: 65–68.

Berloty, B. "Bulletin Scientifique: Sismologie." *Études* 44 (1907): 870–87.

Bernard, F. M. "Erdbebenstudien des Grafen de Montessus de Ballore." *Die Erdbebenwarte* 1 (1902): 129–35.

Bettelheim, Anton, ed. *Biographisches Jahrbuch und deutscher Nekrolog*, vol. 7. Berlin: Georg Reimer, 1903.

Bigourdan, M. G. "Projet de classification bibliographique des matières qui constituent la sismologie actuelle." *Comptes rendus hebdomadaires des séances de l'Académie des Sciences* 144 (1907): 113–19.

Billwiller, R. "Bericht der Erdbeben-Commission, 1899–1900." *VSNG* 83 (1900): 166–68.

Blackmore, John T. *Ernst Mach: His Work, Life, and Influence.* Berkeley: University of California Press, 1972.

Bluestone, Jamie Rae. "Why the Earth Shakes: Pre-Modern Understandings and Modern Earthquake Science." PhD diss., University of Minnesota, 2010.

Boas, Franz. *The Mind of Primitive Man.* New York: Macmillan, 1911.

Bolt, Bruce. "Memorial: Perry Byerly (1897–1978)." *BSSA* 69 (1979): 928–45.

Bourke, Joanna. *Fear: A Cultural History.* Emeryville, CA: Shoemaker & Hoard, 2005.

Bowker, Geoffrey C. "Biodiversity Data Diversity." *Social Studies of Science* 30 (2000): 643–83.

———. *Memory Practices in the Sciences.* Cambridge, MA: MIT Press, 2005.

Branco, Wilhelm. *Wirkungen und Ursachen der Erdbeben.* Berlin: Gustav Schade, 1902.

Branner, John Casper. "Correspondence Relating to the Survey of the Coal Fields of Arkansas." *Science* 24 (1906): 532–37.

———. "Earthquakes and Structural Engineering." *BSSA* 3 (1913): 1–5.

———. "Geology and the Earthquake." In *The California Earthquake of 1906*, edited by David Starr Jordan, 63–78. San Francisco: A. M. Robertson, 1907.

———. "The Policy of the U.S. Geological Survey and Its Bearing upon Science and Education." *Science* 24 (1906): 722–28.

———. "Suggested Organization for Seismologic Work on the Pacific Coast." *BSSA* 1 (1911): 5–8.

Brockmann-Jerosch, Marie, and Arnold and Helene Heim, *Albert Heim: Leben und Forschung.* Basel: Wepf & Co., 1952.

Brown, Robert H. "The 'Demonic' Earthquake: Goethe's Myth of the Lisbon Earthquake and Fear of Modern Change." *German Studies Review* 15 (1992): 475–91.

Brush, Stephen G. "Discovery of the Earth's Core." *American Journal of Physics* 48 (1980): 705–24.

Brussilowski, L. "Beeinflussung der neuropsychischen Sphäre durch das Erdbeben in der Krim 1927." *Zeitschrift für die gesammte Neurologie und Psychiatrie* 116 (1928): 442–70.

Bucholtz, Ferdinand. "Vorwort." In *Der Gefährliche Augenblick: Eine Sammlung von Bildern und Berichten*, edited by Ferdinand Bucholtz. Berlin: Junker und Dünnhaupt, 1931.

Buckle, Henry Thomas. *History of Civilization in England*, vol. 1, 2nd edition. New York: Appleton, 1860.

Burton, Ian, Robert W. Kates, and Gilbert White. *The Environment as Hazard.* 2nd edition. New York: Guilford Press, 1993.

Burton, Captain Richard. "The Volcanic Eruptions of Iceland in 1874 and 1875." *Proceedings of the Royal Society of Edinburgh* 9 (1875–76): 44–58.

Button, Gregory. 2010. *Disaster Culture: Knowledge and Uncertainty in the Wake of Human and Environmental Catastrophe.* Walnut Creek, CA: Left Coast Press, 2010.

Byerly, Perry. "History of the Seismological Society of America." *BSSA* 54 (1964): 1723–41.

———. "Northern California Earthquakes, April 1932–1933." *BSSA* 1934: 115–17.

———. Review of the 1969 edition of the Lawson Report. *BSSA* 60 (1970): 2089–91.

———. "Seismic Intensity Scales." *BSSA* 59 (1969): 1735.

Calhoun, Craig. "The Imperative to Reduce Suffering: Charity, Progress, and Emergencies in the Field of Humanitarian Action." In *Humanitarianism in Question: Power, Politics, Ethics*, edited by Thomas G. Weiss and Michael Barnett, 73–97. Ithaca, NY: Cornell University Press, 2008.

Canaval, Richard. "Das Erdbeben von Gmünd am 5. November 1881." *Sitzungsberichte der kaiserlichen Akademie der Wissenschaften* IIa, vol. 86 (1882): 353–409.

Canguilhem, Georges. "The Living and Its Milieu." Translated by John Savage. *Grey Room* 3 (2001): 6–31.

Cañizares-Esguerra, Jorge. "How Derivative Was Humboldt? Microcosmic Narratives in Early Modern Spanish America and the (Other) Origins of Humboldt's Ecological Sensibilities." In *Nature, Empire, and Nation: Explorations in the History of Science in the Iberian World*, 112–28. Stanford, CA: Stanford University Press, 2006.

Cannon, Susan Faye. *Science in Culture: The Early Victorian Period.* New York: Science History Publications, 1978.

Cannon, Walter Bradford. *The Wisdom of the Body.* New York: Norton, 1932.

Cantor, Geoffrey, and Michael J. S. Hodge, eds. *Conceptions of Ether: Studies in the History of Ether Theories, 1740–1900.* Cambridge: Cambridge University Press, 1981.

Carment, Samuel. *Scenes and Legends of Comrie and Upper Strathearn.* Dundee: James P. Mathew & Co., 1882.

Cecić, Ina, and Roger Musson. "Macroseismic Surveys in Theory and Practice." *Natural Hazards* 31 (2004): 39–61.

Cecić, Ina, Roger M. W. Musson, and Massimiliano Stucchi. "Do Seismologists Agree upon Epicentre Determination from Macroseismic Data." *Annali di Geofisca* 39 (1996): 1013–27.

Cisternas, Armando. "Montessus de Ballore, a Pioneer of Seismology: The Man and His Work." *Physics of the Earth and Planetary Interiors* 175 (2009): 3–7.

Chamberlin, Thomas C. "Letter to George Darwin, June 30, 1906." In *Science in America: A Documentary History 1900–1939*, edited by Nathan Reingold and Ida H. Reingold. Chicago: University of Chicago Press, 1981.

Charcot, Jean-Martin. *Leçons du Mardi à la Salpêtrière*, vol. 1. Paris: Lecrosnier & Babé, 1889.

Charvolin, Florian, André Micoud, and Lynn K. Nyhart, eds. *Des sciences citoyennes? La question de l'amateur dans les sciences naturalists.* Paris: Éditions de l'Aube, 2007.

Clancey, Gregory. *Earthquake Nation: The Cultural Politics of Japanese Seismicity, 1868–1930.* Berkeley: University of California Press, 2006.

Clark, William. *Academic Charisma and the Origins of the Research University.* Chicago: University of Chicago Press, 2005.

Clerke, A. "Earthquakes and the New Seismology." *Edinburgh Review* 201 (1905): 294–312.

Clouzot, E. "Histoire et Météorologie." *Bulletin Historique et Philologique du Comité des Travaux Historiques et Scientifiques* 1906: 117–35.

Coen, Deborah R. "Climate and Circulation in Imperial Austria." *Journal of Modern History* 82, no. 4 (December 2010): 839–75.

———. "Faultlines and Borderlands: Earthquake Science in Imperial Austria." In *The Nationalization of Scientific Knowledge in the Habsburg Empire, 1848–1918*, edited by Mitchell Ash and Jan Surman. Basingstoke: Palgrave, 2012.

———. "Imperial Climatographies from Tyrol to Turkestan." Climate and Cultural Anxiety, edited by James Fleming and Vladimir Jankovic, *Osiris* 26 (2011): 45–65.

———. "Rise, Grubenhund: On Provincializing Kuhn." *Modern Intellectual History* 9 (2012): 109–26.

Cohen, Claudine, and Andre Wakefield. Introduction to Gottfried Wilhelm Leibniz, *Protogaea*, edited by Claudine Cohen and Andre Wakefield, xiii–xlii. Chicago: University of Chicago Press, 2008.

Coit, Daniel Wadsworth. *A Memoir.* Cambridge, MA: Harvard University Press, 1908.

Commenda, Hans. "Aufruf zur Einsendung von Nachrichten über Erdbeben und andere seltene Naturereignisse." *Jahresberichte des Vereins für Naturkunde in Österreich ob der Enns zu Linz* 36 (1907): 3–15.

Conboy, Martin. *Journalism: A Critical History.* London: SAGE, 2004.

Cooper, Alix. *Inventing the Indigenous: Local Knowledge and Natural History in Early Modern Europe.* Cambridge: Cambridge University Press, 2007.

Cornell, C. A. "Engineering Seismic Risk Analysis." *BSSA* 58 (1968): 1583–1606.

Crawford, Elisabeth. *Nationalism and Internationalism in Science, 1880–1939.* Cambridge: Cambridge University Press, 1992.

Crawford, Elisabeth, Terry Shinn, and Sverker Sörlin, eds. *Denationalizing Science: The Contexts of International Scientific Practice.* Dordrecht: Kluwer, 1993.

Cremer, Marielle. *Seismik zu Beginn des 20. Jahrhunderts—Internationalität und Disziplinbildung.* Berlin: ERS-Verlag, 2001.

Darwin, Charles. *The Descent of Man, and Selection in Relation to Sex,* vol. 1. New York: Appleton, 1871.

———. *Voyage of the Beagle.* New York: Penguin, 1989.

"Das Erdbeben vom 26. Mai." *Intelligenzblatt,* 28 May 1910, 6.

Daston, Lorraine, ed. *Biographies of Scientific Objects.* Chicago: University of Chicago Press, 2000.

———. "The Empire of Observation, 1600–1800." In Daston and Lunbeck, *Histories of Scientific Observation,* 81–113.

———. "Life, Chances, and Life Chances." *Daedalus* 137 (2008): 5–14.

Daston, Lorraine, and Peter Galison. "The Image of Objectivity." *Representations* 40 (1992): 81–128.

———. *Objectivity.* Cambridge, MA: Zone Books, 2007.

Daston, Lorraine, and Elizabeth Lunbeck, eds. *Histories of Scientific Observation.* Chicago: University of Chicago Press, 2011.

Daum, Andreas. *Wissenschaftspopularisierung im 19. Jahrhundert.* Munich: R. Oldenbourg Verlag, 1998.

Dauncey, Mrs. Campbell. *The Philippines: An Account of Their People, Progress, and Condition,* vol. 15. Boston: J. B. Millet, 1910.

Davids, Karel. "River Control and the Evolution of Knowledge: A Comparison between Regions in China and Europe, c. 1400–1850." *Journal of Global History* 1 (2006): 59–79.

Davis, Mike. *Ecology of Fear: Los Angeles and the Imagination of Disaster.* New York: Metropolitan Books, 1998.

———. *Late Victorian Holocausts: El Niño Famines and the Making of the Third World.* London: Verso, 2001.

Davis, W. M. "The Work of the Swiss Earth-Quake Commission." *Science* 5 (1885): 196–98.

Davison, Charles. "The Earthquakes of Comrie, in Perthshire." *Knowledge: A Monthly Journal of Science* 6 (1909): 143–46.

———. "Earthquakes in Great Britain (1889–1914)." *Geographical Journal* 46 (1915): 357–74.

———. "The Effects of an Observer's Conditions on His Perception of an Earthquake." *Gerlands Beiträge zur Geophysik* 8 (1907): 68–78.

———. "The Effects of Earthquakes on Human Beings." *Nature* 63 (1900): 165–66.

———. *The Founders of Seismology.* Cambridge: Cambridge University Press, 1927.

———. *The Hereford Earthquake of December 17, 1896.* Birmingham: Cornish Brothers, 1899.

———. *A History of British Earthquakes.* Cambridge: Cambridge University Press, 1924.

————. "On Scales of Seismic Intensity and on the Construction and Use of Isoseismal Lines." *BSSA* 11 (1921): 95–130.

————. "On the British Earthquakes of 1889." *Proceedings of the Royal Society of London* 48 (1890): 275–77.

————. "On the Comrie Earthquake of July 12, 1895, and on the Hade of the Southern Border Fault of the Highlands." *Geological Magazine* 3 (1896): 75–79.

————. "Robinson Crusoe's Earthquake: The Realism of Defoe." *Times* (London), 19 March 1934, 17.

————. "Scales of Seismic Intensity." *Philosophical Magazine* 50 (1900): 44–53.

Dawson, John W. "The Reception of Gödel's Incompleteness Theorems." In *Perspectives on the History of Mathematical Logic,* edited by Thomas Drucker, 84–100. Boston: Birkhäuser, 1991.

Dean, Dennis R. "Robert Mallet and the Founding of Seismology." *Annals of Science* 48 (1991): 39–67.

De la Beche, Henry Thomas. *How to Observe: Geology.* London: Charles Knight, 1836.

Dengler, L. A., and J. W. Dewey. "An Intensity Survey of Households Affected by the Northridge, California, Earthquake of 17 January, 1994." *BSSA* 88 (1998): 441–62.

Dettelbach, Michael. "Alexander von Humboldt between Enlightenment and Romanticism." *Northeastern Naturalist* 8 (2001): 9–20.

————. "The Face of Nature." *Studies in the History and Philosophy of Biological and Biomedical Sciences* 30 (1999): 473–504.

————. "Global Physics and Aesthetic Empire: Humboldt's Physical Portrait of the Tropics." In *Visions of Empire: Voyages, Botany, and Representations of Nature,* edited by David Philip Miller and Peter Hanns Reill, 258–92. Cambridge: Cambridge University Press, 1996.

Dewey, James, and Perry Byerly. "The Early History of Seismometry (to 1900)." *Bulletin of the Seismological Society of America* 59: 183–227.

"Die Chemie der hohen Temperaturen." *Schweizerische Bauzeitung* 27 (1896): 133–35.

"Discussion of President McAdie's Address." *BSSA* 5 (1915): 177–89.

Di Stefano, Diana L. "Disasters, Railway Workers, and the Law in Avalanche Country, 1880–1910." *Environmental History* 14 (2009): 476–501.

Dittmar, Peter. "Mein Kapital: Die Angst!" *Die Welt,* 8 December 2000.

Dombois, Florian. "Über Erdbeben." PhD diss., Humboldt University, 1998.

Dörries, Matthias. "Climate Catastrophes and Fear." *Climate Change* 1 (2010): 885–89.

————. "Krakatau 1883: Die Welt als Labor und Erfahrungsraum." In *Welt-Räume. Geschichte, Geographie und Globalisierung seit 1900,* edited by Iris Schröder and Sabine Höhler, 51–73. Frankfurt am Main: Campus, 2005.

Dror, Otniel. "Seeing the Blush: Feeling Emotions." In Daston and Lunbeck, *Histories of Scientific Observation,* 326–48.

Drummond, James. *The Comrie Earthquakes.* Perth: C. Paton, 1875.

————. "A Table of Shocks of Earthquake, from September 1839 to the End of 1841, Observed at Comrie, Near Crieff." *Philosophical Magazine* 20 (1842): 240–47.

Dunning, Frederick W. "The Story of the Earth: Exhibition at the Geological Museum, London." *Museum* 26 (1974): 99–109.

Dutton, C. E., and Everett Hayden. "Abstract of the Results of the Investigation of the Charleston Earthquake." *Science* 9 (1887): 489–501.

————. *Earthquakes in the Light of the New Seismology.* New York: G. P. Putnam's Sons, 1904.

"The Earthquakes." *Times* (London), 13 January 1885, 10.

Edwards, Paul N. *A Vast Machine: Computer Models, Climate Data, and the Politics of Global Warming.* Cambridge, MA: MIT Press, 2010.

Eliot, George. *Middlemarch: A Study of Provincial Life.* New York: Harper & Brothers, 1876.

Ella, Samuel. "Some Physical Phenomena of the South Pacific Island." In *Report of the Second Meeting of the Australasian Association for the Advancement of Science,* 559–72. Sydney, 1890.

Endersby, Jim. *Imperial Nature: Joseph Hooker and the Practices of Victorian Science.* Chicago: University of Chicago Press, 2008.

Esposito, Fernando. *Mythische Moderne: Aviatik, Faschismus und die Sehnsucht nach Ordnung in Deutschland und Italien.* Munich: Oldenbourg, 2011.

Exner, Franz Serafin. *Der Schlichten Astronomia.* Vienna: self-published, 1908.

"Experiences of Earthquakes in Great Britain." *Chambers's Edinburgh Journal* 9 (1841): 58–59.

Faidiga, Adolf. "Das *Erdbeben* von Sinj am 2. Juli, 1898." *Mitteilungen der Erdbeben-Kommission der Kaiserlichen Akademie der Wissenschaften in Wien* 17 (1903).

Falb, Otto. "Aus dem Leben Rudolf Falbs." *Heimgarten* 29 (1905): 112–21.

Falb, Rudolf. *Das Land der Inca.* Leipzig: J. J. Weber, 1883.

———. *Von den Umwälzungen im Weltall.* Vienna: Hartlebens, 1880 and later editions.

Fan, Fa-ti. "'Collective Monitoring, Collective Defense': Science, Earthquakes, and Politics in Communist China." Witness to Disaster: Earthquakes and Expertise in Comparative Perspective, special issue, edited by Deborah R. Coen, *Science in Context* 25 (2012): 127–54.

———. "Redrawing the Map: Science in Twentieth-Century China." *Isis* 98 (2007): 524–38.

Favre, Louis. "Tremblement de terre observé à Fleurier en 1817." *Musée Neuchatelois* 19 (1883): 131–34.

Felsch, Philipp. *Laborlandschaften: Physiologische Alpenreisen im 19. Jahrhundert.* Göttingen: Wallstein, 2007.

Ferrari, Graziano, and Anita McConnell. "Robert Mallet and the 'Great Neapolitan Earth-quake' of 1857." *Notes and Records of the Royal Society of London* 59 (2005): 45–64.

Fleming, James. *Meteorology in America, 1800–1870.* Baltimore: Johns Hopkins, 1990.

Fleming, Paul. *The Pleasures of Abandonment: Jean Paul and the Life of Humor.* Würzburg: Königshausen & Neumann, 2006.

Fischer, Kuno. *Geschichte der neueren Philosophie,* vol. 3. Heidelberg: F. Bassermann, 1869.

Forel, F. A. "A la Commission du Catalogue de l'Association Sismologique Internation-ale." In *Procès-Verbaux des séances de la Troisième Conférence de la Commission Perma-nente de l'Association Internationale de Sismologie . . . 1909,* edited by R. de Kövesligethy, appendix XII, 190–95. Budapest: V. Hornyánsky, 1919.

———. *Handbuch der Seenkunde. Allgemeine Limnologie.* Stuttgart: Verlag von J. Engelhorn, 1901.

———. *Le Léman: Monographie limnologique.* 3 vols. Lausanne: Librairie de l'Université, 1892, 1895, 1901.

———. "Les tremblements de terre étudiés par la Commission Sismologique Suisse de novembre 1879 à fin décembre 1880." *Archives des sciences physiques et naturelles* 6 (1881): 461–94.

———. "Observations sur les tremblements de terre sentis récemment dans les canons de Vaud et de Neuchâtel." *Archives des sciences physiques et naturelles* 6 (1881): 266–67.

———. "Tremblement de terre du 30 décembre 1879." In *Jahrbücher des Tellurischen Observatoriums zu Bern*, edited by A. Forster, 1–2. Bern, 1880.

———. "Tremblement de terre du 30 décembre 1879." *Jahrbücher des Tellurischen Observatoriums zu Bern*, edited by A. Forster, 14. Bern, 1880.

Forster, Aimé. "Bericht der Erdbeben-Commission, 1878–1879." *VSNG* 62 (1879): 112–18.

———. "Bericht der Erdbeben-Commission, 1879–1880." *VSNG* 63 (1880): 101–3.

———. "Bericht der Erdbebencommission für 1882/83." *VSNG* 66 (1883): 92–94.

———. "Bericht der Erdbebenkommission." *VSNG* 71 (1888): 126–28.

———. *Die Schweizerischen Erdbeben in den Jahren 1884 und 1885*. Bern: Stämpfli, 1887.

Fort, Charles. *The Book of the Damned*. New York: Boni and Liveright, 1919.

Fortun, Kim. "From Bhopal to the Informating of Environmentalism: Risk Communication in Historical Perspective." *Osiris* 19 (2004): 283–96.

Fréchet, J. "Past and Future of Historical Seismicity Studies in France." In Fréchet et al., *Historical Seismology*, 131–46.

Fréchet, J., M. Meghraoui, and M. Stucchi, eds. *Historical Seismology: Interdisciplinary Studies of Past and Recent Earthquakes*. Dordrecht: Springer, 2008.

Friedländer, I. "Beiträge zur Geologie der Samoa-Inseln." *Abhandlungen der Bayerischen Akademie der Wissenschaften* II 24 (1910): 507–41.

Friedman, Robert Marc. *Appropriating the Weather: Vilhelm Bjerknes and the Construction of a Modern Meteorology*. Ithaca, NY: Cornell University Press, 1989.

Frisch, Max. *Tagebuch 1946–1949*. Frankfurt: Suhrkamp, 1950.

Früh, Johannes. "Bericht der Erdbebenkommission für das Jahr 1907/08." *VSNG* 91 (1908): 67–70.

———. "Bericht der Erdbeben-Kommission für das Jahr 1913/14." *VSNG* 96 1914: 117–22.

———. "Ergebnisse fünfundzwanzigjähriger Erdbebenbeobachtungen in der Schweiz 1880–1904." *VSNG* 88 (1905): 144–49.

———. "Ueber die 30-jährige Tätigkeit der Schweizerischen Erdbebenkommission." *Jahresversammlung der Schweizerischen Naturforschenden Gesellschaft*, vol. 1, *Vorträge und Sitzungsprotokolle* 94 (1911): 67–80.

Fuchs, C. W. C. "Bericht über die vulkanische Ereignisse des Jahres 1871." *Neues Jahrbuch für Mineralogie, Geologie und Paläontologie* 1872: 701–19.

Fuller, Steve. *Thomas Kuhn: A Philosophical History for Our Times*. Chicago: University of Chicago Press, 2000.

Für Laibach: Zum besten der durch die Erdbeben-Katastrophe im Frühjahre 1895 schwer betroffenen Einwohner von Laibach und Umgebung. Vienna: Genossenschaft der Bildenden Künstler Wiens, 1895.

Galison, Peter. *Image and Logic: A Material Culture of Microphysics*. Chicago: University of Chicago Press, 1997.

Gardner, Daniel. *The Science of Fear*. New York: Penguin, 2008.

Genthe, Arnold. *As I Remember*. New York: Reynal and Hitchcock, 1936.

Gerland, Georg. *Anthropologische Beiträge*. Halle: Max Niemayer, 1875.

———. "Die Kaiserliche Hauptstation für Erdbebenforschung in Strassburg und die moderne Seismologie." *Gerlands Beiträge zur Geophysik* 4 (1899): 427–72.

———. "Die moderne seismische Forschung." In *Verhandlungen des 7. Internationalen Geographen-Kongresses Berlin 1899*, vol 2. Berlin: Kühl, 1901.

———. Foreword to *Beiträge zur Geophysik* 1 (1887).

———. "Die moderne Erdbebenforschung." *Deutsche Rundschau* 96 (1898): 438–49.

"Immanuel Kant, seine geographischen und anthropologischen Arbeiten." *Kant-Studien* 10 (1905):1–43; 417–547.

———. *Über das Aussterben der Naturvölker.* Leipzig: Friedrich Fleischer, 1868.

Geschwind, Carl-Henry. *California Earthquakes: Science, Risk, and the Politics of Hazard Mitigation.* Baltimore: Johns Hopkins University Press, 2001.

———. "Embracing Science and Research: Early Twentieth-Century Jesuits and Seismology in the United States." *Isis* 89 (1998): 27–49.

Geyer, Martin H. *Die verkehrte Welt: Revolution, Inflation und Moderne: München 1914–1924.* Göttingen: Vandenhoeck & Ruprecht, 1998.

Gilbert, Grove Karl. "The Investigation of the California Earthquake of 1906." In *The California Earthquake of 1906,* edited by David Starr Jordan, 215–56. San Francisco: A. M. Robertson, 1907.

———. Preface to Martin and Tarr, *The Earthquakes at Yakutat Bay.*

Girard, Raymond de. *La théorie sismique du deluge.* Freiburg: Fragnière Frères, 1895.

Gisler, Monika. *Göttliche Natur? Formationen im Erdbebendiskurs der Schweiz des 18. Jahrhunderts.* Zurich: Chronos Verlag, 2007.

Gisler, Monika, Donat Fäh, and Domenico Giardini, eds. *Nachbeben: Eine Geschichte der Erdbeben in der Schweiz.* Bern: Haupt Verlag, 2008.

Gisler, Monika, Jan Kozák, and Jiří Vaněk. "The 1855 Visp (Switzerland) Earthquake: A Milestone in Macroseismic Methodology?" In Fréchet et al., *Historical Seismology,* 231–47.

Gilbert, Grove Karl. "The Investigation of the San Francisco Earthquake." *Popular Science Monthly* (1906): 97–115.

Goldstein, Daniel. "'Yours for Science': The Smithsonian Institution's Correspondents and the Shape of the Scientific Community in Nineteenth-Century America." *Isis* 85 (1994): 573–99.

Golinski, Jan. *British Weather and the Climate of Enlightenment.* Chicago: University of Chicago Press, 2007.

Goltz, James D. "Science Can Save Us: Outreach as Necessity and Strategy." *Seismological Research Letters* 74 (2003): 491–93.

Goodstein, J. R. *Millikan's School: A History of the California Institute of Technology.* New York: Norton, 1991.

———. "Waves in the Earth: Seismology Comes to Southern California." *Historical Studies in the Physical Sciences* 14 (1984): 201–30.

Gould, Stephen Jay. *Time's Arrow, Time's Cycle.* Cambridge, MA: Harvard University Press, 1987.

Graves, John. "On Earth Currents." *Journal of the Society of Telegraph Engineers* 2 (1873): 102–23.

Gray, Malcolm. "The Highland Potato Famine of the 1840s." *Economic History Review* 7 (1955): 357–68.

Greene, Mott. *Geology in the Nineteenth Century: Changing Views of a Changing World.* Ithaca, NY: Cornell University Press, 1982.

Guidoboni, Emanuela, and John E. Eben. *Earthquakes and Tsunamis in the Past: A Guide to Techniques in Historical Seismology.* Cambridge: Cambridge University Press, 2009.

Guinon, Georges. *Les agents provocateurs de l'hystérie.* Paris: De la Haye et Lecrosnier, 1889.

Gutenberg, B., C. F. Richter, and H. O. Wood. "The Earthquake in Santa Monica Bay, California, on August 30, 1930." *BSSA* 22 (1932): 138–54.

Gyorgy, Anna. *No Nukes: Everyone's Guide to Nuclear Power.* Boston: South End Press, 1979.

Hagner, Michael. "Psychophysiologie und Selbsterfahrung. Metamorphosen des Schwindels und der Aufmerksamkeit im 19. Jahrhundert." In *Aufmerksamkeiten. Archäologie der literarischen Kommunikation VII*, edited by Aleida Assmann and Jan Assmann, 241–63. Munich: Fink, 2001.

Hamann, Günther, ed. "Eduard Suess als liberaler Politiker."

———. *Eduard Suess zum Gedenken*. Vienna: Akademie der Wissenschaften, 1983.

Hamblyn, Richard. *The Invention of Clouds: How an Amateur Meteorologist Forged the Language of the Skies*. New York: Farrar, Straus and Giroux, 2001.

Hamlin, Homer. "Earthquakes in Southern California." *BSSA* 8 (1918): 20–24.

Hammerl, Christa, and Wolfgang Lenhardt. *Erdbeben in Österreich*. Vienna: Leykam, 1997.

Hammerl, Christa, et al., eds. *Die Zentralanstalt für Meteorologie und Geodynamik 1851–2001: 150 Jahre Meteorologie und Geophysik in Österreich*. Graz: Leykam, 2001.

"Hans von Hentig, Eighty Years Old." *Journal of Criminal Law, Criminology, and Police Science* 58 (1967): 427.

Hantken, Miksa. *Das Erdbeben von Agram im Jahre 1880*. Budapest: Legrady, 1882.

Harwood, Christopher W. "Human Soul of an Engineer: Andrei Platonov's Struggle with Science and Technology." PhD diss., Columbia University, 2000.

Hecht, Gabrielle. "Introduction: The Power of Nuclear Things." In Hecht, *Being Nuclear: Africans and the Global Uranium Trade*, 1–46. Cambridge, MA: MIT Press, 2012.

Heck, André, ed. *The Multinational History of Strasbourg Astronomical Observatory*. Dordrecht: Springer, 2005.

Heesen, Anke te, ed. *Cut and paste um 1900: Der Zeitungsausschnitt in den Wissenschaften*. Berlin: Vice Versa, 2002.

Heim, Albert. "Bergsturz und Menschenleben." *Der Schweizer Geograph* 8 (1931): 126–28.

———. "Der Bergsturz von Elm." *Zeitschrift der Deutschen Geologischen Gesellschaft* 34 (1882): 74–115.

———. *Die Erdbeben und deren Beobachtung*. Zurich: Zürcher & Furrer, 1879.

———. "Einiges über den Stand der Erdbebenforschung." In *Comptes-Rendus des Séances de la Troisième Réunion de la Commission Permanente de l'Association Internationale de Sismologie Réunie à Zermatt*, edited by R. de Kövesligethy, 146–50. Budapest: Victor Horyánszky, 1910.

———. "Notizen über den Tod durch Absturz." *Jahrbuch des Schweizer Alpenclub* 27 (1892): 327–37.

———. *Sehen und Zeichnen*. Basel: Benno Schwabe, 1894.

———. "Ueber die geologische Voraussicht beim Simplon-Tunnel. Antwort auf die Angriffe des Herrn Nationalrat Ed. Sulzer-Ziegler." *Eclogae Geologicae Helvetiae* 8 (1904): 365–84.

———. "Zur Prophezeiung der Erdbeben." *Vierteljahrsschrift der Naturforschenden Gesellschaft in Zürich* 32 (1887): 129–48.

Heim, Albert, et al. "Étude géologique sur le nouveau projet de tunnel coudé au travers du Simplon." *Bulletin de la Société Vaudoise des Sciences Naturelles* 19 (1883): 1–27.

Heise, Carl Georg. *Persönliche Erinnerungen an Aby Warburg*. Edited by Björn Biester and Hans-Michael Schäfer. Wiesbaden: Harrassowitz, 2005.

Hennig, R. "Die Abhängigkeit des menschlichen Seelenlebens vom Wetter." *Gartenlaube* 37 (1912): 779–80.

Hentig, Hans. "Über die Wirkung von Erdbeben auf Menschen." *Archiv für Psychiatrie* 69 (1923): 546–68.

————. *Ueber den Zusammenhang von kosmischen, biologischen und sozialen Krisen.* Tübingen: J. C. B. Mohr, 1920.

Herbert, Christopher. *Victorian Relativity: Radical Thought and Scientific Discovery.* Chicago: University of Chicago Press, 2001.

Herf, Jeffrey. *Reactionary Modernism: Technology, Culture, and Politics in Weimar and the Third Reich.* Cambridge: Cambridge University Press, 1984.

Herschel, John, ed. *Manual of Scientific Enquiry.* 4th edition. London: John Murray, 1871.

Herscher, Andrew. "Städtebau as Imperial Culture: Camillo Sitte's Urban Plan for Ljubljana." *Journal of the Society of Architectural Historians* 62 (2003): 212–27.

Hesse, Hermann. *Briefe.* Frankfurt: Suhrkamp, 1959.

Heuvel, Charles van den. "Building Society, Constructing Knowledge, Weaving the Web: Otlet's Visualizations of a Global Information Society and His Concept of a Universal Civilization." In *European Modernism,* edited by W. Boyd Rayward, 127–54. Aldershot: Ashgate, 2008.

Hilhorst, Dorothea, and Greg Bankoff. Introduction to *Mapping Vulnerability: Disasters, Development and People,* edited by Greg Bankoff, Georg Frerks and Dorothea Hilhorst. London: Earthscan, 2004.

Hinrichs, Gustavus. *Fifth Biennial Report of the Central Station of the Iowa Weather Service.* Des Moines, IA: Geo. E. Roberts, 1887.

————. *First Annual Report of the Iowa Weather Stations.* Des Moines, IA: R. P. Clarkson, 1877.

————. *Report of the Iowa Weather Service for the Year 1886.* Des Moines, IA: Geo. E. Roberts, 1887.

Hirsch, Francine. 2005. *Empire of Nations: Ethnographic Knowledge and the Making of the Soviet Union.* Ithaca, NY: Cornell University Press.

Hobbs, W. H. Review of Martin, "Alaskan Earthquake of 1899." *BSSA* 2 (1912): 96–97.

Hochstetter, Ferdinand. "Ueber Erdbeben, mit Beziehung auf das Agramer Erdbeben vom 9. November 1880." *Monatsblätter des Wissenschaftlichen Clubs in Wien* 2, no. 3 (1880): 1–14.

Hoefer, H. "Die Erdbeben Kärntens und deren Stosslinien." *Denkschriften der Kaiserlichen Akademie der Wissenschaften* 42 (1880): 1–88.

Hoernes, Rudolf. "Das Erdbeben von Belluno am 29. Juni 1873 und die Falb'sche Erdbeben-Hypothese." *Mittheilungen des naturwissenschaftlichen Vereines für Steiermark* 1877: 33–45.

————. "Das Erdbeben von Saloniki am 5. Juli 1902 und der Zusammenhang der makedonischen Beben mit den tektonischen Vorgängen in der Rhodopemasse." *Mitteilungen der Erdbeben-Kommission der Kaiserlichen Akademie der Wissenschaften in Wien* 13 (1902).

————. *Die Erdbeben-Theorie Rudolf Falbs und ihre wissenschaftliche Grundlage, kritisch erörtert.* Vienna: Broackhausen und Bräuer, 1881.

————. "Erdbeben in Steiermark während des Jahres 1880." *Mittheilungen des naturwissenschaftlichen Vereines für Steiermark* 1880: 65–114.

————. *Erdbebenkunde: Die Erscheinungen und Ursachen der Erdbeben, die Methoden ihrer Beobachtung.* Leipzig: Veit & Co., 1893.

————. "Erdbeben und Stoßlinien Steiermarks." *Mittheilungen der k. k. Erdbebenkommission* Neue Folge 7 (1902): 1–2.

Hoffman, Susanna M., and Anthony Oliver-Smith. "Anthropology and the Angry Earth: An Overview." In Oliver-Smith and Hoffman, *The Angry Earth,* 1–16.

Hofmannsthal, Hugo von. "Der Dichter und diese Zeit." In *Die Prosaischen Schriften Gesammelt,* 1:1–52. Berlin: Fischer, 1907.

Holden, Edward S. *A Catalogue of Earthquakes on the Pacific Coast, 1769–1897.* Washington, DC: Smithsonian Institution, 1898.

———. "Earthquake Observations." *Publications of the Astronomical Society of the Pacific* 2 (1890): 73–74.

———. "Earthquakes and How to Measure Them." *The Century* 47 (1894): 749–59.

———. "Earthquakes in California (1888)." *American Journal of Science* 1889: 392–402.

———. *List of Recorded Earthquakes in California, Lower California, Oregon, and Washington Territory.* Sacramento, CA: J. D. Young, 1887.

———. "Note on Earthquake-Intensity in San Francisco." *American Journal of Science* 35 (1888): 427–31.

———. "A System of Local Warnings against Tornadoes." *Science* 2 (1883): 521–22.

———. "The Work of an Astronomical Society." *Publications of the Astronomical Society of the Pacific* 1 (1889): 9–20.

"Homo nobilis u. bête humaine." *Archiv für Kriminologie* 27 (1906): 362–64.

Hones, Sheila. "Distant Disasters, Local Fears: Volcanoes, Earthquakes, Revolution, and Passion in the Atlantic Monthly, 1880–1884." In *American Disasters,* edited by Steven Biel, 170–96. New York: NYU Press, 2001.

Hösler, Joachim. *Von Krain zu Slowenien.* Munich: Oldenbourg, 2006.

Hough, Susan E. *Richter's Scale: Measure of an Earthquake, Measure of a Man.* Princeton, NJ: Princeton University Press, 2007.

How and Why Stories. Recorded by John C. Branner. New York: Henry Holt, 1921.

Howell, Jr., Benjamin F. *An Introduction to Seismological Research: History and Development.* Cambridge: Cambridge University Press, 1990.

Huber, Walter L. "The San Francisco Earthquakes of 1865 and 1868." *BSSA* 20 (1930): 261–72.

Huler, Scott. *Defining the Wind: The Beaufort Scale, and How a Nineteenth-Century Admiral Turned Science into Poetry.* New York: Random House, 2004.

Humboldt, Alexander von. *Cosmos: A Sketch of a Physical Description of the Universe,* vol. 1. Translated by E. C. Otté. London: Henry G. Bohn, 1848.

———. *Personal Narrative of a Journey to the Equinoctal Regions of the New Continent.* Translated by Jason Wilson. New York: Penguin, 1995.

———. *Personal Narrative of Travels to the Equinoctial Regions of America.* Translated by Thomasina Ross. London: Bohn, 1853.

Hume, David. *A Treatise of Human Nature.* Oxford: Clarendon, 1888.

Humphreys, W. J. "Seismology." *Monthly Weather Review* 42 (1914): 687–89.

Hunt, Bruce. *The Maxwellians.* Ithaca, NY: Cornell University Press, 1991.

Hutchinson, John F. "Disasters and the International Order: Earthquakes, Humanitarians, and the Ciraolo Project." *International History Review* 22 (2000): 1–36.

Huyssen, Andreas. "Fortifying the Heart, Totally: Ernst Jünger's Armored Texts." Special Issue on Ernst Jünger, *New German Critique* 59 (1993): 3–23.

International Nuclear and Radiological Event Scale User's Manual. http://www-pub.iaea .org/MTCD/publications/PDF/INES-2009_web.pdf. Accessed 19 July 2011.

Irwin, Alan. *Citizen Science: A Study of People, Expertise, and Sustainable Development.* New York: Routledge, 1995.

J., H. "Charles Davison." *Quarterly Journal of the Geological Society* 97 (1941): lxxxv–lxxxvi.

Jaehnike, Alfred. "Das Gebäude der Kaiserlichen Hauptstation für Erdbebenforschung zu Strassburg i. E." *Gerlands Beiträge zur Geophysik* 4 (1900): 421–26.

James, William. "On Some Mental Effects of the Earthquake." In *Memories and Studies,* 209–26. New York: Longman, 1911.

Janković, Vladimir. *Confronting the Climate: British Airs and the Making of Environmental Medicine.* New York: Palgrave, 2010.

Janouch, Gustav. *Conversations with Kafka.* Translated by Goronwy Rees. New York: New Directions, 1971.

Janz, Rolf-Peter, Fabian Stoermer, and Andreas Hiepko. "Einleitung." In *Schwindeler-fahrungen: Zur kulturhistorischen Diagnose eines vieldeutigen Symptoms,* 7–45.

Jaspers, Karl. *Allgemeine Psychopathologie.* 9th edition. Berlin: Springer, 1923.

Jennings, Eric. 2006. *Curing the Colonizers: Hydrotherapy, Climatology, and French Colonial Spas.* Durham, NC: Duke University Press.

Jones, E. Lester. *Earthquake Investigation in the United States.* Washington, DC: Government Printing Office, 1925.

Jung, C. G. "Experiences Concerning the Psychic Life of the Child." In *Collected Papers on Analytical Psychology,* edited by C. E. Long, 132–55. New York: Moffat Yard & Co., 1916.

Jünger, Ernst. *Copse 125.* Translated by Basil Creighton. London: Chatto & Windus, 1930.

———. *Strahlungen.* Tübingen: Heliopolis, 1949.

———. "Über die Gefahr." In Bucholtz, *Der Gefährliche Augenblick,* 11–16.

Kaddoura, Abdul-Razzak. "A Tremendous Privilege." In *Sixty Years of Science at UNESCO,* ed. P. Petitjean, V. Zharov, G. Glaser, et al., 487–90. Paris: UNESO, 2006.

Kafka, Alan L. "Public Misconceptions about Faults and Earthquakes in the Eastern United States: Is It Our Own Fault?" *Seismological Research Letters* 71 (2000): 311–12.

Kafka, Alan L., Ellyn A. Schlesinger-Miller, and Noel L. Barstow. "Earthquake Activity in the Greater New York City Area: Magnitudes, Seismicity, and Geologic Structures." *BSSA* 75 (1985): 1285–1300.

Kant, Immanuel. "Geschichte und Naturbeschreibung der merkwürdigen Vorfälle des Erd-bebens. . . ." In *Kants gesammelte Schriften,* 1:431–61. Berlin: Georg Reimer, 1902.

———. *Physische Geographie,* vol. 1, part 1. Mainz: Vollmer, 1801.

Kautz, William. *Opening the Inner Eye: Explorations on the Practical Application of Intuition in Daily Life and Work.* Lincoln, NE: iUniverse, 2003.

Keilhack, Konrad. *Lehrbuch der praktischen Geologie.* Stuttgart: Ferdinand Enke, 1908.

Keller, Susanne B. "Sections and Views: Visual Representation in Eighteenth-Century Earthquake Studies." *British Journal for the History of Science* 31 (1998): 129–59.

Kevles, Daniel J. *The Physicists: The History of a Scientific Community in Modern America.* Cambridge, MA: Harvard University Press, 1995.

Kingsland, Sharon. *Evolution of American Ecology, 1890–2000.* Baltimore: Johns Hopkins University Press, 2005.

Kingsley, Charles. *Madam How and Lady Why.* New York: Macmillan, 1888.

Klotz, Otto. "Present Status of Seismological Work in the Pacific." *BSSA* 10 (1920): 300–310.

Knopoff, Leon. "Beno Gutenberg." *Biographical Memoirs* (National Academy of Sciences) 76: 115–47.

Koch, Julius. "Ursachen des Verfalles der Hochbauten." *Schweizerische Bauzeitung* 20 (1892): 73–78.

Koelsch, William A. "Ben Franklin's Heir: Alexander McAdie and the Experimental Analy-sis and Forecasting of New England Storms, 1884–1892." *New England Quarterly* 59 (1986): 523–43.

Koerner, Lisbet. *Linnaeus: Nature and Nation.* Cambridge, MA: Harvard University Press, 1999.

Komine, A., and A. Maki. "Psychiatric Observations of Earthquake Catastrophe of Tokyo and Vicinity." *Japanese Journal of Neurology and Psychiatry* 1924.

Kövesligethy, Rado von, ed. *Procès-Verbaux des séances de la deuxième conférence de la Commission Permanente et de la première assemblée générale de l'Association Internationale de Sismologie réunie a la Haye du 21 au 25 septembre 1907.* Budapest: V. Hornyanszky, 1908.

Kozák, J. "100-Year Anniversary of the First International Seismological Conference." *Studia Geophysica et Geodaetica* 45 (2001): 200–209.

Kozák, J., and Vladimir Čermák. *The Illustrated History of Natural Disasters.* Dordrecht: Springer, 2010.

Kozák, J., and Axel Plešinger. "Beginnings of Regular Seismic Service and Research in the Austro-Hungarian Monarchy." *Studia Geophysica et Geodaetica* 47 (2003): 99–119, 757–91.

Kozák, J., and J. Van k. "The 1855 Visp (Switzerland) Earthquake: Early Attempts of Earthquake Intensity Classification." *Studia Geophysica et Geodetica* 50 (2006): 147–60.

Krajewski, Markus. *Zettelwirtschaft.* Berlin: Kadmos, 2002.

Kraus, Karl. "Antworten des Herausgebers: Geolog." *Die Fackel* 4, no. 103 (1902): 20–22.

———. "Das Erdbeben." *Die Fackel* 9, no. 245 (1908): 16–24.

———. "The Discovery of the North Pole." In *In These Great Times: A Karl Kraus Reader*, edited by Harry Sohn, translated by Joseph Fabry et al., 48–57. Chicago: University of Chicago Press, 1990.

———. "Erdbeben." *Die Fackel* 10, no. 270 (1909): 45–48.

———. "Erdbeben." *Die Fackel* 12, no. 303 (1910): 4–6.

———. "Messina." *Die Fackel* 10, no. 273 (1909): 1–2.

———. "Nach dem Erdbeben." *Die Fackel* 13, no. 338 (1911): 18–24.

———. "Nestroy und die Nachwelt. Zum 50. Todestag." *Die Fackel* 14, no. 349 (1912): 1–23.

Kraus, Karl, Kurt Wolf, and Friedrich Pfäfflin Zwischen. *Jüngstem Tag und Weltgericht: Karl Kraus und Kurt Wolff Briefwechsel 1912–1921.* Tübingen: Wallstein, 2007.

Kroeber, A. L. "Earthquakes." *Journal of American Folklore* 19 (1906): 322–23.

Krummel, Richard Frank. *Nietzsche und der deutsche Geist*, vol. 2. Berlin: de Gruyter, 1998.

Kuhn, Thomas S. *The Structure of Scientific Revolutions.* 2nd edition. Chicago: University of Chicago Press, 1970.

Lang, Andrew. *The Book of Dreams and Ghosts.* London: Longmans, Green, and Co., 1899.

Lapajne, Janez. "The MSK Intensity Scale and Seismic Risk." *Engineering Geology* 20 (1984): 105–12.

Lasch, Richard. "Die Ursache und Bedeutung der Erdbeben im Volksglauben und Volksbrauch." *Archiv für Religionswissenschaft* 5 (1902): 236–57, 369–83.

Láska, Wenzel. "Die Erdbeben Polens." *Mittheilungen der Erdbeben-Commission der k. k. Akademie der Wissenschaften in Wien* Neue Folge 8 (1902): 1–2.

Latour, Bruno. *We Have Never Been Modern.* Translated by Catherine Porter. Cambridge, MA: Harvard University Press, 1993.

Lauer, Gerhard, and Thorsten Unger. "Angesichts der Katastrophe: Das Erdbeben von Lissabon und der Katastrophendiskurs im 18. Jahrhundert." In *Das Erdbeben von Lissabon und der Katastrophendiskurs im 18. Jahrhundert*, edited by Lauer and Unger, 13–44. Göttingen: Wallstein, 2008.

Lawson, A. C. *California Earthquake of April 18, 1906: Report of the State Earthquake Investigation Commission.* Washington, DC: Carnegie Institution, 1908.

Lecointe, G. "Motion sur la publication annuelle de la bibliographie sismologique." In *Verhandlungen der Internationalen Seismologischen Konferenz 1906*, 181–83. Budapest: V. Hornyanszky, 1907.

Le Bon, Gustave. *La psychologie des foules.* 2nd edition. Paris: F. Alcan, 1896.

Lehner, Martina. *"Und das Unglück ist von Gott gemacht . . .": Geschichte der Naturkatastrophen in Österreich.* Vienna: Praesens, 1995.

Leighly, John. "Methodologic Controversy in Nineteenth-Century German Geography." *Annals of the Association of American Geographers* 28 (1938): 238–58.

"Les Tremblements de terre italiens." *Journal de Genève*, 30 December 1908, 3.

Leuba, J. "Le professeur Hans Schardt: 1858–1931." *Bulletin de la Société Neuchâteloise des Sciences Naturelles* 56 (1931): 106.

Linke, Friedrich. "Bericht über die Arbeiten des Samoa-Observatoriums in den Jahren 1905 und 1906." In Wagner, "Das Samoa Observatorium," 55–70.

Livingstone, David. *Nathanial Southgate Shaler and the Culture of American Science.* Tuscaloosa: University of Alabama Press, 1987.

Locher, Eduard. "Zum Jungfraubahnproject von Oberst Locher." *Schweizerische Bauzeitung* 15 (1890): 149–51.

Locher, Fabien. *Le savant et la tempête: Étudier l'atmosphère et prévoir le temps au XIXe siècle.* Rennes: Presses Universitaires, 2008.

Lombroso, Cesare. *Crime, Its Causes and Remedies.* Translated by Henry P. Horton. Boston: Little, Brown, 1911.

Lombroso, Paola, and Cesare Lombroso. "La Psicologia dei Terremotati." *Archivio di Antropologica Criminale, Psychiatria, Medicina Legale e Scienze Affini* 30 (1909): 122–30.

Long, J. J. "From Das Antlitz des Weltkrieges to Der gefährliche Augenblick: Ernst Jünger, Photography, Autobiography, and Modernity." In *German Life Writing in the Twentieth Century,* edited by Birgit Dahlke, Dennis Tage, and Roger Woods, 54–70. Rochester, NY: Camden House, 2010.

Lovell, J. H. *The Dr ATJ Dollar Papers Held in the National Seismological Archive.* British Geological Survey Technical Report WL/99/15. Edinburgh: British Geological Survey, 1999.

Lowe, E. J. "History of the Earthquake of 1863." *Proceedings of the British Meteorological Society* 2 (1864): 55–99.

Luhmann, Niklas. *Ecological Communication.* Translated by John Bednarz, Jr. Chicago: University of Chicago Press, 1989.

Mach, Ernst. *Contributions to the Analysis of the Sensations.* Translated by C. M. Williams. La Salle, IL: Open Court, 1897.

———. *Lotos* 11 (November 1869): 165.

Maher, Thomas J. "The United States Coast and Geodetic Survey—Its Work in Collecting Earthquake Reports in the State of California." *BSSA* 19 (1929): 77–79.

Makowsky, A. "Ueber die Erdbeben-Theorie Rudolf Falb's im Lichte der Geologie." *Verhandlungen des Naturforschenden Vereines in Brünn* 25 (1886): 54–63.

Mallet, Robert. "Earthquakes." In *A Manual of Scientific Enquiry; Prepared for the Use of Officers in Her Majesty's Navy; and Travellers in General,* edited by John Herschel, 205–36. London: John Murray, 1851.

———. *The First Principles of Observational Seismology,* vol. 1. London: Chapman and Hall, 1862.

———. *The First Principles of Observational Seismology,* vol. 2. London: Chapman and Hall, 1862.

———. "First Report on the Facts of Earthquake Phenomena." In *Report of the Twentieth*

Meeting of the British Association for the Advancement of Science, 1–87. London: John Murray, 1851.

Mallet, Robert, and John William Mallet. *The Earthquake Catalogue of the British Association.* London: Taylor and Francis, 1858.

Markovits, Stefanie. "Rushing into Print: 'Participatory Journalism' during the Crimean War." *Victorian Studies* 50 (2008): 559–86.

Marshall, Michael. "Fukushima Crisis Raised to Level 7, Still No Chernobyl." 12 April 2011. http://www.newscientist.com/blogs/shortsharpscience/2011/04/fukushima-crisis-raised-to-lev.html. Accessed 19 July 2011.

Marshall, Wallace. "Biological Reactions to Earthquakes." *Journal of Abnormal and Social Psychology* 30 (1936): 462–67.

Martin, Daniel S. "Notes upon the Earthquake of December, 1874." *American Journal of Science* 10 (1875): 191–94.

Martin, L., and Ralph S. Tarr. "Alaskan Earthquake of 1899." *Bulletin of the Geological Society of America* 21 (1910): 339–406.

———. *The Earthquakes at Yakutat Bay, Alaska, in September, 1899.* USGS Professional Paper 69 (1919).

———. "Recent Changes of Level in the Yakutat Bay Region, Alaska." *Bulletin of the Geological Society of America* 17 (May 1906): 29–64.

Marvin, C. F. "Foreword: Co-Operative Observers' Department." *Bulletin of the American Meteorological Society* 3 (1922): 72–73.

Matus, Jill. *Shock, Memory, and the Unconscious in Victorian Fiction.* Cambridge: Cambridge University Press, 2009.

Mauch, Christof, ed. *Natural Disasters, Cultural Responses.* Lanham, MD: Lexington Books, 2009.

Mayer-Ahrens, Conrad. "Ueber die Beziehungen des Vulkanismus zur Gesundheit des thierischen Organismus." *Deutsche Klinik* 31 (1857): 293–95, and 329–33.

Mazower, Mark. *Governing the World: The Rise and Fall of an Idea since 1815.* New York: Penguin, 2012.

Mazzotti, Massimo. "The Jesuit on the Roof: Observatory Sciences, Metaphysics, and Nation Building." In Aubin et al, *The Heavens on Earth,* 58–85.

McAdie, Alexander. "Muir of the Mountains." John Muir Memorial, special issue, *Sierra Club Bulletin,* 10, no. 1 (January 1916): 20–22.

McCray, Patrick. *Keep Watching the Skies! The Story of Operation Moonwatch and the Dawn of the Space Age.* Princeton, NJ: Princeton University Press, 2008.

McGee, W. J. "Some Features of the Recent Earthquake." *Science* 8 (24 September 1886): 271–75.

McPhee, John. *Annals of the Former World.* New York: Farrar, Straus, Giroux, 1998.

Meldola, Raphael, and William White. *Report on the East Anglian Earthquake of April 22, 1884.* London: MacMillan & Co. and the Essex Field Club, 1885.

Meltsner, Arnold J. "The Communication of Scientific Information to the Wider Public: The Case of Seismology in California." *Minerva* 17 (1979): 331–54.

———. "Public Support for Seismic Safety: Where Is It in California?" *Mass Emergencies* 3 (1978): 167–84.

"Members of the Seismological Society of America: December 20, 1921." *BSSA* 11 (1921): 205–17.

Menand, Louis. *The Metaphysical Club: A Story of Ideas in America.* New York: Farrar, Straus, & Giroux, 2001.

Mendelsohn, J. Andrew. "The World on a Page: Making a General Observation in the Eighteenth Century." In Daston and Lunbeck, *Histories of Scientific Observation*, 396–420.

Mendenhall, T. C. "On the Intensity of Earthquakes, with Approximate Calculations of the Energy Involved." *Proceedings of the American Association for the Advancement of Science* 37 (1889): 190–95.

Menger, Karl. "The New Logic." *Philosophy of Science* 4 (1937): 336.

Mercalli, Giuseppe. "Sulle modificazioni proposte alla scala sismica De Rossi-Forel." *Bolletino della Società Sismologica Italiana* 8 (1902): 184–91.

Meyer, Theo. "Nietzsche und die klassische Moderne." In *Nietzsche und Schopenhauer: Rezeptionsphänomene der Wendezeiten*, edited by Marta Kopij and Wojciech Kunicki, 13–46. Leipzig: Leipziger Universitätsverlag, 2006.

Micale, Mark. *Hysterical Men: The Hidden History of Male Nervous Illness*. Cambridge, MA: Harvard University Press, 2008.

Middendorf, Ernst W. *Peru: Beobachtungen und Studien über das Land und seine Bewohner*, vol. 1. Berlin: R. Oppenheim, 1893.

Mignani, Vincenzo Domenico. Review of *De effectibus terraemotus in corpore humano*. The *Monthly Review* 72 (1785): 526.

Mileti, Dennis S. "Public Perceptions of Seismic Hazards and Critical Facilities." *BSSA* 72 (1982): 13–18.

Miller, Mary Ashburn. 2011. *A Natural History of Revolution: Violence and Nature in the French Revolutionary Imagination, 1789–1794*. Ithaca, NY: Cornell University Press.

Milne, David. "Notices of Earthquake-Shocks Felt in Great Britain, and Especially in Scotland, with Inferences Suggested by These Notices as to the Causes of Such Shocks." *Edinburgh New Philosophical Journal* 31 (1841): 92–122.

———. *Seismology*. London: Kegan Paul, Trench, Trübner & Co., 1898.

Milne, John. "The Earth in Its Vigour." *Times* (London), 8 September 1883.

———. *Earthquakes and Other Earth Movements*. London: Kegan Paul, Trench, & Co., 1886.

———. "Earthquake Effects, Emotional and Moral." *Seismological Journal of Japan* 11 (1887): 91–113.

———. Letter. *Times* (London), 28 September 1881, 8

———. Letter. *Times* (London), 29 December 1896, 9.

———. "On the Study of Earthquakes in Great Britain" *Nature* 42 (1890): 346–49

———. "Prehistoric Remains from Otaru and Hakodate, Part I." *Transactions of the Asiatic Society of Japan* 8 (1880): 61–73.

———. "Preliminary Notes on Observations Made with a Horizontal Pendulum in the Antarctic Regions." *Proceedings of the Royal Society of London* 76 (1905): 284–95.

———. "Recent Earthquakes." *Nature* 77 (1908): 592–97.

———. "The Recent Earthquake in Essex." *Times* (London), 7 August 1884, 9.

———. "Seismological Observations and Earth Physics." *Geographical Journal* 21 (1903): 1–22.

———. "Seismology at the British Association." *Nature* 88 (1911): 124–25.

Milne-Home, Grace. *Biographical Sketch of David Milne Home*. Edinburgh: David Douglas, 1891.

"Miscellaneous Scientific Intelligence." *American Journal of Science* 28, no. 165 (September 1884): 242.

Mitman, Gregg, Michelle Murphy, and Christopher Sellers. "Introduction: A Cloud over History." Landscapes of Exposure: Knowledge and Illness in Modern Environments, edited by Mitman, Murphy, and Sellers, *Osiris* 19 (2004): 1–17.

Mojsisovics, Edmund von, ed. "Allgemeiner Bericht und Chronik der im Jahre 1900 im Beobachtungsgebiete eingetretenen Erdbeben." *Mittheilungen der Erdbeben-Kommission der Kaiserlichen Akademie der Wissenschaften* Neue Folge 2 (1901).

Monatsblätter des Wissenschaftlichen Clubs in Wien 2, no. 2 (15 November 1880): 13.

Montaigne, Michel de. "Of Cannibals." In *The Complete Essays of Montaigne,* translated by Donald Frame, 150–59. Stanford, CA: Stanford University Press, 1958.

Montessus de Ballore, Fernand. "Earthquake Intensity Scales." *BSSA* 6 (1916): 227–31.

———. *Ethnographie sismique et volcanique.* Paris: E. Champion, 1923.

———. "Introduction à un essai de description sismique de globe et mesure de la sismicité." *Gerlands Beiträge zur Geophysik* 4 (1900): 331–82.

———. "Géosynclinaux et régions à tremblements de terre." *Bulletin de la Société Belge de Géologie, de Paléontologie et d'Hydrologie* 18 (1904): 243–67.

———. "La Sismologia en la Biblia." *Boletin del Servicio Seismológico de Chile* 11 (1915): 27–158.

———. "La théorie sismico-cyclonique du déluge par Suess." *Revue des questions scientifiques* 26 (1902): 577–89.

———. *Les visées de la sismologie moderne.* Louvain: Polleunis & Ceuterick, 1904.

———. "Mémorandum pour la conférence sismologique internationale." *Verhandlungen der ersten internationalen seismologischen Konferenz,* edited by E. Rudolph. *Beiträge zur Geophysik,* supplement (1902): 132–36.

———. "Physique du globe." *Revue Scientifique* 20 (1903): 609–14.

———. "Réponse à la note 'Denkschrift, betreffend die Herstellung der dem Zentralbureau übertragenen makroseismischen Kataloge.'" In *Procès-Verbaux des séances de la Troisième Conférence de la Commission Permanente de l'Association Internationale de Sismologie réunie à Zermatt du 30 août au 2 septembre 1909,* edited by R. de Kövesligethy, 197–98. Budapest: V. Hornyánsky, 1919.

Montgomery, Scott. *Science in Translation: Movements of Knowledge through Cultures and Time.* Chicago: University of Chicago Press, 2000

Morrell, Jack, and Arnold Thackray. *Gentlemen of Science: Early Years of the British Association for the Advancement of Science.* Oxford: Clarendon, 1981.

Mousson, Albert. "Bericht über die Organisation meteorologischer Beobachtungen in der Schweiz." *VSNG* 48 (1864): 196–312.

Muir, John. *Our National Parks.* Boston: Riverside, 1901.

Muir-Wood, Robert. "Robert Mallet and John Milne—Earthquakes Incorporated in Victorian Britain." *Earthquake Engineering and Structural Dynamics* 17 (1988): 107–42.

Muir-Wood, Robert, and Charles Melville. "Who Killed the British Earthquake?" *New Scientist* 100 (20 October 1983): 170–73.

Muir-Wood, Robert, G. Woo, and H. Bungum, "The History of Earthquakes in the Northern North Sea." In *Historical Seismograms and Earthquakes of the World,* edited by W. H. K. Lee, H. Meyers, and K. Shimazaki, 297–306. San Diego: Academic Press, 1988.

Mulholland, Catherine. *William Mulholland and the Rise of Los Angeles.* Berkeley: University of California Press, 2000.

Murphy, Michelle. *Sick-Building Syndrome and the Problem of Uncertainty: Environmental Politics, Technoscience, and Women Workers.* Durham, NC: Duke University Press, 2006.

Musson, R. M. W. "The Comparison of Macroseismic Intensity Scales." *Journal of Seismology* 14 (2010): 413–28.

———. "Comrie: A Historical Scottish Earthquake Swarm and Its Place in the History of Seismology. *Terra Nova* 5 (1993): 477–80.

———. "A Critical History of British Earthquakes." *Annals of Geophysics* 47 (2004): 597–609.

———. "Discovery of a Curious Seismological Monument from 19th Century Scotland." *Terra Nova* 5 (1993): 513.

———. "From Questionnaires to Intensities—Assessing Free-Form Macroseismic Data in the UK." *Physics and Chemistry of the Earth, Part A* 24 (1999): 511–15.

———. "Seismicity: Fatalities in British Earthquakes." *Astronomy and Geophysics* 44 (2003): 14–16.

Musson, R. M. W., and I. Cecić. "Intensity and Intensity Scales." In *New Manual of Seismological Practice*. International Association of Seismology and Physics of the Earth's Interior, 2012.

Musson, R. M. W., Gottfried Grünthal, and Max Stucchi. "The Comparison of Macroseismic Intensity Scales." *Journal of Seismology* 14 (2010): 413–28.

Musson, R. M. W., and P. H. O. Henni. "Methodological Considerations of Probabilistic Seismic Hazard Mapping." *Soil Dynamics and Earthquake Engineering* 21 (2001): 385–403.

Nash, Linda. *Inescapable Ecologies: A History of Environment, Disease, and Knowledge*. Berkeley: University of California Press, 2006.

"Naturlehre." *Der Floh*, 21 April 1895.

Neubauer, Ferdinand J. "A Short History of the Lick Observatory." *Popular Astronomy* 58 (1950): 369–87.

Neumayer, Georg. *Anleitung zu wissenschaftlichen Beobachtungen auf Reisen*. 1st edition, vol. 1, 1875; 2nd edition, vol. 1. Berlin: Robert Oppenheim, 1888.

Neumayr, Melchior. *Erdgeschichte*. Leipzig: Bibliographisches Institut, 1887.

"New Theory of Earthquakes and Volcanoes." *Littell's Living Age* 102 (1869): 387–99.

Nietzsche, Friedrich. "On the Uses and Disadvantages of History for Life." In *Untimely Meditations*, translated by R. J. Hollingdale, 120. Cambridge: Cambridge University Press, 1983.

"1906 Marked the Dawn of the Scientific Revolution." USGS. http://earthquake.usgs .gov/regional/nca/1906/18april/revolution.php. Accessed 5 June 2010.

Nipher, Frances. "The Missouri Weather Service: Shall It Be Sustained as a State Service?" In *Thirty-First Annual Report of the State Horticultural Society of Missouri, 1888*, 340–44. Jefferson City, MO: Tribune, 1889.

Noakes, Richard J. "Telegraphy Is an Occult Art: Cromwell Fleetwood Varley and the Diffusion of Electricity to the Other World." *British Journal for the History of Science* 32 (1999): 421–59.

Nordau, Max. *Degeneration*. New York: Appleton, 1895.

Norris, Kathleen. *Noon: An Autobiographical Sketch*. New York: Doubleday, 1925.

"Notes." *Nature* 26 (4 May 1882): 16–18.

"Notes." *Nature* 37 (3 May 1888): 16.

Nunn, Patrick D. "Fished Up or Thrown Down: The Geography of Pacific Island Origin Myths." *Annals of the Association of American Geographers* 93 (2003): 350–64.

Nyhart, Lynn K. *Modern Nature: The Rise of the Biological Perspective in Germany*. Chicago: University of Chicago Press, 2009..

"Objects of the Society." *Proceedings of the Society for Psychical Research* 1 (1882–83): 3–6.

Oeser, Erhard. "Historical Earthquake Theories." *Konrad Lorenz Institute for Evolution & Cognition Research*. 2001. http://www.univie.ac.at/wissenschaftstheorie/heat/heat .htm. Accessed 17 September 2010.

Oldham, Richard D. "The Depth of Origin of Earthquakes." *Quarterly Journal of the Geological Society* 82 (1926): 67–93.

———. Oldroyd, David, Filomena Amador, Jan Kozák, et al. "The Study of Earthquakes in the Hundred Years Following the Lisbon Earthquake of 1755." *Earth Sciences History* 26 (2007): 321–70.

Oliveira, C. S. "Review of the 1755 Lisbon Earthquake Based on Recent Analyses of Historical Observations." In Fréchet et al., *Historical Seismology*, 261–300.

Oliver-Smith, Anthony, and Susanna M. Hoffman. *The Angry Earth: Disaster in Anthropological Perspective*. New York: Routledge, 1999.

Omori, Fusakichi. "Seismic Experiments on the Fracturing and Overturning of Columns." *Publications of the Earthquake Investigation Committee in Foreign Languages* 4 (1900): 69–141.

Oppenheim, Janet. *The Other World: Spiritualism and Psychical Research in England, 1850–1914*. Cambridge: Cambridge University Press, 1988.

Orange, A. D. "The British Association for the Advancement of Science: The Provincial Background." *Science Studies* 1 (1971): 315–29.

Oreskes, Naomi. *The Rejection of Continental Drift: Theory and Method in American Earth Science*. Oxford: Oxford University Press, 1999.

———. "Weighing the Earth from a Submarine: The Gravity Measuring Cruise of the U.S.S. S-21." In *The Earth, the Heavens and the Carnegie Institution of Washington*, edited by Gregory A. Good, 53–68. Washington, DC: American Geophysical Union, 1994.

Orihara, Minami, and Gregory Clancey. "The Nature of Emergency: The Great Kanto Earthquake and the Crisis of Reason in Late Imperial Japan." Witness to Disaster: Earthquakes and Expertise in Comparative Perspective, special issue, edited by Deborah R. Coen, *Science in Context* 25 (2012): 103–26.

Orr, Jackie. *Panic Diaries: A Genealogy of Panic Disorder*. Durham, NC: Duke University Press, 2005.

Osterbrock, Donald E. "The Rise and Fall of Edward S. Holden: Part 2." *Journal of the History of Astronomy* 15 (1984): 151–76.

Owen, Alex. *The Darkened Room: Women, Power, and Spiritualism in Late Victorian England*. Chicago: University of Chicago Press, 2004.

Palazzo, M. "Projet de réforme du catalogue macrosismique." In *Procès-Verbaux des séances de la Troisième Conférence de la Commission Permanente de l'Association Internationale de Sismologie réunie à Zermatt du 30 août au 2 septembre 1909*, edited by R. de Kövesligethy, appendix XIV, 202–3. Budapest: V. Hornyánsky, 1910.

Palm, Risa, and John Carroll. *Illusions of Safety: Culture and Earthquake Hazard Response in California and Japan*. Boulder, CO: Westview, 1998.

Palmer, Andrew H. "California Earthquakes during 1915." *BSSA* 6 (1915): 8–25.

———. "California Earthquakes during 1916." *BSSA* 7 (1917): 1–17.

———. "California Earthquakes during 1918." *BSSA* 9 (1918): 1–7.

———. "California Earthquakes during 1919." *BSSA* 10 (1920): 1–8.

———. "The Inauguration of Seismological Work in the United States Weather Bureau." *BSSA* 5 (1915): 63–71.

Pfister, Christian, ed. *Am Tag danach. Zur Bewältigung von Naturkatastrophen in der Schweiz 1500–2000*. Bern: Haupt, 2002.

Phillips, Denise. "Friends of Nature: Urban Sociability and Regional Natural History in Dresden, 1800–1850." Science and the City, *Osiris* 18 (2003): 43–59.

Phleps, Eduard. "Psychosen nach Erdbeben." *Jahrbücher für Psychiatrie und Neurologie* 23 (1903): 382–406.

Piccardi, L., and W. B. Masse, eds. *Myth and Geology*. London: Geological Society, 2007.

Pietruska, Jamie. "Looking Forward: Forecasting and the Making of Modern America." Unpublished manuscript.

Pinch, Adela. *Strange Fits of Passion: Epistemologies of Emotion, Hume to Austen.* Stanford, CA: Stanford University Press, 1996.

Poliwoda, Guido. *Aus Katastrophen Lernen: Sachsen im Kampf gegen die Fluten der Elbe 1784 bis 1845.* Cologne: Böhlau, 2007.

Porter, Theodore. "How Science Became Technical." *Isis* 100 (2009): 292–309.

———. *Trust in Numbers: The Pursuit of Objectivity in Science and Public Life.* Princeton, NJ: Princeton University Press, 1995.

Prince, Samuel. *Catastrophe and Social Change: Based upon a Sociological Study of the Halifax Disaster.* New York: Columbia University Press, 1920.

"Prof. Charles G. Rockwood." *Proceedings of the New Jersey Historical Society* (1914): 144–45.

Pyenson, Lewis. *Cultural Imperialism and the Exact Sciences.* New York: Lang, 1985.

Quenet, Grégory. *Les tremblements de terre aux XVIIe et XVIIIe siècles: La naissance d'un risque.* Seyssel: Champ Vallon, 2005.

Quervain, Alfred de. *Die Erdbeben der Schweiz im Jahre 1909.* Zurich: Schweizerischen meteorologischen Zentral-Anstalt, 1910.

———. "Die Erdbeben der Schweiz im Jahre 1910." *Annalen der schweizerischen meteorologischen Centralanstalt* 47 (1910).

———. *Die Erdbeben der Schweiz im Jahre 1912.* Separatabdruck aus den Annalen der schweizerischen meteorologischen Centralanstalt, 1912.

———. "Erdbeben als Folge von Tunnelbau." *VSNG* 96 (1914): 139–42.

———. "Über die Erdbeben von Wallis und der Schweiz und ihre seismographische Erforschung." *VSNG* 104 (1923): 74–95.

Radovanović, S., and J. Mihailović. "Die Erdbeben in Serbien." *Annales Géologiques de la Péninsule Balkanique* 6 (1911): 5–13.

Ratzel, Freidrich. *Völkerkunde,* vol. 1. Leipzig: Bibliographisches Institut, 1885.

Rayward, W. Boyd. Introduction to *European Modernism and the Information Society,* edited by Rayward. Aldershot: Ashgate, 2008.

Reid, Harry. "Records of Seismographs in North America and the Hawaiian Islands." *Terrestrial Magnetism and Atmospheric Electricity* 10 (1905): 177–89.

Reitter, Paul. *The Anti-Journalist: Karl Kraus and Jewish Self-Fashioning in Fin-de-Siècle Europe.* Chicago: University of Chicago Press, 2008.

Report of the Iowa Weather and Crop Service for the Year 1890. Des Moines, IA: Ragsdale, 1891.

Revkin, Andrew C. "On Elephants' Memories, Human Forgetfulness, and Disaster." *New York Times,* 12 August 2008, http://dotearth.blogs.nytimes.com/2008/08/12/on-elephants-memories-human-forgetfulness-and-disaster/?scp=2&sq=wisdom%20disaster&st=cse.

Richards, C. H. "Dr. C. Davison." *Nature,* 25 May 1940, 805.

Richards, Robert J. *Darwin and the Emergence of Evolutionary Theories of Behavior.* Chicago: University of Chicago Press, 1987.

Richter, C. F. Interview with Ann Scheid. 15 February–1 September 1978. Pasadena: Oral History Project, California Institute of Technology Archives. http://resolver.caltech.edu/CaltechOH:OH_Richter_C. Accessed 10 June 2010.

———. "An Instrumental Magnitude Scale." *BSSA* 25 (1935): 1–32.

Richtofen, Ferdinand. *Führer für Forschungsreisende.* Berlin: Robert Oppenheim, 1886.

Robin, Corey. *Fear: The History of a Political Idea.* Oxford: Oxford University Press, 2004.

Rockwood, Charles G. "Japanese Seismology; How an Impetuous Man Fell Down from Up-Stairs." *American Journal of Science* 22 (1881): 468–79.

————. "Notes on American Earthquakes, No. 14." *American Journal of Science* 29 (1885): 425–37.

————. "Notes on American Earthquakes No. 15." *American Journal of Science* 32 (1886): 7–19.

————. "Notices of Recent American Earthquakes" (Article II). *American Journal of Science* 15 (1878): 21–27.

Rohr, Christian. "The Danube Floods and Their Human Response and Perception (14th to 17th C)." *History of Meteorology* 2 (2005): 71–86.

Rolfe, Frank, and A. M. Strong. "The Earthquake of April 21, 1918 in the San Jacinto Mountains." *BSSA* 8 (1918): 63–67.

————. "The Southwest Section of the Seismological Society of America." *BSSA* 11 (1921): 4–5.

Rossi, Count Michele Stefano de. *Programma dell'osservatorio ed archivio centrale geodinamico*. Rome: Pace, 1883.

Rothenberg, Marc. "Organization and Control: Professionals and Amateurs in American Astronomy, 1899–1918." *Social Studies of Science* 11 (1981): 305–25.

Rowlandson, Thomas. *A Treatise on Earthquake Dangers, Causes and Palliatives*. San Francisco: Dewey & Co., 1868.

Rozario, Kevin. *Culture of Calamity: Disaster and the Making of Modern America*. Chicago: University of Chicago Press, 2007.

Rozwadowski, Helen M. "Internationalism, Environmental Necessity, and National Interest: Marine Science and Other Sciences." *Minerva* 42 (2004): 127–49.

Rudolph, Emil, ed. *Verhandlungen der Zweiten Internationalen Seismologischen Konferenz*. Leipzig: Engelmann, 1904.

Rupke, Nicolaas. *Alexander von Humboldt: A Metabiography*. Chicago: University of Chicago Press, 2008.

Sachs, Aaron. *The Humboldt Current: Nineteenth-Century Exploration and the Roots of American Environmentalism*. New York: Penguin, 2007.

Salpeter (pseud.). "Das Erdbeben." *Simplicissimus* 31 (1926): 339.

Sarton, George. "Secret History." *Scribner's Magazine* 67 (1920): 187–92.

Schaffer, Simon. "A Manufactory of Ohms: Late Victorian Metrology and Its Instrumentation." In *Invisible Connections: Instruments, Institutions, and Science*, edited by Robert Bud and Susan Cozzens, 23–56. Bellingham, WA: SPIE Optical Engineering Press, 1992.

Schardt, Hans. "Quelques détails sur les causes du tremblement de terre de Messine." *Bulletin de la Société Neuchâteloise des Sciences Naturelles* 36 (1909): 109–12.

————. "Tremblement du terre du 29 mars 1907." *Bulletin de la Société Neuchâteloise des Sciences Naturelles* 34 (1907): 314–15.

Schivelbusch, Wolfgang. *The Railway Journey: The Industrialization of Time and Space in the 19th Century*. Berkeley: University of California Press, 1977.

Schmidt, J. F. Julius. *Studien über Vulkane und Erdbeben*. Leipzig: C. Alwin Georgi, 1881.

————. "Untersuchung über das Erdbeben am 15. Jänner 1858." *Mittheilungen der k. k. Geographischen Gesellschaft* 2 (1858): 131–203.

Schneider, Jen, and Roel Snieder. "Putting Partnership First: A Dialogue Model for Science and Risk Communication." *GSA Today* 21 (2011): 36–37.

Schröter, Carl. "Prof. Christian Brügger, 1883–1899." *VSNG* 83 (1900): vii–xxiii.

Schütz, Arthur. *Der Grubenhund: Experimente mit der Wahrheit*. Edited by Walter Hömberg. Munich: R. Fischer, 1996.

Schweitzer, Johannes. "The Birth of Modern Seismology in the Nineteenth and Twentieth Centuries." *Earth Sciences History* 26 (2007): 263–80.

———. "Old Seismic Bulletins to 1920: A Collective Heritage from Early Seismologists." *International Handbook of Earthquake and Engineering Seismology* 81B (2003): 1665–1723.

Sedgwick,Eve Kosofsky. *The Weather in Proust*. Edited by Jonathan Goldberg. Durham, NC: Duke University Press, 2011.

Seebach, Karl von. "Erdbebenkunde." In *Anleitung zu wissenschaftlichen Beobachtungen auf Reisen*, edited by G. Neumayer, 309–32. Berlin: Robert Oppenheim, 1875.

"The Seismological Society of America." *Science* 25 (1907): 437.

Seneca, Lucius Annaeus. *Dialogues and Letters*. New York: Penguin, 2005.

Şengör, A. M. Celâl. *The Large-Wavelength Deformations of the Lithosphere: Materials for a History of the Evolution of Thought from the Earliest Times to Plate Tectonics*. Boulder, CO: Geological Society of America, 2003.

Severit, Wilhelm. "Die anthropogeographische Bedeutung der Erdbeben." PhD diss., University of Würzburg, 1928.

Shaler, N. S. "California Earthquakes." *Atlantic Monthly* 25 (1870): 351–60.

———. "The Stability of the Earth." *Scribners'* 1 (1887): 259–79.

———. *The Story of Our Continent: A Reader in the Geography and Geology of North America for the Use of Schools*. Boston: Ginn & Co., 1892.

Shelley, Mary. *Frankenstein, or, The Modern Prometheus, The 1818 Text*. Chicago: University of Chicago Press, 1974.

Shepard, Cola W. "Dear Fellow Co-Ops." *Bulletin of the American Meteorological Society* 3 (1922): 76.

Sieberg, August. *Handbuch der Erdbebenkunde*. Braunschweig: Vieweg & Sohn, 1904.

———. "Methoden der Erdbebenforschung." In *Lehrbuch der praktischen Geologie*, 2nd edition. Stuttgart: Ferdinand Enke, 1908.

———. "Über die makroseismische Bestimmung der Erdbebenstärke." *Gerlands Beiträge zur Geophysik* 11 (1912): 227–39.

Simson, T. "Psychische und psychotische Reaktionen Erwachsener und Kinder bei Erdbeben." *Zeitschrift für die gesammte Neurologie und Psychiatrie* 118 (1929): 130–43.

Skoko, Dragutin, and Josip Mokrović. *Andrija Mohorovičić*. Zagreb: Školska knj., 1982.

Sloterdijk, Peter. *Sphären III—Schäume*. Frankfurth: Suhrkamp, 2004.

Smith, M. Brewster. "Preface." *Journal of Social Issues* 10 (1954): 1.

Smith, N. Kemp. "Fear: Its Nature and Diverse Uses." *Philosophy* 32 (1957): 3–20.

Solnit, Rebecca. *A Paradise Built in Hell: The Extraordinary Communities That Arise in Disaster*. New York: Viking, 2009.

Soulé, Frank, John H. Gihon, and Jim Nisbet. *The Annals of San Francisco*. New York: Appleton, 1855.

Spector, Scott. *Prague Territories: National Conflict and Cultural Innovation in Franz Kafka's Fin de Siècle*. Berkeley: University of California Press, 2000.

Speich, Daniel. "Draining the Marshlands, Disciplining the Masses: The Linth Valley Hydro Engineering Scheme (1807–1823) and the Genesis of Swiss National Unity." *Environment and History* 8 (2002): 429–47.

Spengler, Erich. "Franz Wähner." *Annalen des Naturhistorischen Museums in Wien* 46 (1931): 309–12.

Stearns, Robert E. C. "Dr. John B. Trask, a Pioneer of Science on the West Coast." *Science*, new series, 28 (1908): 240–43.

Stein, Seth. "Approaches to Continental Intraplate Earthquake Issues." In *Continental Intraplate Earthquakes: Science, Hazard, and Policy Issues*, ed. Seth Stein and Stéphane Mazzotti, 1–16. Boulder, CO: Geological Society of America, 2007.

Steinberg, Ted. *Acts of God: The Unnatural History of Natural Disaster in America.* Oxford: Oxford University Press, 2000.

Stierlin, Eduard. "Nervöse und psychische Störungen nach Katastrophen." *Deutsche Medizinische Wochenschrift* 37 (1911): 2028–35.

Strasser, Bruno. "Laboratories, Museums, and the Comparative Perspective: Alan A. Boyden's Quest for Objectivity in Serological Taxonomy, 1924–1962." *Historical Studies in the Natural Sciences* 40 (2010): 149–82.

Strehlau, Jürgen. "' . . . Earthquakes occur on specific points and lines which . . . mostly coincide with traceable fracture lines . . . ': Eduard Sueß' Study of Earthquakes in Lower Austria and Southern Italy . . ." *Berichte des Geologischen Bundesanstalts* 69 (2006): 67–68.

Strickland, Stuart W. "The Ideology of Self-Knowledge and the Practice of Self-Experimentation." *Eighteenth-Century Studies* 31 (1998): 453–71.

Struck, Bernhard, "Zur Kenntnis afrikanischer Erdbebenvorstellungen." *Globus* 95 (1909): 85–90.

Stucchi, Massimiliano. "Do Seismologists Agree upon Epicentre Determination from Macroseismic Data? A Survey of ESC Working Group 'Macroseismology.'" *Annali de Geofisica* 39 (1996): 1013–27.

Suess, Eduard. *Bemerkungen über den naturgeschichtlichen Unterricht an unseren Gymnasien.* Vienna, 1862.

———. *Der Boden der Stadt Wien, nach seiner Bildungsweise, Beschaffenheit, und seinen Beziehungen zum bürgerlichen Leben.* Vienna: Braumüller, 1862.

———. *Das Antlitz der Erde*, vol. 1. Prague: F. Tempsky, 1885.

———. *Die Erdbeben Nieder-Österreichs.* Vienna: k. k. Hof- und Staatsdruckerei, 1873.

———. *Erinnerungen.* Leipzig: G. Hirzel, 1916.

———. Preface to W. H. Hobbs, *On Some Principles of Seismic Geology.* Leipzig: Wilhelm Engelmann, 1907.

Suess, Franz E. *Das Erdbeben von Laibach am 14. April 1895.* Vienna: k. k. Geologische Reichsanstalt, 1897.

Sulzer-Ziegler, Eduard. "Der Bau des Simplon-Tunnels." *VSNG* 87 (1904): 128–71.

Surman, Jan. "Figurationen der Akademia. Galizische Universitäten zwischen Imperialismus und multiplem Nationalismus." In *Galizien—Fragmente eines diskursiven Raums,* ed. Doktoratskolleg "Galizien." Innsbruck: Studienverlag, 2010.

Suter, H. "Prof. Dr. Hans Schardt, 1858–1931." *VSNG* 112 (1931): 411–22.

Taber, Stephen. "The Los Angeles Earthquakes of July, 1920." *BSSA* 11 (1921): 63–79.

———. "Seismic Activity in the Atlantic Coastal Plain near Charleston, South Carolina." *BSSA* 4 (1914): 108–60.

Taber, Stephen, et al. "The Earthquake Problem in Southern California." *BSSA* 10 (1920): 276–99.

Tarr, Ralph S., and Lawrence Martin. "Recent Change of Level in Alaska." *Geographical Journal* 28 (1906): 30–43.

Tatevossian, Ruben. "History of Earthquake Studies in Russia." *Annals of Geophysics* 47 (2004): 811–30.

Taussig, Michael. *Mimesis and Alterity: A Particular History of the Senses.* New York: Routledge, 1993.

Templeton, E. C. "The Central California Earthquake of July 1, 1911." *BSSA* 1 (1911): 167–69.

Tertulliani, A., I. Cecić, and M. Goded. "Unification of Macroseismic Data Collection Procedures: A Pilot Project for Border Earthquakes Assessment." *Natural Hazards* 10 (1999): 221–32. http://www.globalquakemodel.org/about-gem. Accessed 29 July 2011.

Tetens, Otto. "Bericht über die Arbeiten des Samoa-Observatoriums in den Jahren 1904 und 1904." In Wagner, "Das Samoa Observatorium," 27–54.

Thorsheim, Peter. *Inventing Pollution: Coal, Smoke, and Culture in Britain since 1800*. Athens: Ohio University Press, 2006.

Thouvenot, F., and M. Bouchon. "What Is the Lowest Magnitude Threshold at Which an Earthquake Can Be Felt or Heard, or Objects Thrown into the Air?" In Fréchet et al., *Historical Seismology*, 313–26.

Tilley, Helen. *Africa as a Living Laboratory: Empire, Development, and the Problem of Scientific Knowledge, 1870–1950*. Chicago: University of Chicago Press, 2011.

Timms, Edward. *Karl Kraus: Apocalyptic Satirist: Culture and Catastrophe in Habsburg Vienna*. New Haven, CT: Yale University Press, 1986.

"To Foretell 'Quakes." *Washington Post*, 15 May 1910, 7.

Tollmann, Alexander. "Eduard Suess—Geologe und Politiker." In *Eduard Suess zum Gedenken*, edited by Günther Hamann, 27–78. Vienna: Akademie der Wissenschaften, 1983.

"Topics of the Times." *New York Times*, 28 August 1911, 6.

Toula, Franz. "Ueber den gegenwärtigen Stand der Erdbebenfrage." *Schriften des Vereines zur Verbreitung naturwissenschaftlicher Kenntnisse in Wien* 21 (1881): 523–601.

Townley, Sidney D., and Maxwell W. Allen. "Descriptive Catalog of Earthquakes of the Pacific Coast of the United States 1769–1928." *BSSA* 29 (1939): 1–297.

———. "John Casper Branner." *BSSA* 12 (1922): 1–11.

———. "The San Jacinto Earthquake of April 21, 1918." *BSSA* 8 (1918): 45–62.

Trask, John Boardman. "Earthquakes in California, 1812–1855." *Proceedings of the California Academy of Sciences* 1 (1856): 85–89.

———. "Earthquakes in California in 1856." *Science* 23 (1857): 341–45.

———. *Report on the Geology of the Coast Mountains, and Part of the Sierra Nevada: Embracing Their Industrial Resources in Agriculture and Mining*. Sacramento: B. B. Redding, 1854.

Treitel, Corinna. *Science for the Soul: Occultism and the Genesis of the German Modern*. Baltimore: Johns Hopkins University Press, 2004.

Tresch, John. "The Daguerreotype's First Frame: François Arago's Moral Economy of Instruments." *Studies in History and Philosophy of Science* Part A 38 (2007): 445–76.

———. "Even the Tools Will Be Free: Humboldt's Romantic Technologies." In *The Heavens on Earth: Observatories and Astronomy in Nineteenth-Century Science and Culture*, 253–84. Durham, NC: Duke University Press, 2010.

Tribelot, Maurice de. "Notice sur les tremblements de terre ressentis dans le canton de Neuchâtel, du 2 avril au 16 mai 1876." *Bulletin de la Société des Sciences Naturelles de Neuchâtel* 10 (1876): 358–73.

Tributsch, Helmut. *When the Snakes Awake: Animals and Earthquake Prediction*. Translated by Paul Langner. Cambridge, MA: MIT Press, 1982.

Twain, Mark. *Early Tales and Sketches*, vol 2. Edited by Edgar Marquess Branch. Berkeley: University of California Press, 1981.

———. "The Great Earthquake in San Francisco." In *Early Tales and Sketches*, 303–10.

———. "A Page from a Californian Almanac." In *The Celebrated Jumping Frog of Calaveras County, and Other Sketches*, 132–34. London: George Routledge & Sons, 1867.

———. *Roughing It*, vol. 1. New York: Harper Brothers, 1913.

Tylor, Edward B. *Primitive Culture*, vol. 1. London: Murray, 1871.

"Ueber die Geschwindigkeit von Erdbebenstössen." *Schweizerische Bauzeitung* 32 (1898): 100.

"Ursache der Erdbeben." *Aus der Natur* 51 (1870): 550–52.

Valencius, Conevery Bolton. "Accounts of the New Madrid Earthquakes: Personal Narratives across Two Centuries of North American Seismology." Witness to Disaster: Earthquakes and Expertise in Comparative Perspective, special issue, edited by Deborah R. Coen, *Science in Context* 25 (2012): 17–48.

———. *The Health of the Country: How American Settlers Understood Themselves and Their Land.* New York: Basic Books, 2002.

Vaněk, Jiři, and Jan Kozák. "First Macroseismic Map with Geological Background (Composed by L. H. Jeitteles)." *Acta Geophysica* 55 (2007): 594–606.

Verhandlungen des Siebenten Internationalen Geographen-Kongresses. Berlin: Kühn, 1901.

Vetter, Jeremy. Introduction to *Knowing Global Environments.* New Brunswick, NJ: Rutgers, 2010.

———. "Introduction: Lay Participation in the History of Scientific Observation." *Science in Context* 24 (2011): 127–41.

———. "Lay Observers, Telegraph Lines, and Kansas Weather: The Field Network as a Mode of Knowledge Production." *Science in Context* 24 (2011): 259–80.

———. "The Regional Development of Science: Knowledge, Environment, and Field Work in the United States Central Plains and Rocky Mountains, 1860–1920." PhD diss., University of Pennsylvania, 2005.

———. Review of *California Earthquakes,* by Carl-Henry Geschwind. *Environmental History* 7 (2002): 701–2.

Vidal, Fernando. *Piaget before Piaget.* Cambridge, MA: Harvard University Press.

Vidrih, Renato and Jože Mihelič. *Albin Belar: Pozabljen slovenski naravoslovec.* Radovljica: Didakta, 2010.

Volger, G. H. Otto. *Erde und Ewigkeit.* Frankfurt am Main: Verlag von Meidinger Sohn & Co., 1857.

———. *Untersuchungen über das Phänomen der Erdbeben in der Schweiz,* vol. 2. Gotha: Justus Perthes, 1858.

"Vom Tage." *Simplicissimus* 13 (1909): 727.

Voss, Martin. *Symbolische Formen. Grundlagen und Elemente einer Soziologie der Katastrophe.* Bielefeld: transcript, 2006.

Vyleta, Daniel M. *Crime, Jews, and News: Vienna 1895–1914.* New York: Berghahn, 2007.

Wagner, H. "Das Samoa Observatorium, Ergebnisse der Arbeiten des Samoa-Observatoriums I." In *Ergebnisse der Arbeiten des Samoa-Observatoriums der Königlichen Gesellschaft der Wissenschaften zu Göttingen,* vol. 1. Berlin: Weidmannsche Buchhandlung, 1908.

Wähner, Franz. *Das Erdbeben von Agram am 9. November 1880.* Vienna: Akademie der Wissenschaften, 1883.

Walcott, Charles. "Progress of the Smithsonian Institution." *Nature* 85 (1911): 526–28.

Walker, Charles. *Shaky Colonialism: The 1746 Earthquake-Tsunami in Lima, Peru, and Its Long Aftermath.* Durham, NC: Duke University Press, 2008.

Wallace, Alfred Russell. *The Malay Archipelago,* vol. 2. 2nd edition. London: MacMillan & Co., 1869.

Walls, Laura Dassow. *The Passage to Cosmos: Alexander von Humboldt and the Shaping of America.* Chicago: University of Chicago Press, 2009.

Warne, William E. "Earthquake Waves." *BSSA* 23 (1933): 169–71.

Waters, Aaron C. "Memorial to Bailey Willis (1857–1949)." *Geological Society of America Bulletin* 73 (1962): 55–72.

Wegener, Alfred. *Die Entstehung der Kontinente und Ozeane.* Braunschweig: Vieweg & Sohn, 1920.

Wegener, Kurt. "Die seismischen Registrierungen am Samoa-Observatorium der Kgl. Gesellschaft der Wissenschaften zu Göttingen in den Jahren 1909 u. 1910." *Nachrichten der königlichen Gesellschaft der Wissenschaften zu Göttingen*, 1912.

Westermann, Andrea. "Disciplining the Earth: Earthquake Observation in Switzerland and Germany circa 1900." *Environment and History* 16 (2010).

———. "Overcoming the Division of Labor in Global Tectonics: Eduard Suess's The Face of the Earth." Workshop Earth Science—Global Science, 30 September–2 October 2010, York University, Toronto.

White, Gilbert F. "Human Adjustment to Floods." University of Chicago Department of Geography Research Paper No. 29, 1942.

White, Paul. "Darwin, Concepcíon, and the Geological Sublime." Witness to Disaster: Earthquakes and Expertise in Comparative Perspective, special issue, edited by Deborah R. Coen, *Science in Context* 25 (2012): 49–71.

Whitnah, Donald R. *A History of the United States Weather Bureau.* Urbana: University of Illinois Press, 1965.

Whitney, J. D. "Earthquakes." *North American Review* 108 (1869): 578–610.

———. "Owen's Valley Earthquake." *Overland Monthly* 9 (1872): part 1, 131–40, part 2, 266–78.

Wiechert, Emil. "Entwurf einer Denkschrift über seismologische Beobachtungen in den deutschen Kolonien." *Verhandlungen der Ersten Internationalen Seismologischen Konferenz*, 313–18.

Wilke, Jürgen. "Das Erdbeben von Lissabon als Medienereignis." In *Das Erdbeben von Lissabon und der Katastrophendiskurs im 18. Jahrhundert*, edited by Gerhard Lauer and Thorsten Unger. Göttingen: Wallstein, 2008.

Wilkinson, Iain. *Anxiety in a Risk Society.* New York: Routledge 2001.

Williams, G. D. "Greco-Roman Seismology and Seneca on Earthquakes in 'Natural Questions 6.'" *Journal of Roman Studies* 96 (2006): 124–46.

Willis, Bailey. "Coöperation." *BSSA* 13 (1923): 119–23.

———. "Essays on Earthquakes." *BSSA* 16 (1926): 27–40.

———. "The Next Earthquake." *Transactions of the Commonwealth Club of California* 16 (1922): 375–79.

———. "A Study of the Santa Barbara Earthquake of June 29, 1925." *BSSA* 15 (1925): 255–78.

Wilson, Holly. *Kant's Pragmatic Anthropology: Its Origin, Meaning, and Critical Significance.* Albany: SUNY Press, 2006.

Winchester, Simon. *A Crack in the Edge of the World: America and the Great California Earthquake of 1906.* New York: Harper Collins, 2005.

"Wirkung von Naturereignissen auf schwache Gemüter." *Archiv für Kriminologie* 30 (1908): 368.

Wohnlich, M. "Erdbebenprognose und seismisches Risiko." *Schweizerische Bauzeitung* 91 (1973): 1139–48.

Woldřich, J. N. "Předběžná zpráva o zemětřesení v Pošumaví." *České akademie císaře Františka Josefa pro vědy, slovesnost a umění v Praze* II, vol. 6 (1897): 1–6.

Wood, Harry O. "California Earthquakes: A Synthetic Study of Recorded Shocks." *BSSA* 6 (1916): 55–180.

———. "Concerning the Perceptibility of Weak Earthquakes and Their Dynamical Measurement." *BSSA* 4 (1914): 29–38.

———. "The Earthquake Problem in the Western United States." *BSSA* 6 (1916): 197–217.

———. "Earthquake Reports." *BSSA* 14 (1924): 60–68.

_____. "Earthquakes in Southern California, Part 1." *BSSA* 37 (1947): 107–57.

_____. "A Further Note on Seismometric Bookkeeping." *BSSA* 7 (1917): 106–12

_____. "The Observation of Earthquakes: A Guide for the General Observer." *BSSA* 1 (1911): 48–82.

_____. "Preliminary Report on the Long Beach Earthquake." *BSSA* 23 (1933): 43–56.

_____. "'Regional' vs. 'World' Seismology in Relation to the Pacific Basin." In *Proceedings of the First Pan-Pacific Scientific Congress*, 378–91. Honolulu: Honolulu Star-Bulletin, 1921.

_____. "The Seismic Prelude to the 1914 Eruption of Mauna Loa." *BSSA* 5 (1915): 39–51.

_____. "The Study of Earthquakes in Southern California." *BSSA* 8 (1918): 28–33.

Wood, Harry O., and F. Neumann. "Modified Mercalli Scale of 1931." *BSSA* 21 (1931): 2772–83.

Worster, Donald. *Dust Bowl: The Southern Plains in the 1930s.* Oxford: Oxford University Press, 1979.

Yeo, Richard. *Defining Science: William Whewell, Natural Knowledge, and Public Debate in Early Victorian Britain.* Cambridge: Cambridge University Press, 1993.

Zimmer, Oliver. *A Contested Nation: History, Memory and Nationalism in Switzerland, 1761–1891.* Cambridge: Cambridge University Press, 2003.

Zimmerman, Andrew. *Anthropology and Antihumanism in Imperial Germany.* Chicago: University of Chicago Press, 2001.

"Zur Falbschen Theorie." *Humboldt* 8 (1889): 362.